TEMPERATE HORTICULTURE FOR SUSTAINABLE DEVELOPMENT AND ENVIRONMENT

Ecological Aspects

TEMPERATE HORTICULTURE FOR SUSTAINABLE DEVELOPMENT AND ENVIRONMENT

Ecological Aspects

Edited by
Larissa I. Weisfeld, PhD
Anatoly I. Opalko, PhD
Sarra A. Bekuzarova, DSc

Reviewers and editorial board members
Gennady E. Zaikov, DSc
Alexander N. Goloshchapov, PhD

AAP | APPLE ACADEMIC PRESS

Apple Academic Press Inc.
3333 Mistwell Crescent
Oakville, ON L6L 0A2 Canada

Apple Academic Press Inc.
9 Spinnaker Way
Waretown, NJ 08758 USA

© 2019 by Apple Academic Press, Inc.

First issued in paperback 2021

Exclusive worldwide distribution by CRC Press, a member of Taylor & Francis Group

No claim to original U.S. Government works

ISBN 13: 978-1-77463-156-0 (pbk)
ISBN 13: 978-1-77188-668-0 (hbk)

Library and Archives Canada Cataloguing in Publication

Temperate horticulture for sustainable development and environment : ecological aspects / edited by Larissa I. Weisfeld, PhD, Anatoly I. Opalko, PhD, Sarra A. Bekuzarova, DSc ; reviewers and editorial board members, Gennady E. Zaikov, DSc, Alexander N. Goloshchapov, PhD.

Includes bibliographical references and index.
Issued in print and electronic formats.
ISBN 978-1-77188-668-0 (hardcover).--ISBN 978- 978-1-351-24939-3 (PDF)

1. Horticulture. 2. Sustainable agriculture. I. Weisfeld, Larissa I., editor II. Bekuzarova, Sarra A., editor III. Opalko, Anatoly I., editor

SB318.T46 2018	635	C2018-902820-3	C2018-902821-1

Library of Congress Cataloging-in-Publication Data

Names: Weisfeld, Larissa I., editor. | Opalko, Anatoly I., editor. | Bekuzarova, Sarra A., editor.

Title: Temperate horticulture for sustainable development and environment : ecological aspects / editors: Larissa I. Weisfeld, Anatoly I. Opalko, Sarra A. Bekuzarova.

Description: Waretown, NJ : Apple Academic Press, 2018. | Includes bibliographical references and index.

Identifiers: LCCN 2018023160 (print) | LCCN 2018024409 (ebook) | ISBN 9781351249393 (ebook) | ISBN 9781771886680 (hardcover : alk. paper)

Subjects: LCSH: Horticultural crops.

Classification: LCC SB318 (ebook) | LCC SB318 .T46 2018 (print) | DDC 635--dc23

LC record available at https://lccn.loc.gov/2018023160

Apple Academic Press also publishes its books in a variety of electronic formats. Some content that appears in print may not be available in electronic format. For information about Apple Academic Press products, visit our website at **www.appleacademicpress.com** and the CRC Press website at **www.crcpress.com**

DEDICATION

Nikolai Iv. Vavilov
(November 25, 1887 to January 26, 1943)

On November 25, 2017, biologists and agricultural scientists from all over the world celebrated the 130th anniversary of the prominent Russian botanist, plant-breeder, geneticist, geographer, agronomist, selectionist, and organizer of science: Nikolai Iv. Vavilov. The name of Nikolai Vavilov is not widely known among world public. Few people know and even less realize that Nikolai Vavilov was the first in the world to proclaim that in order to save mankind from famine, it is necessary to preserve genetic diversity of cultivated plants by means of creating special "banks of seeds" and to use this diversity for the creation of new plant cultivars.

Nikolai Vavilov, at 29 years old, performed his first expedition to Persia (now called Iran) for the collection of seeds in 1916 in the heat of the First World War. Since this time, the scientist has devoted his whole life to the study and collection of plants. For their conservation, Nikolai Vavilov reached to create the first bank of seeds in the world in 1924, in Saint Petersburg. It is practically impossible to count how many thousands of cultivars were created using Vavilov's worldwide collection or the many modern gene banks; banks of seeds, pollen, tissues; and cryobanks that are replenished by the collections of field gene pools all over the world. The life of Nikolai Vavilov and his contribution to biological science are well known and will remain in the history of science through the ages.

The authors of the dedication to Nikolai Iv. Vavilov did not have the luck of personal communication with this elegant and innovative personality because in 1943 the scientist went to a better world at the age of 56. It is assumed that Nikolai Vavilov died in a Saratov prison from starvation during the époque of Stalinism.

"The great scientist, genius, and pride of the home science the academician Vavilov did not perish—he croaked!!! He croaked as a dog in the prison in Saratov!!!"—cried Soviet geneticist Vladimir Efroimson to the scientific community. In 1987, in the lobby of the 5th Congress of the N. I. Vavilov All-Union Society of Geneticists and Selectionists devoted to the 100th anniversary of the scientist, the terrible history of his death was discussed many times. In spite of the whole crying injustice and para-doxicality of early death of the scientist who did not manage to achieve many of his plans, it is necessary to note that even what Nikolai Vavilov managed to do is so weighty, so significant, that it is simply beyond the strength of many venerable scientists who lived for many years longer than Nikolai Vavilov.

Let us only think over a short cold lines of his biography:

Nikolai Iv. Vavilov was born in Moscow 1887, studied with William Bateson (first director of John Innes Institute) in 1913/1914; was a professor at Moscow University 1914, in Saratov 1917; was President of the Lenin Academy of Agricultural Sciences and Director of Institute of Applied Botany in Leningrad (St. Petersburg) 1921; established more than 400 research institutes with a staff of 20,000; 1921–1934: organized expe-ditions to about 180 regions in 52 different countries around the world; was elected foreign member of the Royal Society of Great Britain 1942.

In this biography of the highest achievements of the successful scien-tist, the line about the main, maybe the single mistake of the genius is missing: in 1935 he was involved in the science of his enemy, the enemy of genetics, and the agricultural and biological science in general, Trofim D. Lysenko. Namely, Nikolai Vavilov was one of the first scientists who started to complement and stimulate the work of Lysenko. The error of the genius braked Soviet biological, medicine, and agricultural science for more than half of century and braked the life to Nikolai Iv. Vavilov himself.

Soon, Trofim D. Lysenko, who not only denied the existence of genes and DNA, and who tried to discredit natural selection and revise Darwin's evolution theory—but also successfully used repressive apparatus of special services for ideological pressing and physical elimination of oppo-nents, became a favorite of Stalin and was proclaimed "the engine of Soviet genetics." Scientific polemics was replaced with accusations, denuncia-tions, and prohibitions to perform research, and those who infringed in many cases were exposed to arrest, exile, prison, and physical elimination.

In 1929–1932, Trofim D. Lysenko, Vasiliy R. Vilyams, Isaak Iz. Prezent, and other "people's promoted workers" initiated pointed polemics about the hereditability of obtained characters' issue and the reality of "heredity substance" (genes). Soon many advanced scientific schools, of which works became later the pride of Soviet science, were proclaimed bourgeois, idealistic, and anti-Marxist. Gradually, trends were evolved (called later the sociopolitical evolution of the Union of Soviet Socialist Republics [USSR]), which later led to the phenomenon of "Michurin genetics and Soviet creative Darwinism." However, in spite of arrests of possible opponents, the decisive session of the All-Union Academy of Agricultural Sciences, held in December 1936, did not follow Lysenko's scenario. The resistance of geneticists was not broken at this session; that is why the further wave of actions aimed to suppress genetics was transformed into the ideological channel and was finished by broad public discussion conducted in autumn 1939 under the aegis of the editors of the magazine *Under the Banner of Marxism*. The transition from scientific to ideological discussion approved by the leaders of the party and the government dramatically decreased the possibilities of Trofim D. Lysenko's opponents.

In August 5, 1940, the decree of People's Commissariat for Internal Affairs USSR for arrests, where it was, in particular, noted: "It was ascertained that a number of departments of VIR (Vavilov Research Institute of Plant Industry), following the instructions of Nikolai Iv. Vavilov, in order to disprove new theories in the domain of vernalization and genetics, advanced by Soviet scientists Lysenko and Michurin, performed special work aimed to discredit the theories advanced by Lysenko and Michurin."

In August 6, 1940, Lavrentiy P. Beria confirmed the decree, and the same day the scientist was arrested in Chernivtsi city (Western Ukraine). On August 20, 1955, the Military Collegium of the Supreme Court of the USSR abolished the sentence of July 9, 1941, and terminated the case against Nikolai Iv. Vavilov for lack of *corpus delicti* (facts and circumstances constituting a breach of a law). However, the revival of domestic genetics began only after a decade. After a more than 30-years-long domination of Lysenko's inquisition, genetics started gradually to become a theoretical base of selection, overcoming fierce resistance of Lysenko's supporters.

It is not at all usual that we, representatives of the generation burnt by the Second World War, became selectionists—geneticists of the new generation. The interest in genetics, which was forbidden until still very recently, was strengthened by the information on the sacrificial activity of Nikolai Iv. Vavilov becoming available.

Our life and especially our approach to the science were formed under the great influence of the ideas of Nikolai Vavilov, which were transmitted to us by our teachers, Vavilov pupils Joseph A. Rapoport, Yuriy P. Miryuta, Lev N. Delone, Sergei M. Gerschenzon, Peter K. Shkvarnikov, and Aleksandr P. Ivanov.

That is why we, together with our colleagues, the authors of the papers in this volume, devote this book to Nikolai Iv. Vavilov in memory of his great contribution to science and society and for the entire personal charm of this great man. We wish to complete our humble dedication with the words of Nikolai Iv. Vavilov: "We will go to fire, we will be burned, but we will not refuse our beliefs!"

We sincerely hope that this volume as a whole will meet the needs of students, young and venerable scientists, involved in studies, teaching, research, and spreading of temperate horticultural crops with the aim to serve science and society.

—Anatoly I. Opalko
Sarra A. Bekuzarova
Larissa I. Weisfeld

ABOUT THE EDITORS

Larissa I. Weisfeld, PhD

Larissa I. Weisfeld, PhD, is a chief specialist at the Emanuel Institute of Biochemical Physics, Russian Academy of Sciences, Moscow, Russia. She is the author of about 300 publications in scientific journals as well as several patents and conference proceedings; and the co-author of publications on three new cultivars of winter wheat, which are included in the State Register of the Russian Federation. She is a member of the Vavilov Society of Genetics and Breeders. Her main field of interest concerns basic problems of chemical mutagenesis, mutational selection, and the mechanisms of action of para-aminobensoic acid. She has worked as a scientific editor at the publishing house Nauka ("Science") (Moscow) and an editor of the journals *Genetics* and *Ontogenesis*. She has co-edited several books, including *Ecological Consequences of Increasing Crop Productivity: Plant Breeding and Biotic Diversity; Biological Systems, Biodiversity, and Stability of Plant Communities; Temperate Crop Science and Breeding: Ecological and Genetic Study; Heavy Metals and Other Pollutants in the Environment: Biological Aspects;* and *Chemistry and Technology of Plant Substances: Chemical and Biochemical Aspects.*

Anatoly I. Opalko, PhD

Anatoly I. Opalko, PhD in Agriculture, is Leading Researcher of the Genetics, Plant Breeding and Reproductive Biology Division at the National Dendrological Park "Sofiyivka" of the National Academy of Sciences of Ukraine. He is also Professor of the Genetics, Plant Breeding and Biotechnology Chair in Uman National University of Horticulture and Head of the Cherkassy Regional Branch of the Vavilov Society of Geneticists and Breeders of Ukraine. He is a prolific author, researcher, and lecturer. He has received several awards for his work, including the badge of honor, "Excellence in Agricultural Education," and the badge of honor of the National Academy of Sciences of Ukraine "for professional achievement." He has also received the Nikolai Cholodny Prize in Botany and Plant Physiology. He is member of many professional organizations and is on the editorial boards of the Ukrainian Biological and Agricultural Science Journals. He is the author and co-editor of the books *Ecological Consequences of Increasing Crop Productivity: Plant Breeding and Biotic Diversity; Biological Systems, Biodiversity, and Stability of Plant Communities;* and *Temperate Crop Science and Breeding.* He was also a reviewer and advisory board member for the book *Heavy Metals and Other Pollutants in the Environment: Biological Aspects.*

Sarra A. Bekuzarova, DSc

Sarra A. Bekuzarova, DSc in Agriculture, is a Professor and Head of the Laboratory of Plant Breeding and Feed Crops of Long-term Seed Grower of Fodder Crops in North Caucasus Institute of Mountain and Foothill Agriculture of the Republic of North Ossetia-Alania, Russia. She is also a Professor at the Gorsky State University of Agriculture, Vladikavkaz, Republic of North Ossetia-Alania, Russia, as well as Professor at L. N. Kosta Khetagurov North Ossetia State University, Vladikavkaz, Republic of North Ossetia-Alania, Russia. Dr. Bekuzarova has been named a Deserved Inventor of Russia and has been awarded the Medal of Popova. She is a prolific author, researcher, and lecturer. She is a corresponding member of the Russian Academy of Natural Sciences and a member of the International Academy of Authors of the Scientific Discoveries and Inventions, the International Academy of Sciences of Ecology; Safety of Man and Nature; the All Russian Academy of Nontraditional and Rare Plants; and the International Academy of Agrarian Education. Dr. Bekuzarova is also a member of the editorial boards of several scientific journals. She is the author and a coeditor of several books, including *Ecological Consequences of Increasing Crop Productivity: Plant Breading and Biotic Diversity*; *Biological Systems, Biodiversity and Stability of Plant Communities*; *Temperate Crop Science and Breeding: Ecological and Genetic Studies;* and *Heavy Metals and Other Pollutants in the Environment; Biological Aspects.*

CONTENTS

LIST OF CONTRIBUTORS

Rafail A. Afanas'ev, DSc
Professor, Pryanishnikov All-Russian Scientific Research Institute of Agrochemistry, Project Leader, 31A, Pryanishnikov St., Moscow, Russia 127550, Tel. +79191040585, E-mail: rafail-afanasev@mail.ru

Boris V. Anisimov, DSc
A. G. Lorch All-Rissian Research Institute of Potato Farming, Head of Department, 23 Lorch St., Lyubertsy district, Kraskovo, Moscow Region, Russia 140051, Tel. +79057440494, +74955571018, E-mail: anisimov.bv@gmail.com

Zhamal H. Bakuyev, DSc
Associate professor, North Caucasian Research Institute of Mountain and Foothill Gardening, Director for Science, 23, Shardanov St., Nal'chik, Kabardino-Balkarsk Republic, 360004, Russia, Tel. +78662722733, E-mail: kbrapple@mail.ru

Alla V. Balabak, PhD
Associated Professor, Uman National University of Horticulture, 1 Instytutska Str., Uman, Cherkasy region, 20305, Ukraine, Tel. +380983419167, E-mail: a.v.balabak@ukr.net

Oleksandr A. Balabak, PhD
Senior Scientist, National Dendrological Park "Sofiyivka" of The National Academy of Sciences of Ukraine, Head of the Department of Genetics, Plant Breeding and Reproductive Biology of Plants, 12-a Kyivska Str., Uman, Cherkasy region, 20300, Ukraine, Tel. +380672516383, E-mail: o.a.balabak@ukr.net

Ludmila O. Barabash, PhD
Senior Scientist, Institute of Horticulture of National Academy of Agriculture of Ukraine, Head of Department, 23 Sadova St., Kyiv, 03027, Ukraine, Tel. +380445267144, E-mail: labeko@rambler.ru

Soltan S. Basiev, DSc
Professor, Head of Plant-Growing Department, Gorsky State Agrarian University, Head of Plant-Growing Department, 37 Kirov St., Vladikavkaz, Republic of North Ossetia Alania, Russia 362040, Tel. +79194286525, E-mail: basiev_s@mail.ru

Alina S. Basieva
Graduate student, Gorsky State Agrarian University, Plant-Growing Department, 37, Kirov St., Vladikavkaz, Republic of North Ossetia Alania, Russia 362040, Tel. +79897441113

Sarra A. Bekuzarova, DSc
Honored the Inventor of the Russian Federation, Professor, Gorsky State Agrarian University, 37, Kirov St., Vladikavkaz, Republic of North Ossetia Alania, Russia 362040, Tel. +79618259796, E-mail: bekos37@mail.ru

Nina V. Biserova
Municipal Autonomous Comprehensive Educational Institution Secondary Comprehensive School № 9 with the profound studying of local lore, 15 Shishkov St., Tyumen, Russia 625031, Tel. +79829714356, E-mail: nwnag@mail.ru

Lyubov D. Boldyzheva, PhD
Senior Scientist, Institute of Horticulture of National Academy of Agriculture of Ukraine,
Senior Scientist, 23 Sadova St., Kyiv, 03027, Ukraine, Tel. +380445266548

Zarema A. Bolieva, PhD
Gorsky State Agrarian University, Laboratory of Potato Selection and Seeds Growing,
Senior Scientist, 12 Williams Street 1, village Mikhaylovskoye, Republic of North Ossetia Alania,
Russia 363110, Tel. +79194274439

Akexander Ya. Bome, PhD
Senior Research Associate, the Tyumen Basing Point of Federal Research Center N. I. Vavilov
All-Russian Institute of Plant Genetic Resources, 42–44, Bol'shaya Morskaya St., Saint Petersburg,
Russia 190000, Tel.: 78123142234, E-mail: office@vir.nv.ru

Nina A. Bome, DSc
Professor, Head of the Department of Botany, Biotechnology and Landscape Architecture, Institute
of Biology of the Tyumen State University, 10 Semakov St., Tyumen, Russia 625003,
Tel.: 73452464061, +79129236177, E-mail: bomena@mail.ru

Mykola O. Bublyk, DSc
Professor, Institute of Horticulture of National Academy of Agriculture of Ukraine, First Deputy
Director, 23 Sadova St., Kyiv, 03027, Ukraine, Tel. +380445266548, E-mail: mbublyk@ukr.net

Alex E. Burakov
Engineer-Scientist, Gas Analysis Laboratory and Environmental Toxicometry, Emanuel Institute of
Biochemical Physics, Russian Academy of Sciences, 4 Kosygin St., Moscow, Russia 119334

Tatiana L. Egoshina, DSc
Professor, Prof. B. M. Zhitkov Russian Research Institute of Game Management and Fur Farming,
Head of Department of Ecology and Plant Resources, 79 Preobrazhenskaya St., Kirov,
Russia 610020, Tel. +79097166866, E-mail: etl@inbox.ru

Lyudmyla A. Fryziuk
Researcher, Institute of Horticulture of National Academy of Agriculture of Ukraine, 23 Sadova St.,
Kyiv, 03027, Ukraine, Tel. +380445266542, E-mail: lufri@ukr.net

Yulia V. Gudovskikh
Prof. B. M. Zhitkov Russian Research Institute of Game Management and Fur Farming,
Department of Ecology and Plant Resources, Junior Research Fellow, 79 Preobrazhenskaya St.,
Kirov, Russia 610020, Tel. +78332353715, E-mail: etl@inbox.ru

Natalya V. Kapustina
Prof. B. M. Zhitkov Russian Research Institute of Game Management and Fur Farming, Department
of Ecology and Plant Resources, Junior Scientist, 79 Preobrazhenskaya St., Kirov, Russia 610020,
Tel. +79127156368, E-mail: natalika.vasil@yandex.ru

Oleg S. Khutinaev, PhD
A. G. Lorch All-Rissian Research Institute of Potato Farming, Leading Scientist, 23 Lorch St.,
Lyubertsy district, Kraskovo, Moscow region, Russia 140051, Tel. +79057440494, +74955571018,
E-mail: okhutina@mail.com

Anastasya V. Kislitsyna
Prof. B. M. Zhitkov Russian Research Institute of Game Management and Fur Farming,
Department of Ecology and Plant Resources, Junior Research Fellow, 79 Preobrazhenskaya St.,
Kirov, Russia 610020, Tel. +78332353715, E-mail: etl@inbox.ru

Konstantin P. Korolev, PhD
Tyumen State University, Institute of Biology, Department of Botany, Biotechnology and Landscape Architecture, Researcher, 10 Semakov St., Tyumen, Russia 625003, Tel. +79829722770, E-mail: corolev.konstantin2016@yandex.ru

Ivan S. Kosenko, DSc
Professor, Director of the National Dendrological Park "Sofiyivka" of National Academy of Science of Ukraine, 12-a, Kyivska St., Uman, Cherkassy region, 20300, Ukraine, E-mail: sofievka@ck.ukrtel.net

Peter V. Lapshin, PhD
Timiryazev Institute of Plant Physiology, Russian Academy of Sciences, Laboratory of Phenolic Metabolism, Researcher, 35 Botanicheskaya St., Moscow, Russia 127276, Tel. +74999779433, E-mail: p.lapshin@mail.ru

Ekaterina A. Luginina
Prof. B. M. Zhitkov Russian Research Institute of Game Management and Fur Farming, Department of Ecology and Plant Resources, Senior Research Fellow, 79 Preobrazhenskaya St., Kirov, Russia 610020, Tel. +78332353715, E-mail: e.luginina@gmail.com

Genrietta E. Merzlaya, DSc
Professor, Pryanishnikov All-Russian Scientific Research Institute of Agrochemistry, Head of laboratory, 31A Pryanishnikov St., Moscow, Russia 127550, Tel. +79623694197, E-mail: lab.organic@mail.ru

Volodymyr M. Mezhenskyj, DSc
Professor, National University of Life and Environmental Sciences of Ukraine, prof. V. L. Symyrenko Department of Horticulture, Senior Scientist, 15 Heroiv Oborony Street, Kyiv, 03041 Ukraine. Tel. +380502972739. E-mail: mezh1956@gmail.com

Anatoly I. Opalko, PhD
Full professor, Leading Researcher of the Genetics, plant Breeding and Reproductive Biology Division in National Dendrological Park "Sofiyivka" of National Academy of Science of Ukraine, 12-a Kyivska Str., Uman, Cherkassy region, 20300, Ukraine; and Professor of the Genetics, Plant Breeding and Biotechnology Chair in Uman National University of Horticulture, 1 Instytutska St., Uman, Cherkassy region, 20305, Ukraine. Tel. +380506116881. E-mail: opalko_a@ukr.net

Olga A. Opalko, PhD
Associate Professor, Senior Scientist of the Physiology, Genetics, Plant Breeding and Biotechnology Division in National Dendrological Park "Sofiyivka" of National Academy of Science of Ukraine, 12-a, Kyivska St., Uman, Cherkassy region, 20300, Ukraine. Tel.: +380664569116. E-mail: opalko_o@ukr.net

Natalia N. Sazhina, PhD
Senior Scientist, Emanuel Institute of Biochemical Physics Russian Academy of Sciences, Laboratory of Complex Antioxidant Assessment, 4 Kosygin Street, Moscow, Russia 119334. Tel. +74959397056, +79160879868. E-mail: Natnik48s@yandex.ru

Marina V. Semenova, PhD
Tyumen State University, Institute of Biology, Department of Botany, Biotechnology and Landscape Architecture, Associated Professor, 10, Semakov St., Tyumen, Russia 625003. Tel. +79129925333. E-mail: ssmmvv@list.ru

Nizam E. Shabanov, PhD

Lorch All-Rissian Research Institute of Potato Farming, Senior Scientist, 23 Lorch St., Lyubertsy district, Kraskovo, Moscow region, Russia 140051. Tel. +79057440494 +74955571018. E-mail: shaban-sky@mail.com

Michail O. Smirnov, PhD

Senior Scientist, Pryanishnikov All-Russian Scientific Research Institute of Agrochemistry, Pryanishnikov All-Russian Scientific Research Institute of Agrochemistry, 31A Pryanishnikov St., Moscow, Russia 127550. Tel. +79057966323. E-mail: User53530@yandex.ru

Fyodor A. Tatarinov, PhD

Scientist, A. N. Severtsov Institute of Ecology and Evolution, Russian Academy of Sciences, V. N. Sukachev Laboratory of Biogeocenology, 33 Leninskij Prosp., Moscow, Russia 119071, Tel. +79163881561, E-mail: f.tatarinov@gmail.com

Larissa I. Weisfeld, PhD

Chief Specialist, N. M. Emanuel Institute of Biochemical Physics of Russian Academy of Sciences, 4 Kosygin St., Moscow, Russia 119334, Tel.: +79162278685, E-mail: liv11@yandex.ru

Natalia V. Zagoskina, DSc

Professor, Timiryazev Institute of Plant Physiology of Russian Academy of Sciences, Laboratory of Phenolic Metabolism, Leading Scientist, 35 Botanicheskaya St., Moscow, Russia 127276, Tel. +74999779433, E-mail: nzagoskina@mail.ru

Tat'yana V. Zhidyokhina, PhD

Head of the Department of Berry Crops, I. V. Michurin All-Russia Research Institute for Horticulture, Russian Academy of Sciences, Deputy Director for Science, Associated Professor, 30 Michurin St., Michurinsk, Tambov Region, Russia 393774, Tel./Fax +74754520761, E-mail: berrys-m@mail.ru

LIST OF ABBREVIATIONS

a	is the value of the character under consideration under salinization
A	retinol
A_1	provitamin A
AD	Anno Domini (used to indicate that a date comes the specified number of years after the traditional date of Christ's birth)
ADSR	Artemivsk Nursery Experimental Station
AOA	antioxidant activity
$(AP^{2-})^*$	aminophthalate anion in excited state (free radical)
As	arsenic
b	is the value of the character under consideration under salinization
B	borum
B_1	thiamine
B_2	riboflavin
B_3	niacin
B_5	pantothenic acid
B_6	pyridoxine
B_7	biotin
B_9	folacin, folic acid
B_{10}	para-aminibenzoic acid
BC	before Christ (used to indicate that a date is before the Christian era)
Belgium	Kingdom of Belgium
BS	biological stock
°C	degrees Celsius
Ca	calcium
$CaCl_2$	calcium chloride
$CaCO_3$	calcium carbonate
$CaSO_4$	calcium sulphate
Cd	cadmium
CJSC	Closed Joint Stock Company

CL	chemiluminescense metod
CL-	gramm curve of chemiluminescense
Cl−	ion-chloride
cm	centimeter is a measure of length
C:N	carbon–nitrogen ratio
Co	Cobaltum
Cr	Chromium
Cu	Cuprum
cv.	cultivar
CV	coefficient of variation
cvs.	cultivars
cwt	center, hundred weight; the American centner is equal to 45.359 kg, the English centner is 40.802 kg, Russian is 100 kg
D	calciferol, vitamin
DISoft	Russian Software Developer
E	tocopherol, vitamin
e	base of natural logarithms, mathematical constant, irrational and transcendental number. The approximate value is 2.71828
EC	electrical conductivity
EDTA	ethylenediaminetetraacetic acid
equiv.	equivalent
ES	exploitation stock
et	and (from Latin)
F_1	first generation by crossing a selection process
FAO	The International Food and Agriculture Organization of the United Nations
FAOSTAT	FAO Statistics Division
Fe	ferrum
Fe-EDTA	ferrum chelate of ethylenediamintetraacetic acid
FSC	Federal Scientific Center
g	gram
GOMF	granulated organic-mineral fertilizer
GOST	Sate Standard "Rosstandard"
GOST-12,038–84	methods for determining germination of agricultural seeds
gram/pot	gram per pot

H	hydrigen
H^+	hydrogen ion
ha	hectare
Hb	hemoglobin
(Hb ($^{\bullet+}$)–Fe^{4+}=O)	ferril-radical of hemoglobin
Hg	mercury
HNO_3	nitric acid, azotic acid
H_2O_2	hydrogen peroxide
HHLC	high-performance liquid chromatography
HPUFA	higher polyunsaturated fatty acids
HTC	hydrothermal coefficient
I	iodium
IBA	indole-3-butyric acid
ICNCP	scientific breeding or farmer's selection methods: after International Code of Nomenclature for Cultivated Plants
I.e.	Id est
IH of NAAS	Institute of Horticulture of the National Academy of Agrarian Sciences
ISO	International Organization for Standardization
K	potassium
K	vitamin of blood coagulation
kcal	kilocalorie
KCl	potassium chloride
kg	kilogram
kg/ha	kilogram per hectare
kg/thous. ha	kilogram per thousand hectare
km	kilometer
km^2	square kilometers
K_2O	potassium oxide
kW	kilowatt
l	liter
LSD	least significant difference
Lum-5373	brand of ultraviolet light source
M	manure
m	meter
m^2	square meter
Mg	magnesium

mg-equiv.	miligram-equivalent
MgO	magnesium oxide (burnt magnesia, periclase)
mg/kg	milligram per kilogram
$MgSO_4$	magnesium sulfate
Mln	million
mM	millimol
mm	millimeter
Mn	manganum, manganesium
Mo	molybdenum
mPa	megapascal
MPL	minimum permissible level
Mo	molybdenum
MS	Microsoft
n	mathematical symbol indicating the number of observations
n	haploid number of chromosomes
N	nitrogen is element V of the group Periodic. Systems, at. N. 7, at. M. 14,0067; colorless gas, odorless and tasteless
2n	double set of chromosomes
NaCl	sodium chloride
NAS	National Academy of Sciences
NDP	National Dendrological Park
Ni	nickel is element of the tenth (in obsolete short-period form - the eighth) group, the fourth period of the periodic system of chemical elements of DI Mendeleyev, with atomic number 28.
NO_3	petre(obsolete name) saltpetre, nitre (modern terminology)
NPK	complex soil fertilizer, containing the main nutrient elements for plants: nitrogen, phosphorus and potassium. In Russian-language everyday life is called *azophoska*.
*n*1 *P*	the number of germs in salt solution.
OH	hydroxyl group
OH*	free radical
O—H	chemical bond between molecules of oxygen and hydrogen

o-pyrocatechic	orto-pyrocatechic
P	phosphorus
P	sample salt resistance
P	active substances, whose name begins with the letter P
Pb	lead, plumbum
pH	hydrogen ion concentration
PK	phosphorus and potassium
PP	nicotinic acid, vitamin B_3, nicotinamide, niacin
PABA	para-amino benzoic acid
P_2O_5	phosphorus oxide
R	a general notation for any substance
r	correlation coefficient
RCH	Research Centre of Horticulture
R-OH	natural phenolic antioxidants
RSH	Research Stations of Horticulture
RSNP	Research Station of Nursery Practice
S	sulfur
SAI	sugars to titrated acids
SAR	sugar-acid ratio
SAT	sugars to titrated acids
spp.	reduction from the Latin *"speciales"* (species)
sq. m	square meter
sq. cms	square centimeters
SRN	sanitary regulations and norms
subsp.	subspecies
Sx	error of arithmetic mean
t	metric ton
t°	temperature
T°C	temperature in degrees Celsius
t/ha	tons per hectare
thous. ha	thousand metric hectares
thous. tons	thousand metric tons
TSO-1/80 SPU	thermostat electric dray-aero cooling
U	potato crop
UK	United Kingdom
USSR	Union of Soviet Socialist Republics
var.	variety
VIR	Vavilov Research Institute of Plant Industry

VPHOS	volatile phyto-organic substances
WHO	World Health Organization
Zn	zinc
X	parameter under study
x	monoploid chromosome number
X–axis	is a horizontal line, abscise axis
Y–axis	is a vertical line, ordinate axis
μM	micromole—one-millionth of a mole
μl	microliter—one-millionth of a liter

PREFACE

World vegetation of our planet numbers above 300 thousand plant species. Among them, more than 1200 species are used as vegetables and above 1000 species are used as fruit and berry plants, belonging to 78 (vegetables) and 40 (fruit and berry plants) different families. In post-Soviet countries, traditionally not more than 60 species of vegetables and 25 species of fruit and berry plants are cultivated, although the state registers of plant cultivars recommended for implementation in particular states numbers two times more of them. In the same time, the natural potential of many regions of East-European countries, especially southern and west-southern ones, is quite favorable for the production of a much wider assortment of horticultural products in the amounts sufficient not only for the full coverage of internal demand but also for the development of export options.

Global trends in the world market of horticultural production are characterized by stable growth of unsatisfied demand for the various organic fruit and vegetable production, which was formed by the increase of the consumption of vegetables, fruits, and berries, first of all in the countries of European Union (EU), North America, Japan, and some other developed countries of northern hemisphere with high level of population incomes. During recent years, substantial changes in the structure of nutrition in favor of fruits and berries took place in these countries. Now, fruits and berries are more and more used not only in salads, in former times traditionally made from vegetables, but they also are added to soups, meat, and fish dishes and used in the baking of not only cakes and muffins but also rye bread. That is why the countries of EU, North America, and Japan, in spite of high domestic production, belong to the world's largest importers of vegetables, fruits, and berries.

Similar regularities are also observed for decorative plants, which demand increases in parallel with the growth of the population and for its welfare. Decorative vegetation improves the architectural view of cities, increases their diversity, creates attractive space silhouettes of urban areas, and provides pleasure to herald in the countryside, giving a special charm to big suburban parks as well as to small homestead gardens. Decorative cultivars and forms of practically all plants growing on our planet can be used in gardening and landscape design. The experience of China and Japan, where a broad assortment of ornamental plants has been cultivated since ancient times, shows it. Ornamental plants were known in Egypt for more than thousand years ago; they were used in the Middle East since Biblical times and described the technology of transplantation and cultivation of ornamental plants in Greece. It may be assumed that from Greece they spread in the gardens of Europe and further to North and South America. Ornamental plants play an important role in the life of human society, religion, and science. The diversity of ornamental plant species now makes this group of plants more and more important because the plants are being used as model plants in a series of physiological, biochemical, genetic, and biotechnological studies.

The study of the dynamics of demand in the world food market, as well as the analysis of changes in the food preferences of the population in the countries that have different living standards, shows that as the initial stage of population welfare increases, the consumption of meat products increases. However, with further improvement of population welfare, the consumption of vegetables, fruits, and berries permanently increases in the increasingly diverse assortment.

It seems evident that during the domestication of animals and consumption of some groups of wild plants as food and their selected cultivation, there were fruits and leaves suitable for consumption as fresh fruits and vegetables, which saved populations from hunger after unsuccessful hunting. Taking into account that primitives were hunters and consumed mainly meat, first, the animals helping in the hunting were domesticated. If examining closely the feeding regime of carnivorous animals, one can notice that they periodically eat some green leaves, young shoots, fruits, and berries in different degree of ripeness, roots, and edible roots. The most actively carnivorous animals eat greens in the case of a disease.

Probably, like animals, prehistoric humans, being sated with meat, begun to eat wild fresh plants. With time self-sown crops from the best of harvested plants appeared near the cave (or hut). After centuries of comprehended cultivation of plants for food as well as for curing started. With time cereals started to be cultivated, and even later, with the development of settled livestock breeding the cultivation of forage plants begun. The appearance of technical cultures coincides with the development of industry, where they were used as raw materials. It was noticed that geographically the locations of ancient agricultural civilizations correspond to the centers of the origin of fruit plants or to adjacent territories. This period did not leave anything except the petrified remains of fruits, seeds, and their schematic images found during excavations of the earliest human settlements. It is possible to approximate date the age of these finds: they are 10–15 thousand years old. According to Nikolay Vavilov (1926), two conditions were necessary for the rise of the center of cultivated flora: (1) richness of local flora by plants suitable for cultivation and (2) presence of ancient agricultural civilization. It is hard not to agree with this hypothesis (almost axiom), but it is possible to discuss cause-and-effect relations. Probably, the progress in agriculture did not contribute to the cultivation of fruit and berry plants and their improvement, but the richness of local flora by edible plants contributed to the survival and multiplication of humankind, and further to its formation and development. At the same time, humans did not linger for a long time in the regions depleted by edible plants. If they lingered here for different reasons, it was related to the depletion of hunting grounds as a result of uneconomical hunting, with further unavoidable decline.

Modern data about life duration in different regions of our planet with similar levels of food calorie content but different food traditions testify to the primacy of natural conditions contributing to the local flora enrichment by edible plants and the multiplication of wild animals suitable for hunting, as well as for survival and development of humans. It was proved that insufficient consumption of fruits, berries, and vegetables is one of the main factors of premature aging; and congenital malformations of development, mental and physical deficiency; weakened immune system, blindness; and increased mortality. The World Health Organization (WHO) attributes about 3 millions deaths per year to the diseases related to insufficient consumption of fruits, berries, and vegetables.

This sorrowful statistic did not miss post-Soviet stats, where more than half of untimely death (before 65 years) cases for men and women is conditioned by diseases considerably related to irrational nutrition. Thus, the investigation of households of towns' people and countryside people showed that the latter consume almost half the amount of fruits, berries, nuts, and grapes. At the same time, countryside households consume 1.5 times more potatoes, bread, and baking products. The population of many countries of the European region also does not follow the norms of consumption of fruits, berries, and vegetables recommended by WHO experts (400 g per day). In particular, one human in a South European country consumes on in average 272 g of fruits, berries, and vegetables daily; in Baltic countries this value amounts to 374 g; 254 g in Azerbaijan, Moldova and Ukraine; and only 199 g in Kazakhstan, Kyrgyzstan and Uzbekistan. At the same time the amount of calories consumed with food daily by one human in the an European region countries is 3301.6 kcal in particular, EU and post-Soviet country is 3301.6, 3501.6, and 2944 kcal, respectively, that is, around norm.

The deficit of vitamins and minerals, including microelements, conditioned by refusing to consume a sufficient amount of fruits, berries, and vegetables is presently considered as a factor of risk almost as high as the use of tobacco or practicing of sex without condoms. The measure of mutagenic activity of different compounds in food products depends on the usefulness of the diet: under long-term deficit of methionine, arginine, lysine, and vitamins A, E, C, the amount of cells with chromosome mutations leading to hereditary diseases and malignant growths increases considerably. By overcoming the deficit of fruits, berries, and vegetables in the diet, it is possible not only avoid general disorders of health, but also to decrease the risk of cardiovascular diseases, which belong to the leading causes of mortality all over the world, as well as to prevent the development of many varieties of cancer. Products with elevated content of vitamins C, E, A combine with promoters (activating substances) and hamper, in particular, steps of cancer development. The representatives of leading international organizations, including the Food and Agriculture Organization (FAO), and national organizations of public health of developed countries actively cooperate in order to remove all obstacles and reach the increase of the consumption of fruits, berries, and vegetables by all categories of population.

Different reasons preventing balanced diet of humans can be divided into two groups: mental-psychological and material-industrial. Traditionally, the display of these obstacles is closely related to the general population welfare and country development. Rudiments of rather slighting relation to the consumption of fruits, berries, and vegetables remain in post-Soviet countries until now, not only among socially weak categories of population. That is why in order to improve the culture of nutrition it is necessary not only to reach the increase of production of fruits, berries, and vegetables and corresponding increase of purchasing capacity of potential consumers, but also to create corresponding stereotypes of behavior. In the assortment of gifts, it is necessary to replace traditional and, rather expensive but almost inedible cakes and sweets with festively decorated basket of fruits and grapes. It should become more prestigious to entertain guests with apples and pears, peaches, and apricots than with sprats, sausages, or different smoked products...

The nutritional, therapeutic, and preventive values of vegetables, fruits, and berries are determined by the content of easily digestible carbohydrates and organic acids of 4.5–23.0% and 0.1–3.8%, respectively, by high content of vitamins C, A, B1, B2, B6, P, PP, E, and so forth, as well as phenolic compounds, aromatic and tanning substances and mineral salts containing above 50 chemical elements, including iron, potassium, calcium, magnesium, boron, molybdenum, and others. Walnuts, hazelnuts, pistachios, almonds contain up to 22% of proteins and 60–80% of fats. The biochemical composition is the main indicator of vegetables quality. They contain up to 96−97% of water and, however, they have enormous importance in human nutrition. This is because a small amount of dry substances of vegetables contain a lot of biologically important compounds, which are necessary for normal functioning of organism. Vegetables also contain organic acids: citric, malic, and tartaric, oxalic and others. They improve taste properties of vegetables and contribute to their better assimilation. Volatile oils of vegetables (onion, parsley, dill) have properties of volatile phytoorganic substances (VPHOS) influencing an abiotic effect. It is known that VPHOS are being used in medicine since from old times: they protect humans from many infectious diseases. Leaves of parsley, peas, onions, cabbage, parsnips are rich in phosphorus; leafy vegetables and root vegetables are rich in potassium; salad, spinach, beets, cucumbers, tomatoes contain high amount of iron; cauliflower, lettuce, spinach are rich in calcium. Vitamins from the K group are necessary for humans for normal fibrillation. They are contained in spinach, cauliflower and white cabbage, tomatoes.

Mineral substances are formed during digestion compounds with alkalinity. Vegetable food contributes to the maintenance of alkalescent reaction of blood and neutralizes harmful effect of acidic substances contained in meat, bread, and fats. Inclusion of vegetables, fruits, and berries in the diet makes it more harmonized, positively affects metabolism, prevents the origination of gastrointestinal diseases, and contributes general resistance against pathogens of many diseases.

Flowers and ornamental plants, which decorate our earth and fill our souls with joy, are useful as well. They intrigue us and inspire to good, the saturation with positive emotions, provided by ornamental plants, relieve fatigue and contribute to the strengthening of physical and mental health.

The authors of this book, united by common understanding of the necessity to increase the consumption of traditional vegetable, fruit, and berry plants, as well as by a need for a broad introduction of vegetable, fruit, and berry plants not in current use, and also by the necessity to use flowers and ornamental plants in recreational purposes, have focused not only on the current problems of horticulture but also have designed key directions of horticulture development in different regions. The maintenance of standards concerning the conservation of biodiversity as the base of evolutionary adaptability of wild vegetable, fruit and berry plants, and the source of initial material for the selection of new ones, was defined as the leading principle of stable horticulture development.

—**Anatoly I. Opalko, Larissa I. Weisfeld**

INTRODUCTION

The reader is presented with a new volume from a series of monographs on the problems of conservation of biotic diversity in horticulture and related sciences. In this book, the experimental work of the leading specialists of each in its own field of science is collected. These studies have been conducted in differing and often contrasting soil-geographic conditions: in Russia these are North Caucasus, North East, Chernozem Region, and also in different climatic zones of Ukraine. A wide range of horticultural crops was covered: vegetables, fruit, berries, and flowers.

At present, even among the scientific community, the misconception is widespread that any impact on the chromosome apparatus causes the synthesis of substances harmful to the human body. The millennial history of plant and animal breeding is based namely on the modification of hereditary material and the choice of useful properties of breeds or varieties used by man. More than 3000 cultivars of different cultures have been created in the world only by the method of experimental mutagenesis.

The editorial board invited scientific workers to publish in this book their achievements in the fields of topics of gardening, including gardening of vegetable, fruit, and ornamental horticulture. The works also cover the successes of introduction of crops into production, expansion of a set of cultivars in conditions of different environment. The chapters published here are written by specialists on the basis of their own original experiments and supplemented by reviews of their own previous publications and studies by other authors. The book is illustrated by a large number of drawings, maps, and photographs.

The chapters were selected by taking into account the requirements of the editorial board and the publishing houses.

With the development of scientific knowledge, the complexity, diversity, and interrelation of ecological and biological modern scientific knowledge, whether it is chemistry, biology, agriculture, etc., became apparent.

Ecological and genetic control of plant resistance to unfavorable environmental influences is being carried out all over the world, new cultivars and hybrids are being created, the assortment of source material has expanded considerably, and increasingly new methods are being created.

The studies presented in this book are united by a common way of searching and eliminating painful points of development of cultivated plants in modern conditions of geographical disasters, depletion of biota, and pollution of the environment.

The book is dedicated to the memory of the great scientist—geneticist, breeder, inventor Nikolai Ivanovich Vavilov.

Part I is titled "INNOVATIONS IN VEGETABLE GROWING" and consists of four chapters focusing completely on new methods increasing the yield of potatoes and cucumber. These methods are also applicable to other cultures.

A number of new technologies for obtaining of vegetable products are proposed: "snow technology" for processing potato sprouts before planting in the field (Chapter 3); a method of accelerated the development of cucumber seedlings with the use of extracts from conifers (Chapter 4); for potato cultivation additionally to organic fertilizer manure, bird droppings, application of fertilizers based on sewage sludge—compost and organic-mineral granular fertilizer (Chapter 1).

An original method for growing an increased number of minitubers in the open space is described in Chapter 2, without the costs of additional lighting and ventilation that are characteristic of enclosed spaces.

Part II is titled "ARCTIC BERRIES: ECOLOGY, BIOCHEMISTRY, AND USEFUL PROPERTIES." Here are presented data about phytocenotic properties of wild-growing and cultivated berry plants (three chapters): of arctic raspberry and blueberry in natural populations of the taiga zone (Chapters 5–7). The authors studied berry crops, cranberry, Arctic bramble, blueberry, Arctic raspberry, cowberry, growing on the boggy soil and peat lands in the taiga part of the Kirov region.

The analysis of biochemical composition of Arctic raspberry (*Rubus arcticus* L.), cloudberry and cranberry revealed that wild berries are sources of the most valuable carbohydrates (fructose, cellulose, pectin substances), vitamins, organic acids and other substances, useful for people living in the northern latitudes. Arctic raspberry was grown on the experimental plot in the conditions of the taiga zone of the Kirov region. Plants successfully were developed on peat bogs and were characterized by winter hardiness and high yield; it is a promising species for growing in the southern taiga sub zone of the European part of Russia.

The productivity of the short-bush blueberry, *Vaccinium uliginosum* L., at cultivating on cutover peatlands in the southern taiga zone was studied. The winter hardiness and high productivity of that species were shown in these growing conditions. Genus *Vaccinium* L. belongs to the family Ericacea, which includes a large number of species of blueberries. The ways of their cultivation are also discussed here.

Part III is titled. "BREEDING AND BIOCHEMISTRY OF DECORATIVE PLANTS" (three chapters), has provided an overview of winter gardens plants and shows the ways of their successful cultivation (in Chapter 8); the study the range of resistance to salinization and summated other stress of ornamental plants growing (in Chapter 9) was carried out biochemical analyses of biological active compounds—antioxidants among various species of the genus *Aloe* (in Chapter 10).

In Chapter 8 the possibility of increasing the tolerance of indoor plants in the winter gardens of the Baltschug Kempinski Hotel in winter (Moscow, Russia) and in the garden of the marble monument in the House of Scientists (Uman, Cherkassy region, Ukraine) is shown. To preserve and grow room in the winter gardens, leaves were treated with the solution of para-aminobenzoic acid, and the roots and rootstocks watered with solutions of kornevin, zikron, and a complex of microelements. Such a combined effect without the use of pesticides positively influenced the growth of plants, the development of the root system, and the resistance of plants to pathogens and parasites. Descriptions of the most common houseplants are given.

The problem of salinization of cultural layers of soil led to the need to analyze the salt tolerance of plants. In Chapter 9, resistance to salinization in 17 studied cultivars of annual flower plants from the collection of Tyumen State University is provided. The cultivars belong to the five species: *Eschscholzia californica*, common flax, and three species from the family Asteraceae. These species are used in the floral compositions in Tyumen and other settlements of Tyumen region. The study was performed under controlled laboratory conditions using NaCl solutions. Cultivars differed significantly according to their reaction to salinization stress, namely according to germination capacity in vitro and main characters of sprouts (length and biomass). The heterogeneity of cultivars was determined by the reaction to soil solvation. The cultivars were divided into three groups according to the complex of characters, considered as indicators: cultivars with high, medium, and low resistance.

In Chapter 10, the total antioxidant activity of 15 various species of *Aloe* (*Aloe* L.) extracts were measured by ammetric and chemilumines-cence methods for the purpose of searching for the biologically active components. The essential distinctions in accumulation of different classes of polyphenols and their antioxidant activity were shown. *Aloe* species with antioxidant activity, which can be used in pharmacology and medi-cine, are determined.

Part IV is titled "FRUIT GROWING AND BREEDING" (five chapters). Here, the review of various technologies of cultivation of perspective cultivars of an apple-tree, in the creation of new cultivars of hazelnut, overview of data about breeding rare fruit crop, about breeding of black currant cultivars are given.

Chapter 11 considers the historical information about the apple produc-tion in the country since the middle of the 20th century to the present day and determines the main regions for industrial apple growing. Descrip-tions of the main and promising apple cultivars and rootstocks as well as the technologies for that fruit growing in the scientific institutions of Ukraine are presented. The total area of apple orchards in Ukraine now is 145.6 thousand ha, about 41% of which lie on the most suitable lands in the Western Forest-Steppe and Prydnistrovia. The average annual apple output in the mentioned period was about 1.1 million t; among them almost three-quarters, mostly in individual holdings, meet their own requirements for fresh fruits. Six research stations in different natural areas are presented here.

Chapter 12 presents an overview of previously published works on the productivity of an apple tree in intensive orchards on the slope lands in the foothills of the North Caucasus. Results on the productivity of intensive gardens of an apple-tree are given in conditions such as the terraced and gentle slopes of North Caucasus foothills. In the conditions of vertical zoning, it is preferable to develop the gardens: winter cultivars of apples—in the foothills and forest and mountain areas, in the steppe zone—autumn cultivars of apples, in the mountain Steppe Zone—summer and autumn cultivars.

Particular interest in a new hazelnut breeding scheme was designed by Ivan S. Kosenko and colleagues at the National Dendrological Park "Sofiyivka" of National Academy of Sciences (NAS) of Ukraine (in Chapter 13). This breeding scheme helped increase the passing of breeding material through the stages of a breeding scheme for 5–8 years; that is, at the growing stage of F_1 seedlings for 3–4 years, in a hybrid orchard—for 1–2 years, and at the propagation stage of the best seedlings with layering—for 1–2 years. The whole number of cultivars, the best of which were prepared for submission to the State Veterinary and Phytosanitary Service of Ukraine and were determined in the garden of the primary cultivation, in particular, new hazelnut cultivar Sofiyivsky 15, is characterized by spherical fruits that are higher in comparison with Turkish and Azerbaijan cultivars, winter hardiness and drought resistance, and the lack of periodicity of fruiting.

Chapter 14 gives a detailed analysis of a large number of rare fruit-crop breeding in Ukraine. The natural conditions of this country allow cultivating a wide range of fruit plants. In the second half of the 20th century, zoning and registration of rare fruit crops cultivars began. Today, there are 158 registered cultivars of rare fruit crops of Ukrainian breeding.

Chapter 15 presents a comprehensive assessment of the adaptive and productive potentials of black currant cultivars, which were derived by I. V. Michurina at the All-Russian Scientific Research Institute of Certification. The cultivars were characterized by a standard good and excellent plant condition (five cvs.). Development of cultivars was strongly affected by plant age ($r=0.63$), but not by the sum of the temperatures of the previous period ($r=0.35$). Three cultivars of early flowering, medium-ripening 10 cultivars, the medium-late flowering of six cultivars are distinguished. Cultivars with complex resistance to fungal diseases of four cultivars and six types of mites were obtained. The most plastic cultivars are Chernavka, the Little Prince, and the Green Haze. During the period of research 2000–2016, they were characterized by high productivity.

An extensive glossary is placed at the end of the book, which was compiled by editors of the book, Sarra A. Bekusarova and Larissa I. Weisfeld.

The book will be useful for the scientific community, ecologists, geneticists, breeders, and for young people coming into science, as well as for businessmen interested in using science for practical applications in production and for market access in the post-Soviet states.

—Larissa I. Weisfeld, Gennady E. Zaikov

PART I
Innovations in Vegetable Growing

CHAPTER 1

POTATO PRODUCTION IN VARIOUS SOILS AND CLIMATIC CONDITIONS

RAFAIL A. AFANAS'EV, GENRIETTA E. MERZLAYA*, and MICHAIL O. SMIRNOV

Pryanishnikov All-Russian Scientific Research Institute of Agrochemistry, 31A, Pryanishnikov St., Moscow, Russia 127550, E-mail: rafail-afanasev@mail.ru, User53530@yandex.ru

*Corresponding author. E-mail: lab.organic@mail.ru

CONTENTS

ABSTRACT

Studies in the field experiments showed that in the technology of potato culti-vation in different regions of Russia the most important factor in the formation of a stable yield and high quality of tubers is a science-based joint application of organic and mineral fertilizers. At the sod-podzolic light-textured loamy soil, when using organic fertilizer system comprising 20 t of manure per year per 1 ha and 45 kg/ha of active ingredient of mineral nitrogen, phosphate, and potash fertilizers, you could achieve stable potato productivity with a potato starch containing 14.4% at the level of 36.6 t/ha, which is 40% higher than the control without fertilization. When potatoes were fertilized we observed the efficiency of the use of traditional organic fertilizers—manure, chicken manure, and fertilizers on the basis of sewage sludge—compost and granulated organic-mineral fertilizer (GOMF). At the sod-podzolic middle-textured loamy soil by applying the GOMF at a dose of 4 t/ha per year, as aftereffect the yield 22.4 t/ha of potato tubers was obtained, which exceeded the control at 3.7 t/ha or 20%. The content of heavy metals, arsenic, and nitrates in potato tubers was at the level of the control values, indicating the environmental safety of plant production. In the northern region (Yakutia) at the permafrost taiga soils, we established, while cultivating potatoes, the usefulness of chicken manure and cattle manure in a pure form, as well as peat-litter compost at a ratio of 1:1 at a dose of 300 kg/ha of nitrogen, which ensured the yield of tubers 26.5–26.6 t/ha or 29–30% above control.

1.1   INTRODUCTION

Potato (*Solanum tuberosum* L.) is distributed almost everywhere and belongs to cultures of universal use—for food, fodder, and technical purposes. World potato production per year is about 320 million t, the landing area—about 19 million ha. In foreign countries, due to the high culture of agriculture yields of potatoes in 2002–2008 years were: in the Netherlands 40–47, in the United States of America 39–44 t/ha, in Belgium 38–47, in France 40–45, in Denmark 35–42, in the United Kingdom 40–44, and the world average was 15–16 t/ha. The yield of potatoes in the Russian Federation was 13 t/ha in households in 2009, 9.6 in 2010, 13 t/ha in 2011-14; 2012 - 13 t/ha, in farms of all categories—14, 10, 15, and 13 t/ha, respectively.[1,2] Thus, the yield of potato tubers in Russia was about three times lower than in the developed countries. The potato has a high demand for nutrients. Removal of 1 t of main potato production in view of an accessory production is 5.7 kg N, 1.6 kg P_2O_5, and 7.9 kg K_2O.

According to reports,[3–7] the yield of potato tubers and their quality are directly dependent on fertilizer systems, whose main role belongs to the manure. Potatoes responded favorably to direct application of organic matter in manure and other organic fertilizers, and to its aftereffect.[8] A loose soil is also a mandatory condition for the formation of high-class potato tubers. Tubers, which are thickened underground stem shoots, to a lesser extent than, for example, roots of root crops can push the soil particles, so on loose soils, rich in organic matter, to create the most favorable conditions for the development of potato plants.[9] The impact of various types of organic fertilizers is of interest, in particular, composts on the basis of various organic materials and also organic-mineral fertilizers for the productivity of potatoes.

In recent years, in agricultural technique sewage sludge produced at wastewater treatment plants in cities and other human settlements is becoming increasingly important. With its application, the soil receives a large amount of organic matter, a lot of macro-and micronutrients as well as plant growth stimulants. Sludge is applied to the soil in pure form, as compost with peat, wood waste and other fillers and in the form of organic-mineral fertilizers.[10] In this regard, we performed the experimental verification of the efficiency of fertilizers on the basis of urban sewage sludge when cultivating potato for seed purposes.

Given the fact that many questions of potato fertilizing are poorly investigated, the task of our research was to establish when cultivating this crop in various soil and climatic conditions action of organic and mineral fertilizers in various combinations and doses in the soil-plant system.

1.2   MATERIALS AND METHODOLOGY

Investigations were carried out in the field experiments on sod-podzolic soils in the west and northwest chernozem zone of Russia, as well as on permafrost taiga pale yellow soil in Yakutia.

We experienced the impact of mineral and different types of organic fertilizers: litter manure, chicken manure, compost from cattle manure, compost from sewage sludge and peat and also organic-mineral fertilizer based on sewage sludge on yield and quality of potatoes.

Field experience with a wide range of doses and combinations of litter manure and fertilizer for potatoes was carried out on the experimental field of Pryanishnikov All-Russian Scientific Research Institute of Agro-chemistry and Smolensk Scientific Research Institute of Agriculture, in the village of Olsha (Smolensk region). The experience was lengthy; it was laid in the year 1978. Field experiment had a factorial scheme and presented a selection of $1/27$ $(6 \times 6 \times 6 \times 6)$, included 48 variants.

The sod-podzolic light-textured loamy cultivated soil was used in the experiment. Before lying of an experiment in the 0–20 cm layer the soil contained 1.3% of humus, 110 mg/kg of mobile phosphorus (P_2O_5), and 115 mg/kg of potassium (K_2O); and pH was 5.5. As an organic fertilizer, we used cattle manure with a small amount of litter, 70% moisture, which contained 0.46% total nitrogen, 0.2% phosphorus (P_2O_5), 0.66% potassium (K_2O), and 59% organic matter. The ratio of C:N=19, and heavy metal content of the manure was low, that is, Cd—0.1, Cr—1.0, Ni—1.0, Cu—0.6, and Zn—7.0 mg/kg.

In the experiment, we studied four factors: manure litter, nitrogen, phosphorus, and potassium fertilizers. The area of the experimental plot was 112 m². The repetition of experience was done three times. Each of the four factors was represented by 6 gradations (0, 1, 2, 3, 4, and 5 of unit doses). Every single dose at potato cultivation was adopted by the following amount of fertilizer: 20 t/ha of manure and 45 kg of nitrogen, phosphorus (P_2O_5) and potassium (K_2O).

Crop rotation was grain-grass-tilled crop in the first three rotations. Alternation of crops in the first rotation (1979–1989): potatoes, barley, winter rye, oats, pea-oats mixture, winter wheat, barley, perennial grass of the first and second year use, winter rye, oats; the second (1990–1995) and third (1996–2001) rotations: potatoes, barley, perennial grass of first and second year use, winter wheat, oats.

Studies of fertilizer on the basis of sewage sludge in a field experiment were carried out in the North-West of Russia, in the Vologda region using the sod-podzolic middle loamy soil.[11] Potato cv. Elizabeth was cultivated in a link of a field rotation: fiber flax, potatoes, barley.

As a fertilizer, we applied compost on the basis of sewage sludge of the city Vologda and granulated organic-mineral fertilizer (GOMF) generated on the basis of the dewatered sewage sludge with the addition of mineral nitrogen and potash fertilizers by the technology of the firm Closed Joint Stock Company, "Twin Trading Company." Organic-mineral fertilizer was produced in the form of granules of size 14×20 mm and 84 kg NPK (complex nitrogen, phosphorus, and potassium) characterized by a neutral soil reaction and contained in 1 ton 225 kg of organic matter.

The concentration of heavy metals and arsenic in GOMF and also in other fertilizers used in this field experiment was below the permitted by State Standard of the Russian Federation R 17.4.3.07–2001.[12]

The scheme of experiment presented below in Table 1.2 consists of a control and 6 variants with fertilizers, including 3 variants with an organic system, that is, with increasing doses of compost (2, 4, and 6 t/h), a variant with mineral system (NPK, equivalent to 4 t/ha compost), variant with organic-mineral system (compost 2 t/ha + NPK, equivalent to 2 t/ha compost), and a variant with organic fertilizer at a dose of 4 t/ha.

All fertilizers were applied to the first crop of rotation link—fiber flax. The second crop of rotation link (potatoes) was cultivated on the aftereffect of fertilizers made under fiber flax.

Under the conditions of permafrost taiga pale yellow soil in Central Yakutia, we studied the effect and aftereffect of different kinds of organic fertilizer: chicken manure, cattle manure, and compost from manure and peat on the yield of potatoes in the region. For composting we used chicken manure from Yakut poultry farm at 75% moisture. Compost mixture was laid after the fieldwork in the second decade of June. Finished compost was applied to the soil in the autumn plowing.

1.3 RESULTS AND DISCUSSION

The research results in the long field experience are reflected in the regression Equations 1.1 and 1.2, and showed that the yield of potato tubers in the first and second rotations of crop rotation was largely dependent on the application of nitrogen fertilizer. In the second and third rotations of crop rotation, litter manure and potash fertilizers were of great importance in the formation of potato yield Equations 1.2 and 1.3). The yield calculation was carried out according to the following methods:[13,14]

Potatoes cv. 'Loshitsky', in 1979

$$y = 37.77 + 4.75N^{0.5} - 1.02N + 0.01M - 0.75\,(NP)^{0.5}$$
$$-1.66\,(NP)^{0.5} + 0.48\,(PK)^{0.5},\, R = 0.80 \tag{1.1}$$

Potatoes cv. 'Nevsky', in 1990

$$y = 20.07 + 13.293N^{0.5} - 3.685N + 1.705P^{0.5} + 6.41K^{0.5} + 9.182M^{0.5}$$
$$-3.554\,(MK)^{0.5},\, R = 0.93 \tag{1.2}$$

Potatoes cv. 'Nevsky', in 1996

$$y = 19.481 + 3.8K^{0.5} + 6.473M^{0.5} - 2.258 \left(NM\right)^{0.5} - 2.96 \left(KN\right)^{0.5}, R = 0.62 \quad (1.3)$$

According to Table 1.1 and Figure 1.1, the greatest impact on the productivity of a potato provided mineral fertilizer system which achieved an average 3-year yield increase of 60% as compared to control. Organic-mineral fertilizer systems also were characterized by the high efficiency. Thus, 37.97 t/ha tubers, or in addition to control 45.9%, were obtained in the form of triple doses.

TABLE 1.1   Yield and Quality of Potato Tubers Depending on Fertilizers on Sod-Podzolic Middle-Textured Loamy Soil in West of the Non-chernozemic Zone (The Average for 1979, 1990, and 1996).

Fertilizer	Yield of potato tubers (t/ha)	Yield increase		Starch content in tubers (%)	Nitrate content in tubers, raw matter (mg/kg)
		(t/ha)	(%)		
Control without fertilizers	26.03	–	–	15.2	143
N135	30.03	4.00	15.4	13.8	214
P135	27.07	1.04	4.0	14.5	Not determined
K135	31.97	5.94	22.9	13.9	146
N135 P135 K135	41.67	15.64	60.1	12.6	190
Manure, 60 t/ha	36.80	10.77	41.4	14.8	179
N45 P45 K45 + manure, 20 t/ha	36.50	10.47	40.3	14.4	170
N90 P90 K90 + manure, 40 t/ha	37.77	1.74	45.1	13.3	142
N135 P135 K135 + manure, 60 t/ha	37.97	11.94	45.9	13.1	250
N180 P180 K180 + manure, 80 t/ha	36.30	10.27	39.5	11.9	301
N225 P225 K225 + manure, 100 t/ha	34.53	8.50	32.7	11.9	282
LSD$_{05}$	—	9.74	—	—	—

LSD least significant difference

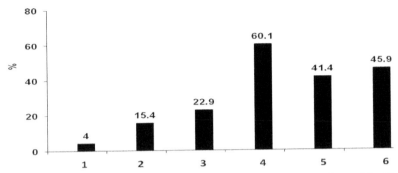

FIGURE 1.1 The potato tuber yield increases, depending on the options of fertilizers in the cultivation on the sod-podzolic light-textured loamy soil: 1—P135; 2—N135; 3—K135; 4—N135 P135 K135; 5—manure 60 t/ha; 6—N135 P135 K135 + manure 60 t/ha.

It is important to note that a further increase in doses of fertilizers in organic-mineral options was not accompanied by an increase of potato yield (Fig. 1.2).

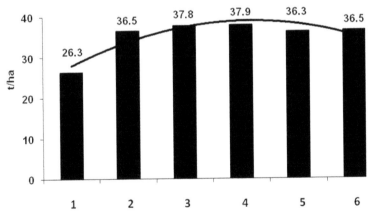

FIGURE 1.2 Effect of increasing doses of fertilizers in organic-mineral systems on the yield of potato tubers on the sod-podzolic light-textured loamy soil: 1—control; 2—N45 P45 K45 + manure 20 t/ha; 3—N90 P90 K90 + manure 40 t/ha; 4—N135 P135 K135 + manure 60 t/ta; 5—N180 P180 K180 + manure 80 t/ha; 6—N225 P225 K225 + manure 100 t/ha.

The experiment had a discernible effect of litter manure, and potato yield in this variant was 36.8 t/ha, and yield increase against the control was 10.77 t/ha, that is, it was significant, exceeding the control option on 41.4%.

With the separated application of nitrogen and potassium fertilizers, potato yield increases were unreliable—15.4 and 22.9%, respectively.

Inefficient was the use of only phosphate fertilizers, where excess tuber yield over the control was at the level of 4%.

The content of starch and nitrates in potato tubers for options experience was determined for their quality characteristics.

As a result of studies, it was found that the starch content in tubers was higher in control without fertilization. In other variants of the field experiment, there was a decrease of this indicator which when observed the higher value of the potato yield was obviously connected with growth dilution. On the starch content of tubers positive impact had manure and phosphorus and organic-mineral fertilizer at low doses, increasing the starch content to 14.4–14.8%.

The nitrate content in tubers in control was 143 mg/kg (Fig. 1.3). The use of fertilizers was usually accompanied by an increase in nitrate content in tubers, especially when applying high doses. However, exceeding the hygienic standards for this indicator, made in Russia, that is, 250 mg/kg NO_3, was observed only in cases with high 4- and 5-fold doses of fertilizers in combination with manure.

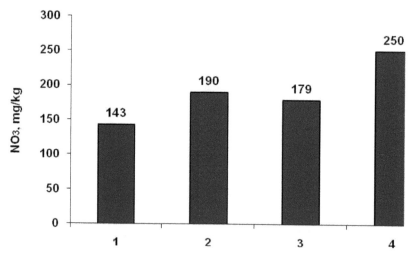

FIGURE 1.3 Nitrate content in potato tubers, depending on the options of fertilizers on the sod-podzolic light-textured loamy soil: 1—control, 2—complex nitrogen, phosphorus, and potassium (NPK), 3—manure, 4—manure + NPK.

The impact of different types of organic fertilizers on the potato productivity is of interest, primarily composts on the basis of different organic materials such as sewage sludge, as well as organic-mineral fertilizers.

In the experiment, conducted for 3 years (2011–2013) in conditions of the Vologda region, we studied the aftereffects of composts based on sewage sludge and organic-mineral fertilizer on potato yield. According to the information received, a positive aftereffect of the fertilizers on potato cv. Elizabeth was detected (Table 1.2). Significant yield increment of potato crop was obtained in all experimental variants, except mineral fertilizers. The highest yield of tubers—22.4 t/ha was when applying GOMF where significant yield increase was 3.7 t/ha, or 19.7% relative to the control.

TABLE 1.2   The Yield of Potato Tubers Depending on Fertilizers on the Sod-Podzolic Middle-Textured Soil in the Vologda Region (2011–2013).

Fertilizer	Tuber yield(t/ha)	Yield increase	
		(t/ha)	(%)
Control	18.7	–	–
Compost, 2 t/ha	19.5	0.8	4.3
Compost, 4 t/ha	19.8	1.1	5.8
Compost, 6 t/ha	21.3	2.6	3.9
Complex nitrogen, phosphorus, and potassium (NPK), equivalent to 4 t/ha of compost	19.4	0.7	3.7
Compost, 2 t/ha+NPK, equivalent to 2 t/ha of compost	20.1	1.4	7.5
GOMF, 4 t/ha	22.4	3.7	19.7
LSD_{05}	0.8	–	–

GOMF granulated organic-mineral fertilizer, LSD least significant difference

Compost aftereffect increasingly manifested itself when we used the highest dose 6 t/ha in the experiment.

The use of sewage sludge generated in waste treatment plants of cities, towns, and other urban settlements in agriculture, is related to the possible risks of pollution of the environment.by pollutants. Nevertheless, the use of sewage sludge as fertilizer should be considered as an important agricultural method in the technologies of cultivation of agricultural crops, that is, they contain large amounts of organic matter, essential nutrients (NPK) and trace elements, as well as growth promoters.

Our attention, while studying, was paid to the research of cultivated plant product quality, primarily to a change in its content of heavy metals and arsenic under the influence of fertilizers based on sewage sludge (Table 1.3).

According to the research, in the tubers of potato content of heavy metals was at a level of control in fertilizer variants of sewage sludge, or even lower than the control values, indicating the environmental safety of crop production obtained.[15]

The data in Table 1.4 show that composts based on sewage sludge and GOMF had a great impact on the agrochemical properties of the soil. In the end of the experiment, soil mobile phosphorus content increased in the variants with composts at doses 4 and 6 t/h, and when applying GOMF. At the same time, mobile potassium content in the soil decreased in all variants of the experiment, which was associated with the low supply of sewage sludge with the element, having in mind that sewage sludge was part of fertilizer used in the experiment. An exception in this respect was the only option with GOMF, in the production of which as a concomitant fertilizer were used potassium fertilizers (5% active ingredient).

The use of fertilizers based on sewage sludge did not cause pollution of sod-podzolic soil with heavy metals and arsenic. This is evidenced by data on their gross content that obtained at the end of the link rotation after harvest of the final crop—spring barley. In all variants of the experiment, there were no exceeding the maximum permissible (roughly allowable) concentrations for the sod-podzolic loamy soils, and the total value of the index of soil contamination was in the range of 0.2–1.2, which was significantly below the permissible level equal to 16.

It is important that a sufficiently high efficiency of different types of organic fertilizers—composts based on chicken manure and peat was obtained in the cultivation of potatoes in the conditions of permafrost taiga pale yellow soil in Central Yakutia.

According to Pryanishnikov All-Russian Scientific Research Institute of Agrochemistry and Safronov Yakut Research Institute of Agriculture,[15,16] the use of chicken manure and cattle manure in a pure form, as well as peat-litter compost (at a ratio of 1:1) at a dose of 300 kg/ha of nitrogen for potatoes under irrigation provided tuber yields 26.5–26.6 t/ha, that is, almost equal in both types of fertilizer and yield increase to control was 29–30% (Table 1.5).

TABLE 1.3 The Content of Heavy Metals and Arsenic in the Potato Tubers According to Fertilizers Based on Sewage Sludge, mg/kg of Dry Matter.

Fertilizer	Cu	Zn	Pb	Cd	Ni	Cr	Mn	Co	Hg	As
Control	1.4	7.1	0.4	0.03	0.21	0.15	4.2	0.08	0.01	0.02
Compost, 2 t/ha	1.2	7.5	0.3	0.03	0.16	0.12	4.4	0.05	0.01	0.03
Compost, 4 t/ha	1.2	7.5	0.3	0.02	0.22	0.14	5.0	0.04	0.01	0.02
Compost, 6 t/ha	1.2	5.8	0.3	0.02	0.09	0.16	4.6	0.04	0.01	0.02
NPK, equivalent to 4 t/ha of compost	1.4	8.2	0.3	0.02	0.18	0.18	6.5	0.03	0.01	0.02
Compost, 2 t/ha+NPK, equivalent to 2 t/ha of compost	1.1	6.6	0.4	0.02	0.20	0.17	5.5	0.05	0.01	0.02
Organic-mineral fertilizer—GOMF,4 t/ha	1.2	8.4	0.3	0.03	0.25	0.13	4.5	0.05	0.01	0.02
MPL 123-4/281–87	30.0	50.0	5.0	0.3	3.0	0.5	—	1.00	0.05	0.5
SRN 2.3.2.1078–01	—	—	0.5	0.03	—	—	—	—	0.02	0.2

MPL the minimum permissible level, SRN sanitary regulations and norms

TABLE 1.4 Agrochemical Properties of Sod-Podzolic Middle-Textured Soil Depending on Fertilizers (2012).

Fertilizer	Humus (%)	pH	P_2O_5	K_2O	Hydrolyticacidity	Ca	Mg
			(mg/kg)		mg-equiv./100 g		
Control without fertilizers	3.1	5.9	213	92	2.2	8.7	3.5
Compost, 2 t/ha	3.5	5.9	226	90	1.8	9.2	3.1
Compost, 4 t/ha	3.5	5.9	227	92	2.0	8.8	3.4
Compost, 6 t/ha	3.6	5.8	231	97	2.0	8.8	2.8
NPK, equivalent to 4 t/ha of compost	3.2	5.5	219	87	2.6	9.5	2.4
Compost, 2 t/ha + NPK, equivalent to 2 t/ha of compost	3.3	5.4	214	108	2.8	9.4	3.0
GOMF 4 t/ha	3.4	5.3	251	106	2.8	8.5	2.8

GOMF granulated organic-mineral fertilizer

TABLE 1.5 Effect of Organic Fertilizers on Yield and Quality of Potato Tubers in Central Yakutia.

Fertilizer	Yields of tubers on the average for 4 years (1988–1991) (t/ha)	Yield increase (%)	The nitrate content in tubers, NO3 (mg/kg) of raw matter	
			1st year of effect (1988)	on average over 3 years of aftereffect (1989–1991)
Control without fertilizers	20.6	–	114	108
Chicken manure (N300)	26.5	29	228	175
Cattle manure (N300)	26.5	29	182	144
Compost from manure and peat (N150)	24.9	21	135	126
Compost from manure and peat (N300)	26.6	30	156	144
Compost from manure and peat (N450)	28.2	37	177	162
LSD$_{05}$	4.6	–	–	–

Composting manure with peat allowed obtaining tubers with a lower content of nitrates both in the year of the fertilizer and during their aftereffect.

1.4 CONCLUSION

Studies have shown for successful that successful potato cultivation, in different regions of Russia, was an important scientifically grounded application of organic and mineral fertilizers by taking into account the optimization of their doses and combinations. This ensured the preservation of soil fertility, and steady productivity of potatoes reached the level of 25–36 t/ha. While cultivating potatoes we detected high efficiency of traditional organic fertilizers: manure, chicken manure and also fertilizers on the basis of sewage sludge—compost and GOMF.

It was proved that in the technologies of potato cultivation in different regions of Russia that cultivation should be done combined with the application of organic and mineral fertilizers, in which optimization of combinations and doses had a positive influence on the formation of tuber yield and potato quality.

In the sod-podzolic light-textured loamy soil in the west chernozem zone of Russia when applying organic-mineral system, consisting of 20 t/ha of manure and 45 kg/ha of active ingredient of mineral nitrogen, phosphate and potash fertilizers, we achieved stable potato productivity at the level of 36.6 t/ha with a starch content 14.4%, which was 40% higher than the control without fertilization. The use of variants of fertilizers not balanced nutritionally, that is, nitrogen alone, phosphorus alone, potash alone fertilizers, did not provide a reliable yielding increase of potato tubers, while the application of only manure at 60 t/ha was accompanied by a significantly increased yield (10.77 t/ha, or 41.4%), where yield amounted to 36.8 t/ha in an average of years of research. Thus, tubers contained 14.8% of starch at low amounts of nitrate (179 mg/kg fresh weight).

We discovered the high efficiency of fertilizer for potatoes based on sewage sludge—compost and GOMF. At the sod-podzolic middle-textured loamy soil in the northwest (Vologda region) in a link of crop rotation: fiber flax, potatoes, barley—from GOMFat dose of 4 t/ha, we obtained in the aftereffect 22.4 t/ha of potato tubers, which was 20% greater than control. The content of heavy metals, arsenic, and nitrates in potato tubers in this variant was at the level of control values, indicating the environmental safety of plant products.

In the northeast region, Yakutia in the permafrost taiga pale yellow soil, it is advisable to use chicken manure and cattle manure in a pure form for potatoes, as well as peat-litter compost at a dose of 300 kg/ha of nitrogen, which provides the yield of tubers 26.5–26.6 t/ha under irrigation, that is, at 29–30% higher than the control without fertilizers.

KEYWORDS

- fertilizers
- nitrates
- heavy metals
- manure

REFERENCES

1. Agroindustrial Complex Russia in 2012. The Printers Rosinformagroteh: Moscow, 2013; p 604 (In Russian).
2. Derzhavin, L. M.; Merzlaya, G. E.; Khaidukov, K. P. *Integrated Use of Organic and Mineral Fertilizers in the Resource-Saving Agro-Technologies of Potato Production;* Pryanishnikov All-Russian Scientific Research Institute of Agrochemistry: Moscow, 2015; p 376 (In Russian).
3. Kidin, V. V. *Fertilizer System*; Publishing House Russian State Agrarian University, Moscow Agricultural Academy: Moscow, 2012; p 534 (in Russian).
4. Prianishnikov, D. N. *Selected Works. Private Farming (Field Crop Plants).* Kolos (Ear): Moscow, 1965, Vol. II; p 708 (In Russian).
5. Cook, D. U. *Fertilizer Systems for Maximum Yields.* [Trans. from English. N.V. Gadelia, E.I. Shkonde, Ed.] Kolos ("Ear" in Eng.): Moscow, 1975; p 416 (In Russian).
6. Methodological Guide for the Design of Fertilizer Application in Technologies of Adaptive-Landscape Agriculture. Ivanov, A. L., Derzhavin, L. M., Eds.; The Ministry of Agriculture of the Russian Federation, Academy of Agricultural Sciences: Moscow, 2008; p 392 (In Russian).
7. Fedotova, L. S. Efficiency of Fertilizers in Intensive Crop Rotation with Potatoes. Thesis abstract for the degree of DSc of agriculturic, Moscow, 2003; p 51 (In Russian).
8. Lukin, S. M.; Eskov, V. I. The Duration of Effect of Organic Fertilizers. *Fertility* **2004,** *1,* 15–17 (in Russian).
9. Iakushkin, I. V. Crop. Selkhozgiz: Moscow, 1947; p 680 (In Russian).

10. Strategy for the Use of Sewage Sludge and Compost on Their Base in the Agricultural Technique. Milaschenko, N. Z., Ed.; Pryanishnikov All-Russian Scientific Research Institute of Agrochemistry, Agroconsult: Moscow, 2002; p 140 (In Russian).

11. Vlasova, O. A. Agroecological Efficiency of Composts Application Based on Sewage Sludge on Sod-Podzolic Soil Under North-West of Chernozemic Zone. Thesis abstract for the degree of PhD of agriculture, Moscow, 2014 25p (In Russian).

12. State Standard of the Russian Federation R 17.4.3.07–2001 (2001). Protection of Nature. Soils. The Properties of Sludge Requirements When They are Used as Fertilizers. State Standard of Russia: Moscow, 2001 (In Russian).

13. Lakin, G. F. Biometrics. *Textbook for Universities and Pedagogical Universities;* Higher School: Moscow, 1973; p 343 (In Russian).

14. Peregudov, V. N. Planning of Multifactorial Field Experiments with Fertilizers and Mathematical Processing of Their Results. Kolos: Moscow, 1978; p 182 (In Russian).

15. Sanitary Regulations and Norms. 2.3.2.1078–01. Hygienic Requirements for Safety and Nutritional Value of Food Products. www.dioxin.ru/doc/sanpin2.3.2.1078–01.pdf (in Russian).

16. Stepanov, A. I.; Okhlopkova, P. P.; Merzlaya, G. E.; Gavrilova, V. A. Manufacturing Peat-Litter Compost and Their Use for the Potatoes in the Conditions of Yakutia. Recommendation. Siberian Branch of Russian Academy of Apicultural Sciences: Novosibirsk, 1993; p 16 (In Russian).

17. Stepanov, A. I. Agroecological Bases of Production and Use of Organic Fertilizers on Frozen Soils of Yakutia. Thesis abstract for the degree of DSc of agriculture, Moscow, 2016; p 50 (In Russian).

CHAPTER 2

INNOVATIONS IN THE SYSTEM OF ACCELERATED PROPAGATION OF POTATO MINITUBERS

OLEG S. KHUTINAEV[*], BORIS V. ANISIMOV, and NIZAM E. SHABANOV

A. G. Lorch All-Russian Research Institute of Potato Farming, 23, Lorch St., Settlement Kraskovo, Lyubertsy District, Moscow Region, Russia 140051

[*]*Corresponding author. E-mail: okosk@mail.ru*

CONTENTS

ABSTRACT

The chapter discusses a method of accelerated reproduction of potato minitubers on the aerohydroponic module in an open area in the natural conditions of the environment. The results showed characteristics of plant development and tuber formation. We studied quantitative and qualitative characteristics of the harvest of minitubers obtained by using the aerohydroponic method of growing in natural conditions of the environment.

The average quantity of standard minitubers counting per plant was 57 pieces. The total quantity obtained from 60 plants amounted to 3467. Less than 1% of minitubers were fraction from 40 mm and above, more than 7% of minitubers were fraction from 30 to 40 mm, more than 75% of minitubers were optimal fraction from 20 to 30 mm, more than 9% of minitubers were fraction from 15 to 20 mm, and approximately 7% of minitubers were fraction from 10 to 15 mm. Minitubers of a size less than 10 mm were not considered.

2.1 INTRODUCTION

With an annual cultivation of potatoes, there is a possibility of the gradual accumulation of viral and bacterial diseases.[1] This leads to decrease in the quality of minitubers, despite compliance with all safety measures for conservation plantings from pests and diseases. To preserve the varietal characteristics, specialized companies are taking steps to improve methods of producing minitubers on the basis of biotechnological methods of reproduction.

For these purposes, a few cells from potato sprout are isolated and planted in a nutrient medium in test tubes or containers under sterile conditions, which then grow plants that are free from viruses and diseases.[2]

Once they have reached a sufficient development, they cut into cuttings and seated into new tubes containing nutrient substrate. This operation is performed several times to get as much of the original in vitro microplants. Upon receipt of the required quantity of original plants, they are planted in greenhouses to get a healthy generation of original minitubers, free of viruses and diseases. From minitubers planted in the field, receive the harvest of the first generation, next year—super-super elite, super elite, elite, the first and second reproduction. The super-super elite is the seeded (or planting) material obtained from the first field generation grown from the microtubers. Super elite is the material obtained from the super-super elite and corresponds to the modern notion of certified seeds. Elite is the seeded material obtained from the super elite and corresponds to the concept of basic seeds.

Currently, the method for year-round growing of minitubers based on traditional technologies in greenhouses is used. It requires the high financial costs of climate maintenance and lighting. Currently, it significantly limits widespread use of this method for the reproduction of original potato minitubers. In addition, the number of tubers usually is eight units on average per plant.

Modern innovations in the system of micropropagation have allowed to significantly improving the methods of producing in vitro microtubers[3] and effectively using them for the cultivation of minitubers in indoors of various types and designs.[4] Improving methods of vegetative propagation by increasing the density of crops and the use of different substrates allowed to significantly increase the number of minitubers per unit area.[5] The cultivation of potato plants in the biocontainer allows getting with one microplant up to two in vitro micro-tubers of sizes ranging from 5 to 10 mm.[6]

On the basis of the technology of production in greenhouses, about 80% minitubers are produced. However, in recent years, interest in the use of advanced technologies has significantly increased, based on the use of hydroponic[7] and aeroponic[8] methods of minitubers producing.

These technologies are becoming increasingly popular, especially for accelerated breeding of new and scarce varieties. Research in the field of aeroponic technologies proved a definite advantage over traditional technology in terms of cost savings and increased quantitative and quality indicators of the final product. Aeroponic technology for the production of minitubers allows to increase the productivity up to 10 times higher[9] than with the conventional method, while significantly reducing the degree of contamination due to the lack of soil substrates. While comparing aeroponic and hydroponic cultivation methods, it was revealed that aeroponic method allows you to obtain more than 2.5 times more minitubers per plant than by hydroponic method. [10]

The aim of our researches was to optimize the conditions for the formation of vegetative mass and intensive development of root and tuber. The use of aerohydroponic method of growing minitubers greatly simplifies the production process and gives the potential to reduce costs and substantially reduce the cost of the final product. The use of the aerohy-droponic method eliminates the risk of crop losses during a power emergency shutdown in relationship with the introduction of passive systems of feeding plants.

On the basis of above method, an inexpensive and effective technology for the production of high-quality minitubers was developed that reduces production costs and simplifies the technological process of minituber production, which is an urgent task of scientific research.

2.2 MATERIALS AND METHODOLOGY

In the studies, the method of aerohydroponic growing of minitubers on the aerohydroponic module is used with a pump of 100 W, 12 V. The module was located in an open area in the natural growing conditions. To protect plants from insects, the module was covered with a special covering material (lutrasil). Lutrasil is a protective fibrous material for covering the vegetable bed. The module is equipped with two different feeding systems, which operate independently from each other. The main system of roots nutrition is presented in the aeroponic part of the module, which performs the functions of active feeding plants. The auxiliary system of the root nutrition is presented in the hydroponic part of the module, which performs the functions of passive nutrition of plants. This system is used for continuous feeding roots and does not require electrical equipment. The passive (hydroponic) system of plant nutrition allows eliminating the risk of the plant's death if in the event of an accident, aeroponic (active) system of plant nutrition shuts down.

The module is a special hydroponic container enclosed in a metal frame, inside of which are mounted aeroponic (active) and hydroponic (passive) supply systems of nutrient solution (Fig. 2.1).

Hydroponic part of the module consists of water tank *1* with a displacement of 150 l in which the wetting of lower parts of plants occurs. The water tank is used to store the nutrient solution and is a part of the hydroponic system of the module. It is equipped with filters and valves for draining. In case of connecting two or more modules into one integrated block, each of the water tanks is connected through these valves. The aeroponic system of feeding is located under transplanter cover *6* and consists of a pipe with a finely divided built-in sprinkler *7*, which is connected to the water pump through filter. While spraying the plant's roots, there is an active saturation of the nutrient solution with oxygen from the air. In addition, the liquid after wetting the roots is drained in the water tank, where it mixes with nutrient solution, and prepared for further recycling. The liquid in the water tank is constantly circulated; therefore, it does not reach stagnation of the nutrient solution. To provide an easily accessible monitoring of the root system, the module is equipped with doors 4. All modules are made in such a way so as to be able to connect several modules in one complex block. The module's block can be used when it is necessary to multiply large quantities of minitubers of the same variety. Feeding system of each module can be easily interfaced with feeding systems of other modules and can form the unified system of feeding plants.

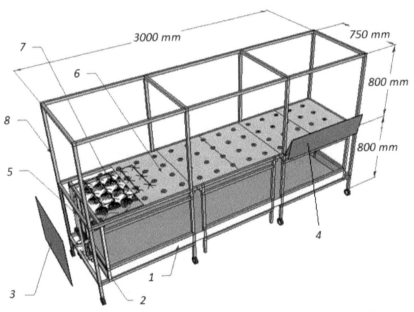

FIGURE 2.1   The universal aerohydroponic module. 1—water tank; 2—filter; 3—side cover; 4—door; 5—pipes with sprinklers; 6—transplanter cover; 7—sprinklers; and 8—rack.

The water tank is covered with transplanting panel with cells for planting out plants, which is placed according to the scheme 190 × 190 mm. The total area of the module for the landing is 2.28 m². The racks 8 of frame serve for fastening the retaining meshes for plants, which prevents lodging of the stems. Timers that are configured on the relevant regimes control the consumption of the nutrient solution. Measurement of acidity (pH) and electrical conductivity (EC) of the nutrient solution are held two times a week. The acidity of the solution is maintained from 5.5 to 6.5. The EC of the nutrient solution is maintained at a level of 0.7–1.3, depending on the phase of growth and development during the growing season. The replacement nutrients solution was performed with the change of the phenological phases of growth and development of plants. Replacement of the nutrient solution is carried out with the change of the phenological phases of growth and development of plants. For fertilizing plants, three different compositions of the nutrient solution applied at different stages of cultivation are used.

2.3   RESULTS AND DISCUSSION

The main features of the technological process of production of minitubers by using the aerohydroponic module are as follows:

- Nutrition of microplants on the device is accomplished simultaneously while operating active and passive methods of feeding.
- Increasing the density of plants per unit area can significantly increase the growth of minitubers per square meter.
- This technology allows us to hold events to initiate and stimulate the processes of growth at different stages of growth and development of plants and to use a differentiated method of harvest by visual control of tuberization.
- The module has a device for fixing of plants, which helps in keeping the plants upright during ontogenesis.
- The module is compact and mobile, and can work in any environment under natural or artificial lighting.
- The modules can be gathered in groups and work as one unit.
- The modules can be equipped with lamps to illuminate plants for growing indoors.
- Retrofitting of the module is provided by its own power supply (solar panels) for the implementation of the method in offline mode, in any terrain conditions.
- The developed method allows to simplify the technology of growing minitubers, reduce the cost of the technological process and significantly reduce the cost of the final product.

To study the effectiveness of aerohydroponic method of growing, one of the most promising varieties was chosen, which is popular and widely used in potato cultivation in different regions of Russia. The choice was made in favor of 'Zhukovskij rannij', which gives highest yields, very early matures, and enjoys great demand among the population. The offered module was placed in an open area under a slight shelter from lutrasil. When the plants grow up to 10–15 cm, the restraint mesh is installed on top of them (Fig. 2.2).

FIGURE 2.2 The initial development of plants in the aerohydroponic module.

Such retaining mesh is installed up to three times with the development of plants. Dense placement of plants on the aerohydroponic device does not affect the development of plants, if you provide sufficient feeding and good illumination (Fig. 2.3).

During the research, it was established that the use of aerohydroponic method of growing has been given with one plant over 50 minitubers. In cultivation process of potatoes, natural lighting conditions are used and supply method of the nutrient solution is differentiated. The spacious enough chamber for the root system provides full visual monitoring and easy access to the root system, and it allows to avoid the risk of damage to the roots while multiple harvesting of minitubers.

FIGURE 2.3   The 45-day development of potato plants on the aerohydroponic device.[11]

During operation on the aerohydroponic devices, the cycle of events starts with transplanting of microplants directly on the module without prior adaptation. Before planting, the plants are carefully washed from the remnants of agar-agar, to prevent it from falling into the aeroponic system.[11] At the initial stage of plant development, special attention is paid for the creation of conditions for the development of the large number of stolons (Fig. 2.4).

Successful carrying of this event directly affects the increase in the number of minitubers grown from one plant. Stimulation of the process formation of stolons directly affects the increase in the number of minitubers grown from a single plant. After beginning the process of the tuber formation, the basic operations on the care of plants are directed to increase the process of tuberization (Fig. 2.5).

FIGURE 2.4 The formation of stolons in the aerohydroponic module.

FIGURE 2.5 The process of tuberization on the aerohydroponic module.

The main part of minitubers obtained on the aerohydroponic module was the tubers of size from 15 to 30 mm (Fig. 2.6).

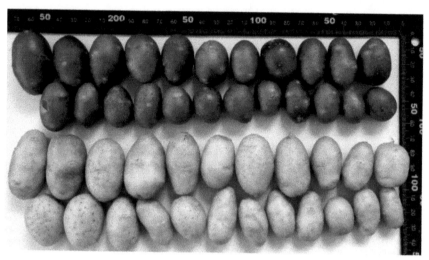

FIGURE 2.6 Size fractions of minitubers (mm), grown in the aerohydroponic module. Cultivars 'Red Scarlet' (above) and 'Innovator' (below).

For each phase of plant development, the different compositions of nutrient solutions were used, which best met the requirements of the technological process for the production of minitubers at the different stages of plant development. The first phase of plants development starts from planting until the beginning of budding and lasts up to 30 days. The second phase of plants development starts from the beginning of budding to the beginning of tuber formation. The third phase of plants development lasts from beginning of tuberization to the end of the growing season.

For the first phase of plant development, the nutrient solution with the following of nutrients content (mg/l) was used: the contents of nutrient elements in the starting nutrient solution were: N 195, P 45, K 185, Ca 170, Mg 35, pH 5.8–6.0, and the water EC 0.8; in the second phase of plants development, the nutrient solution for stimulating process of tuberization with following nutrients content was used: N 150, P 45, K 270, Ca 150, Mg 35, pH 5.8–6.0, and EC 0.7; for the third phase of plant development, the solution with nutrients content N 70, P 45, K 200, Ca 60, Mg 25, pH 5.8–6.0, and EC 1.2 was applied; the content of salts in solution (mg/l) was Fe–EDTA 0.8, B 0.5, Mn 0.5, Zn 0.1, Cu 0.05, I 0.63, Co 0.006, and Mo 0.1. The monitoring and adjustment of the acidity of solution (pH) and the water EC were carried out every 2–3 days. A complete change of the solution was carried out once a month.

The feeding mode of the nutrient solution at day and night time of the growing season was carried out according to the scheme explained in subsequent text.

The first irrigation mode was lasted for 30 days from 6:00 to 22:00 (1 min work and 9 min break) and at night from 22:00 to 6:00 (1 min work and 29 min break). Within 30 days, the pump worked for 56 h. Total energy consumption amounted to 5.6 kW. The second irrigation mode was lasted for 30 days from 6:00 to 22:00 (1 min work and 19 min break) and at night from 22:00 to 6:00 (1 min work and 29 min break). Within 30 days, the pump worked for 33 h. Total electricity energy consumption amounted to 3.3 kW. The third irrigation mode was lasted for 30 days from 6:00 to 22:00 (1 min work 29 min break) and at night from 22:00 to 6:00 (1 min work and 59 min break). Within 30 days, the pump worked for 18 h. Total energy consumption amounted to 1.8 kW.

The first month used the starting nutrient solution. After 1 month of use, the solution was replaced, and stimulation of plants in the next 2 weeks was carried out. After that, before the end of the growing season, the third nutrient solution was applied. At the stage of tuber formation, the process of harvesting minitubers was held by the differential method, which means a collection of minitubers over several stages during the process of tuber formation. Minitubers were harvested when they grow to the size 25–28 mm in diameter. The interval between the picking of minitubers was 7 days. To prevent diseases of minitubers after harvesting, the tubers were treated with 0.1% sodium hypochlorite solution followed by rinsing in water. The harvest minitubers were dried at high humidity for 3 days, and then the tubers were greened at room temperature for 3–5 days. The further operation to prepare for long-term storage was carried out according to the traditional technology of storage.

Research data showed that 1 m^2 of the useful area can get more than 1500 pieces of minitubers. Of the 60 plants planted in the area of 2.28 m^2, 3467 pieces of minitubers were obtained. An average quantity of minitubers per plant is 57. Of them, the minitubers of a size from 20 to 30 mm in diameter accounted for more than 75%; the minitubers of a large fraction (>30 mm) was approximately 7%; and the minitubers from 15 to 20 mm was about 9%. The fraction of small tubers from 10 to 15 mm did not exceed 7%. Tubers of a size less than 10 mm in the calculation are not accepted (Table 2.1).

TABLE 2.1 Quantitative Yield of Various Fractions of Minitubers Grown in the Aerohydroponic Module, cultivar 'Zhukovskij rannij'. The Total Number of Harvested Tubers was 3467.[11]

Fraction (mm)	Average weight of a tuber (g)	Quantity (pcs)	Percentage (%)
10–15	1.3–3	248	7.15
15–20	3–7.5	334	9.63
20–25	7.5–12	1543	44.50
25–28	12–14	724	20.88
28–30	14–25	354	10.21
30–35	25–30	197	5.68
Less than 35 mm	More than 30 g	67	1.94

Within 90 days of the growing season for the production of 3467 minitubers, the total consumption of energy was 10.7 kW. The energy consumption for receiving one minituber amounted to 3.08 W.

Along with the above method of producing minitubers on the aero-hydroponic module, the method of producing minitubers, obtained by repeated cuttings, was studied, which is also carried on the aerohydroponic module. The method consists of the cutting of the original plants followed by planting cuttings directly on the module. Further, when they grow up, they are also cut, and the newly obtained cuttings are also planted immediately on the module. During the period of cultivation, it can be managed to hold up to four cutting. With the development of plants, physiological maturity occurs and there comes a time when they begin to form minitubers (Fig. 2.7).

The growing of plants by the method of repeated graftage makes it possible to place on 1 m² of the usable area up to 675 cuttings, which are capable of producing minitubers in size from 0.6 to 20 mm (Fig. 2.8).

The overall yield of minitubers grown by the method of repeated cuttings can reach 877 and more pieces from 1 m² (Table 2.2).

FIGURE 2.7   The formation of minitubers in potato miniplants, grown by the method of repeated cuttings on the aerohydroponic module. 'Zhukovskij rannij'.

FIGURE 2.8   The process of producing of minitubers by the method of repeated cuttings on the aerohydroponic module. 'Zhukovskij rannij'.

TABLE 2.2 The Structure of the Harvest of Minitubers Obtained from 1 m², Grown by the Method of Repeated Cuttings in the Aerohydroponic Module, Cultivar 'Zhukovskij rannij'.

Number of planted cuttings (pcs) per square meter	Total number of harvested minitubers (pcs) per square meter	Distribution of minitubers by fractions (mm)			Weight of a tuber (g)
		0.5–10	10–15	15–20	
675	877	5%	75%	20%	From 0.6 to 7.5

While growing minitubers, the method of repeated cuttings is used; the bulk of the tuber has a size of from 10 to 15 mm and is 75% of the total quantity of minitubers (Fig. 2.9).

FIGURE 2.9 Minitubers obtained from potato miniplants, grown by the method of repeated cuttings on the aerohydroponic module. Cultivar 'Zhukovskij rannij'.

On the basis of research results, it can be concluded that aerohydroponic growing of minitubers in natural conditions of the environment has advantages in comparison with the traditional method of producing of minitubers under greenhouse conditions.

The aerohydroponic method of plant growing in natural conditions of the environment from spring to early autumn allows to avoid high-energy costs required under artificial lighting in indoors or when the autumn-winter growing in greenhouses.

The selection of the optimal nutrient solution for the plant's growth on the different phenological phases in indoors is also a very promising direction of investigations to create the most favorable conditions for the cultivation of minitubers.

The combination of active (aeroponic) and passive (hydroponic) methods allows eliminating the risk of plant doom in the case of emergency shutdown of the pump since the passive feeding of plants occurs all the time.

ACKNOWLEDGMENT

The author expresses his sincere gratitude to Sarra A. Bekuzarova, professor of Gorsky State Agrarian University, for the initiation of this chapter and the full support in its preparation.

KEYWORDS

- aerohydroponic device
- nutrient solution
- virus-free potato tubers

REFERENCES

1. Simakov, E. A.; Anisimov, B. V.; Jurlova, S. M.; Uskov, A. I.; Zeyruk, V. N.; Chugunov, V. S.; Mityushkin, A. V.; Khutinaev, O. S. The Technological Process of

Production of Original, Elite and Reproduction Seed Potatoes. All-Russian Research Institute of Potato RAAS: Moscow, 2011; p 32 (In Russian).

2. Anisimov, B. V.; Smolegovets, D. V.; Shatilova, O. N. Recommendations on Technology of Cultivation in vitro Micro-Tubers and Their Use in Original Seed. All-Russian Research Institute of Potato RAAS: Moscow, 2009; p 21 (In Russian).

3. Oves, E. V.; Kolesova, O. S.; Fenina, N. A. Cultivation in vitro Micro-Tubers with the Use of Container Technology. Status and Prospects of Development. *Materials of the VI Interregional Scientific-Practical Conference*, Cheboksary, 2014; 111–115 (In Russian).

4. Ranalli, P. Innovative Propagation Methods in Seed Tuber Multiplication Programs. *Potato Res.* **1997,** *40,* 439–453.

5. Wiersema, S. G.; Cabello, R.; Tovar, P.; Dodds, J. H. Rapid Seed Multiplication by Planting into Beds Microtubers and in vitro Plants. *Potato Res.* **1987,** *30,* 117–120.

6. FAO. The Potato Micropropagation Armstatehydromet and Ministry of Agriculture Yerevan, March 2012 Rolot Jean-Louis, 2012. www.cra.wallonie.be

7. Boersig, M. R.; Wagner, S. A. Hydroponic System for Production of Seed Tubers. *Am. J. Potato Res.* **1988,** *65,* 470–471.

8. Farran, I.; Mingo-Castel, A. M. Potato Minituber Production Using Aeroponics: Effect of Plant Density and Harvesting Intervals. *Am. J. Potato Res.* **2006,** *83*(1), 47–53.

9. Otazu, V. *Manual on Quality Seed Potato Production Aeroponics;* International potato Centre (CIP): Lima. Peru, 2010; p 44.

10. Ritter, E.; Angulo, B.; Riga, P.; Ritter, E.; Relloso, J.; San Jose, M. Comparison of Hydroponic and Aeroponic Cultivation Systems for the Production of Potato Minitubers. *Potato Res.* **2001,** *44*(2), 127–135.

11. Khutinaev, O. S.; Anisimov, B. V.; Yurlova, S. M.; Meleshin, A. A. Mini-Tubers by Aero- and Hydroponic Method. *Potato Veg.* **2016,** *11,* 12–14 (In Russian).

CHAPTER 3

SNOW TECHNOLOGY OF POTATO CULTIVATION

SARRA A. BEKUZAROVA*, ZAREMA A. BOLIEVA,
SOLTAN S. BASIEV, and ALINA S. BASIEVA

Gorsky State Agrarian University, 37, Kirov St., Vladikavkaz, the Republic of North Ossetia-Alania, Russia 362040

*Corresponding author. E-mail: bekos37@mail.ru

CONTENTS

ABSTRACT

"Snow technology" of cultivation potato lies in the fact that the tubers are dipped in the snow. For this, tubers were covered to a depth of 1–2 cm layer of snow in a light room at a temperature of 10–12°C for 1.5–2 months before planting. Such measures are carried out every 7–8 days. This method avoids disease of the seed potatoes and damage of plants by Colorado potato beetle.

3.1   INTRODUCTION

New strategic principles of agriculture development must be economically justified, environmentally safe, and socially acceptable in the short and long term. The practical realization of these principles requires, first and foremost, the effective use of "gratuitous" forces of nature and resources recovery that properly corresponds to the essence of plant growing. Fundamentally different relation to resource and energy supply of productive and environmental processes in the agrophytocenosis is the essence of one of the main differences between adaptive-intensive and predominantly technologically intensive strategies of agriculture intensification. The use of natural, environmentally safe resources is in dependence of time.

It is widely known application of natural sources of mineral and thermal waters.[1] Zeolite clays[2] provide high yields by reducing the plant's diseases and increasing quality indicators.

These approaches are applied to cereals and forage crops, potato tubers, and fruit plants.[3]

Natural zeolites are used for treating the seeds for sowing.[4] Some kilograms of zeolite clay per 1 ha is sufficient to increase the ability of nitrogen-fixing bacteria and the amount of biological nitrogen in the root zone.

These nontraditional sources of raw materials increases soil fertility and rehabilitate degraded lands. As an illustration can be used that shows the progress in the agrophysical soil properties, its water, and nutritive regimes on the basis of natural bischofite.[5] One of the most effective means of biological correction of the crop production is the surface treatment of crops seeds with solutions of humic substances. Their effect facilitates the movement of nutrients in plants, intensification of photosynthesis, better works when introduced into the soil fertilizers.[6]

The modern technological processes deal with issues of using natural methods to increase plant growth and development. The pre-sowing seed treatments are thermal, [7,8] electromagnetic, [9] microwaves exposure, [10] laser exposure [11] ultraviolet light [12,13] irradiations, and exposure to alternating electromagnetic field. [14]

There are a number of works relating to chemical effect on seeds germination, growth, and seedlings development. [15] These are humic preparations, plant extracts, zeolite clays, mineral waters, and others that increase the energy of germination. The seed hardening and frost resistance increase occur at the same time. Such plants blossom and produce crops earlier. The seeds are exposed to low temperatures (-1 to $-3°C$ for 2–5 days in cold rooms).

The biology of technologies helps to reduce the dependence of agro-ecosystems on uncontrolled environmental factors (frosts, light frosts, droughts, hot winds, and so forth), to improve the quality of agricultural products, reduce anthropogenic energy costs, transportation, storage, and processing of agricultural products.

However, the known methods do not provide plant hardening and their resistance to sudden change in the environmental factors, especially low temperatures.

For example, early-planted potato tubers are exposed to light frosts and when the temperature falls from -3 to $-5°C$, germinated plants die. It shows that the resistance to low temperatures can be formed at the appropriate hardening using biostimulants of natural raw materials. [16]

During pre-seeding treatment I cryoprotectants—substances that increase plants resistance to cold, for example, hydrogen peroxide, selenium, and others are used. [17] Additional means used to increase the plants adaptability increase the production cost.

Natural impact sources, such as snow and ice, on biological objects are widely used in agriculture and medicine. [18] To improve the adaptability of potato tubers to cold spring period, it is necessary to carry out "hardening" when the planting material in containers is exposed to the conditions of negative temperatures in certain time intervals. [19]

Elements of snow technology are developed and protected by patents for inventions by the authors and their workers. In particular, "snow technology" suggests fourfold or fivefold treatment with snow spaced 5–8 days apart combined with nutrient supply at low temperatures and tubers planting to 5–6 cm depth. At this potato planting is done by hand, mixture introduced with the snow micronutrients significantly increase costs. During the process of mechanical planting, the sprouts are being destroyed, which reduces the method efficiency.[20] We propose a new technology that differs from the known ones.

In this technology, tubers, cereals and legumes, and fruit seedlings of early spring are treated with the mixture of natural sources of raw materials, and to strengthen the appeared sprouts, they are treated with nutrient solutions prepared from wastes of the food industry (starch production, distillery stillage, and so forth). These components ensure the safety of sprouts at low temperatures in early spring. The formation of shell on the sprouts ensures the durability of seeds allowing mechanized planting. The method reduces the production costs due to recycling wastes of food and distillery industry.

The attempts to use wastes of food industry were made earlier during seed treatment.[21] However, in the known experiments, there was no seed hardening that reduced the effectiveness of the proposed methods.

A very important direction is to obtain plants adaptiveness to cold conditions, providing the improvement of their immune system, quality indicators (increase in the content of starch, sugars, protein), and the rise of the seeds standard features as well. The preservation of planting material is one of the main cost-effective methods of snow technology. The method reduces the growing season by 30 days and for intents and purposes to refuse all herbicides, pesticides, and tedious Colorado beetle control.

The analysis of known technologies concludes that the integrated use of snow, with the addition of natural sources of raw materials having sufficient content of macro and micronutrients provide high-quality yield.

Linear chemical–technological approach to intensification of plant-growing ignoring concomitant factors, is fraught obviously with the danger of total catastrophe for humanity disruption of the biosphere ecological balance.

For this reason, in recent years, more attention is paid to the environment where living organisms, including plants, are found. The attention of scientists to industrial factors in the environment is quite grounded because when you save existing objects of the industrial emissions already in a relatively short time (100–200 years), a problem of preserving life on our planet will be acute.

In connection with global environmental problems, it is important to search for natural, low-cost resources that stimulate the growth and development of plants and reduce their morbidity.

The successes achieved in recent years in the study of low-temperature physics contributed significantly to the wide use of low temperature in biology and medicine. The result was a new science Cryobiology.

The cryobiological research ongoing in Russia and abroad, testify to the relevance of this program to solve some most important economic problems.[22]

Significant progress has been made in studying the cold resistance of plants. It is established that for most crops, low temperatures close to zero and even below are not harmful.[23] To characterize cold resistance of plants, it used the concept of the temperature minimum at which plant growth stops. These studies concentrate on vegetative plants, where under the influence of cold destructive processes in biomembranes occur.

The other thing is when plant material is exposed to the cold prior to the beginning of the vegetative processes. Research in the field of cryo-phytobiology is intended to creating low-temperature genomes of germ and somatic cells of plants that are under threat.

Requirements for the conservation of rare and endangered species are comprehensively formulated and summarized in the Red Book of International Union for Conservation of Nature.[19] The issues of low temperatures impact on plants are still empirical.

The main achievement of cryophytobiology in recent years can be considered the fact—finding of the hydration influence on the functional activity of organisms and the resistance of plants to low temperatures.[24] Many authors[25,26] initiated the study of the mechanism of temperatures impact on plant objects.

To analyze the mechanism of ultralow temperatures impact on the plant, special attention is paid to the cooperative nonspecific adaptation and reactivity of the organism. Plants hardening with various methods are significant to increase the resistance to the effect of extreme factors and stresses.[27] Possibly, plants that have been intensively hardened can be resistant to low temperatures.

It is impossible that the origin of processes involving resistance to low temperatures (cryoresistance) only by adapting to slight negative temperatures under natural conditions. Only slow temperature fall allows increasing the tissue resistance to such an extent that they have the ability to withstand the snow layer effect. Our proposed method of batch treatment of potato tubers and sprouts with snow layer prepares seed material for adaptation to low temperatures of cultivation.

Modern research on the plant adaptation suggests that seeds, tubers, and fruit crops can successfully survive a short period in low temperatures.

The batch treatment of potato seeds with 2–3 cm snow layer accelerates tubers' vernalization. Additionally, zeolite clays irlits that have the ability to retain moisture for a long time were added while treating potato tubers with snow. Under the combined effect of snow and irlits, there is total influence of abiotic factors, causing hydrolysis, degradation of oxidation intensity, contributing to the reduction of the content of nitrate and nitrite compounds in tubers.

3.2   MATERIAL AND METHODOLOGY

The survey target in the given work was zoned potato cultivars (Volzhanin, Yubileyny Ossetiya, Vladikavkazsky). The experiments were started in mid-January. The tubers were laid out in a single layer in a light greenhouse room at the temperature above 10–12°C. After 10–15 days (after seedling emergence), tubers were put over 1–2 cm snow layer. After every 5–6 days, seeds were treated with snow. While preparing snow mass nutrients were injected, mainly food processing wastes: distillery stillage, corn steep extract, as well as zeolite clay of local (North Ossetian) field—leskenit.

Leskenit is alkaline (pH of 8.46) in nature and has a high content of calcium (30–35%), macro and microelements in plant-available forms, as well as iodine that is absent in other species. During last treatment of planting material, leskenit (1–2 kg per 10 l of water) is introduced into meltwater. Simultaneously, 0.01% solution of para-aminobenzoic acid (PABA) was mixed with leskenit as a growth-promoting factor.

3.3   RESULTS AND DISCUSSION

The results show that the used technological scheme of potato tubers preplanting treatment can contribute to a significant increase in potato yield (Table 3.1).

Data in Table 3.1 show that the best variant is processing with snow + leskenit + PABA. The yield thus amounted to 28.2 t/ha that is above all other variants by 1.6 and 8.4 t/ha. This increase in yield is due to lower incidence of late blight of potato plants at 10–24%, high adaptation to early spring cold temperature extremes.

TABLE 3.1 The Influence of Snow Technologies on Potato Yield and Tubers Affection by Buckeye Rot (Cultivar Volzhanin).

Method of treatment	Sprouts length (cm)	Potato yield (t/ha)	Sprouts affection with buckeye rot
Treatment by snow without the use of stimulants—control	10.2	25. 0	23
Treatment by snow + PABA	9.8	19. 8	36
Treatment by snow + leskenit + PABA	13.2	28. 2	12
Processing of snow + leskenit	11.6	26. 6	22

This treatment method has contributed not only to an increase in the harvest but also reduce the growing season all the studied cultivars (Table 3.2).

TABLE 3.2 The Phenological Phases of Development of Cultivars After Snow Treatment of Sprouts Potato of Cultivars Volzhanin, Yubileyny Osetiyi, and Vladikavkazsky.

Method of treatment	Number of days between the stages of development				
	Planting— complete sprouts	Sprouts— budding	Budding— flowering	Flowering— tops drying	Length of the growing season
			Volzhanin		
Without treatment—control	25	25	18	39	107
Snow treatment	21	19	14	32	86
			Yubileyny Osetiyi		
Without treatment—control	26	21	14	42	103
Snow treatment	19	15	11	33	78
			Vladikavkazsky		
Without treatment—control	23	26	18	37	104
Snow treatment	17	19	14	30	80

The given data in Table 3.2 indicate that snow and growth-promoting substances on germinated tubers treatment are reducing the vegetation period of the three tested cultivars and lead to a reduction of the growing season days by 21–25 days.

At the application of snow in the first hour, the temperature rises in the cells by 1–5°. The main role is played by stress proteins that are induced in plants under the influence of cold.[11] During cold stress, this protein shunts the flow in the distribution chain, mitochondria, switching it to end of the chain.[24] The path of the electrons becomes shorter. This releases energy, some thermogenesis occurs, without ATP synthesis, which is very important in the first moment of low-temperature impact on the cell. In local areas of the cell, temperature is even higher than in the environment. At the same time, during such electrons distribution, reactive oxygen—containing substances quite toxic for cells inhibit a series of reactions and lead to cells destruction.[26,27]

Therefore, stress proteins perform two functions—create local thermogenesis and stop the oxidizing process. Stress proteins that are formed in response to low-temperature impact can be used as markers of plant resistance to low temperatures.

In snow technology, the reaction of various cultivars has some changes depending on the genotype for cold responsiveness. Cold shock stimulates the growth intensity, development, and leads to the formation of high-quality yield;[27] especially, according to these indicators, cultivar Vladikavkazsky is isolated, where the starch content reached more than 16%, vitamin C—4.9 mg%, and dry matter—more than 22% (Table 3.3).

TABLE 3.3 The Biochemical Composition of Sprouted Potato Tubers Treated with Snow of Cultivars Volzhanin, Yubileyny Osetiyi, and Vladikavkazsky.

Method of tubers treatment	Starch (%)	Vitamin C (mg%)	Dry matter (%)
Volzhanin			
Without treatment—control	12.2	3.28	17.9
Treatment with snow	14.8	3.83	20.4
Yubileyny Osetiyi			
Without treatment—control	13.8	3.44	19.5
Treatment with snow	15.9	4.25	21.8
Vladikavkazsky			
Without treatment	14.5	3.47	20.2
Treatment with snow	16.4	4.92	22.2

Data in Table 3.3 confirm that the treated tubers have significantly more starch, dry matter, and vitamin C. All studied cultivars have overall trends: increase in the qualitative indicators owing to preplanting treatment of sprouted tubers with snow.

The main advantage of the snow technology is achieving high marketability of tubers while in-parallel reducing morbidity. In our studies, the percentage of marketable tubers achieved more than 90%, and the average weight of one tuber—60–70 g that is optimal when making seed material (Table 3.4).

TABLE 3.4   The Productivity of Tubers when Treated with Snow Technology.

Cultivar	Marketable tubers (%)		Marketable weight of a commodity tuber	
	Non-treated	Treated	Non-treated	Treated
Volzhanin	78.3	89.7	62.0	70.2
Yubileyny Osetiyi	80.1	92.3	58.4	60.0
Vladikavkazsky	81.2	93.1	61.4	69.4

Data in Table 3.4 show that snow treatment of the sprouted tubers before planting ensures the increase in yield by 10–12%, tubers marketability, the weight of each tuber by 2–8 g. At the same time, varietal differences of low temperatures impact are revealed. Maximum indices are found in the local cultivar Vladikavkazsky.

3.4   CONCLUSION

Snow treatment of the sprouted tubers before planting ensures an increase in yields of potatoes, resistance to disease, tubers marketability, and the weight of each tuber.

In this study, an improved material was received, which when cultured for 2–3 years in the field conditions does not require additional processing due to the cold effect of snow on tubers and seedlings, and decreasing morbidity and the presence of sufficient nutrients quantitative. This method can reduce the frequency of the Colorado potato beetle infection due to early planting.

KEYWORDS

- potato tubers
- cryophytobiology
- para-aminobenzoic acid
- leskenit

REFERENCES

1. Bekuzarova, S. A.; Basiev, S. S.; Unezhev, KhM.; Shogenov, M. K. Stimulating Method for Legumes Germination. Ru Patent 2,242,108, December 20, 2004, IPC A01C1/00, Bul. # 17 (In Russian). 2004.

2. Bekuzarova, S. A.; Kochieva, Z. S.; Tsogoeva, I. B. Method of Preparing Winter Wheat Seeds for Sowing. Ru Patent 2,178,962, February 10, 2002, IPC A01C 1/00 (In Russian). 2012.

3. Albegov, Z. B.; Tsogoev, V. B.; Bekuzarova, S. A. Method of Long-Term Plantations on Stony Ground. Ru Patent 2,107,422, March 27, 1996, IPC A01C1/ (In Russian). 1998.

4. Zakharov, P. Ya.; Pshonko, L. N.; Golman, N. A.; Ustimenko, F. N. Method of Presowing Treatment of Forage Perennial Plants Seeds. Ru Patent 2,268,569, January 27, 2006 (In Russian).

5. Yashin, E. A. The Efficiency of Using Diatomite and its Mixtures with Chicken Manure as the Fertilizer for Agricultural Crops on Leached Chernozem of the Middle Volga Region. Ph. D. Thesis, Saransk, 2004, p 21 (In Russian).

6. Aniskin, V. I.; Gubiev, YuK.; Erkinbaeva, R. K.; Naleev, O. N. The Method of Grain Heat Treatment. Ru Patent 2,031,585, March 27, 1995, A 23 B 9/04 (In Russian). 1992.

7. Saldaev, A. M.; Rogachev, A. F. Method of Presowing Cornseed treatment. Ru Patent 2,268,571, January 27, 2006, IPC A01C1/00 (In Russian). 2006.

8. Statsenko, A. P. Stimulation of Seed Germination Big Seeds and Crops. *Adv. Sci. Technol. APC.* **2012,** *12,* 34–35 (In Russian).

9. Ermakov, E. I.; Popov, A. I. Stimulation Strategy of Plant Production Process when Spatial Habitat Heterogeneity. *Herald Russ. Acad. Agric. Sci.* **2005,** *6,* 4–7 (In Russian).

10. Krishchenko, V. P.; Ushakova, T. F.; Vereshchak, G. V.; Fomina, L. G. Method of Determining Seeds Viability. Ru Patent 2,025,928, January 9, 1995, IPC A01C1/02 (In Russian). 1990.

11. Luki, A. L.; Kotov, V. V.; Lukina, E. A.; Rymar, V. T.; Gvozdev, N. V.; Slavgorodsky, S. V. Method of Presowing Seeds Treatment. Ru Patent 2,270,547, February 27, 2006, IPC A01C1/00 (In Russian). 2006.

12. Puzanina, T. I.; Prudnikov, I. S.; Yakushina, N. I. Selenium Effect on Hormone Balance and Photosynthetic Activity of Potato Plants. *Rep. Russ. Acad. Agric. Sci.* **2005,** *6,* 7–9 (In Russian).

13. Popov, A. I.; Sukhanova, P. A. Humic Preparations are the Effective Means of Biological Correction of Crops Mineral Nutrition, Their Growth and Development. Agro–Pilot. Information Sheet of St.–Petersburg, 2002; 18–19 (In Russian).

14. Apasheva, L. M.; Duskov, V. Yu.; Glinyanov, V. S.; Komissarov, G. G. Means to Increase the Plants Winter–Hardiness. Ru Patent 2,264,070. November 20, 2005, IPC A01C1/00 (In Russian). 2006.

15. Maeno, E. *The Science of Ice;* Mir: Moscow, 1988; p 312 (In Russian).

16. Pisarev, B. A. *Production of Early Potatoes;* Rosselkhozizdat: Moscow, 1986; p 286 (In Russian).

17. Albegov, Z. B.; Bekuzarova, S. A.; Dzutsev, R. S.; Arsagova, L. P. Method of Potato Cultivation. Ru Patent 2,120,716, October 27, 1998 (In Russian). 1998.

18. Samokhin, V. T. Trace Elements Deficiency in the Body—The Most Important Environmental Factor. Moscow. *Agrar. Russ.* **2000,** *5,* 69–72 (In Russian).

19. Pushkar, N. S.; Belous, A. M. *Introduction to Cryobiology;* Naukova Dumka: Kiev, 1975; p 342 (In Ukraine).

20. Tsutsaeva, A. A.; Popov, V. G.; Sytnik, K. M. *Cryobiology and Biotechnology;* Tsutsaeva, A. A., Ed.; Naukova Dumka: Kiev, 1987; p 216 (In Russian).

21. Tumanov, I. I. *Physiological Bases of Plants Winter Hardiness;* Sel'khozgiz: Moscow, 1970; p 365 (In Russian).

22. Sitnik, K. M.; Manuilsky, V. F. The Problem of Cryophytophysiology. *Herald Acad. Sci. Ukr. SSR* **1978,** *9,* 23–28 (In Ukraine).

23. Manuilsky, V. F. Cryoresistance of Membranes At Different Levels of Plant Cells Hydration. *Rep. Acad. Sci. Ukr. SSR. Ser. B* **1982,** *4,* 62–65 (In Ukraine).

24. Belous, A. M.; Bondarenko, M. A. Structural Changes of Biological Membranes While Cooling; Naukova Dumka: Kiev, 1982; p 254 (In Ukraine).

25. Udovenko, G. V. Mechanisms to Stress Adaptation. *Physiol. Biochem. Cultiv. Plants* **1979,** *11,* 99–107 (In Ukraine).

26. Isayenko, V. V. The Method of Determining the Reliability of Ecological Conditions of Grapes Wintering. International Patent CA1,132,851 A1, 1985.

27. Shulgin, L. S. Method for Rooting of Vine Cuttings. International Patent 518,184, US 29/219:434, 1970.

CHAPTER 4

THE USE OF WOOD WASTES OF CONIFEROUS WHEN GROWING CUCUMBER SEEDLINGS

GENRIETTA E. MERZLAYA[1,*], R. A. AFANAS'EV[1],
MICHAIL O. SMIRNOV[1], and SARRA A. BEKUZAROVA[2]

[1]*Pryanishnikov All–Russian Scientific Research Institute of Agrochemistry, 31A, Pryanishnikov St., Moscow, Russia 127550, E-mail: rafail–afanasev@mail.ru, User53530@yandex.ru*

[2]*Gorsky State Agrarian University, 37, Kirov St., Vladikavkaz, North Ossetia Alania, Russia 362000, Tel. +786725408776, E-mail: bekos.37@mail.ru*

Corresponding author. E-mail: lab.organic@mail.ru

CONTENTS

ABSTRACT

The studies on the effect of extract of spruce and pine cones on growth and development of cucumber plants grown seedlings were carried out by us in the pot experiment in phytotron of Pryanishnikov All–Russian Scientific Research Institute of Agrochemistry. We tested variants watering cucumber plants with extract from the crushed cones of spruce and pine trees and with water such as (1) control with water irrigation; (2) watering 0.2% extract from crushed cones; (3) applying chicken manure at the rate of 100 kg/ha of nitrogen in combination with water irrigation; (4) applying chicken manure at the rate of 100 kg/ha of nitrogen in combination with irrigation of 0.2% extract of crushed cones. Based on the studies, we found that when growing cucumber seedlings the most efficient was the use of chicken manure in combination with coniferous cones extract irrigation, in which—an increase of aboveground biomass in relation to watering with tap water was 81.6% and increase of the root biomass—97.1%.

4.1   INTRODUCTION

In recent years, due to the intensification of the most important branches of industry, the development of utilities and agricultural production increase volumes of wastes, including organic, that during processing may be sources of organic matter and nutrients needed for crops. This are wastes of timber and woodworking industry—bark and sawdust of conifers and other trees, lignin, lignosulphonates, and so forth; agricultural wastes—liquid manure and chicken manure; cities and other settlements wastes—sewage sludge, municipal solid wastes, and others.

The significant amounts of organic trade and domestic wastes are not used and are accumulated in the dumps, slime storage, on the silt playgrounds of treatment facilities, whereas agricultural wastes, especially, livestock wastes and poultry wastes in the form of manure and humid chicken manure are often concentrated in the depots unequipped waterproofing, or lagoons for a long time, often for decades, which increases the risks of environmental degradation of surrounding areas and water bodies.[1–4]

At the same time, the country's agriculture is undergoing an acute shortage of organic fertilizers. In recent years (2012–2014) in Russia, the amount of organic fertilizers was 54–62 million tons as compared with 390 million tons in 1990. The provision of 1 ha canopy with organic fertilizers in terms of litter manure is 1.1–1.3 m, or at 15%.[5] In these circumstances, it is advisable better to use all possible sources of organic matter, including various nontoxic wastes, in particular wood, agricultural, and other.

The wood wastes in the form of sawdust, bark, pine needles; cones in combination with nitrogen-containing additives (fertilizers, manure, and chicken manure) can serve as useful components in the production of organic and organic mineral fertilizers. The application of fertilizers using wood wastes improves the water physical properties of the soil, creating loose topsoil, keeps soil density, increases the content of humus in the soil, and increases the yield of crops.[6]

On the basis of wood wastes in the Arkhangelsk region high-performance fertilizer was obtained.[7] The mixture in this case, included the bark of coniferous trees in the amount of 13%, hot lignin 11%, and liquid manure and chicken manure 38%. During the fermentation of mixture, we used urea (1.5% nitrogen content) and superphosphate (0.5% P_2O_5) as minerals. The completed product after the fermentation contains, in terms of raw material 30% organic substance, 0.5–0.7% nitrogen, 0.3–0.5% P_2O_5, and 0.3–0.4% K_2O at pH 7.4–7.5. When we tested the fertilizer on the Arkhangelsk experimental ameliorative station, it is not inferior manure, increasing the yield of potatoes by two times in comparison with a control without fertilization.

Recycling organic wastes which accumulate in the forest sector, including wastes produced annually in the form of cones of coniferous tree species (spruce and pine) also is an important environmental issue. Their use as agricultural plant growth stimulants can serve a way of utilization of such wastes considering they contain organic substances. At the same time by achieving the growth and increase of plant organism biomass, the agronomic problem could be solved, and by reducing the amount of wastes—ecological, environmental protection.[8–11]

Cones of spruce and pine contain lipids, and tannins, polyterpene bioflavonoids, oleic acids, resins, carotenes, and essential tannins. Essential oils contain a lot of macro and micronutrients, terpene series compounds, biologically active substances (iron, chromium, and copper), and volatile. Together, these ingredients have antimicrobial action, provide protection of plants from diseases and pests (powdery mildew, anthracnose, bacteriose, and others).

There are reports of the stimulating and fertigation effect of products derived from the needles of pine and spruce, as well as their protective properties against pests and diseases when growing different crops such as lettuces, cauliflower, radish, and potatoes. According to the literature,[12–15] the growth stimulants have a great influence on the physiological, formative, and growth processes that occur in plants, provide increased productivity and quality of plant products, accelerated maturation, improving fruit set, increased drought and frost resistance, improved vegetative propagation, increase of immunity, reduction in nitrate content, radionuclides, easier harvesting, and improving the safety of products. The low rates of expenditure and the ability to control the processes of growth and development of plants determine the prospects of wide application of the biological products in agriculture.

Mainly, Russian developers and manufacturers form the market of the plant growth regulators in Russia. The volume of their use is still small, because their effectiveness largely depends on the compliance with all the necessary technical measures, including the use of fertilizers and pesticides, compliance with consumption rates, timing and use of technology.

As previously mentioned, to environmental contaminants except wastes wood, boffins often referred poultry wastes, in the first instance humid chicken manure, which is particularly dangerous under excessive accumulation of liquid and semiliquid form, as well as when applying the litter mass per field in high doses.[16,17] The unregulated storage of humid chicken manure in unequipped storage facilities, which often takes place at poultry farms, and the application of nonnormalized form to agricultural fields is a serious threat to the natural environment. Moreover, the large number of substances from the economic and the biological cycle are derived and irrevocably lost for agricultural production that sharply breaks biological balance. However, the combination of these wastes in defined doses and ratios may be useful for their application as fertilizers. In connection with the above, it is important to determine the possibility of joint use of wood and poultry wastes in the soil–plant system to achieve a certain agro-economic and environmental effects.

At modern environmental pollution of agricultural production, an important work is to find the environmentally friendly ways to improve the survival and productivity of crops. One way is to find an economically justified application of natural remedies that help to avoid or reduce the harm of chemical management of the celiac farm. For this purpose, we have studied the feasibility of extraction of wastes wood from spruce and pine cones for watering cucumber seedlings grown in soil fertilized with chicken manure.

The research task was to study the effect of irrigation of cucumber seedlings with aqueous extract from spruce (first) and pine cones, that are deemed to be waste, are in combination with chicken manure, on accumulation of aboveground and root biomass of cucumber.

4.2 MATERIALS AND METHODS

The study of effect of the wood wastes extract from fallen cones of coniferous trees cones was conducted on cucumber seedlings. Cucumber seedlings were treated with an extract of wood wastes from pine cones. The test culture was cucumber cultivar "Salt coating."

The seedlings were grown in a pottery experiment in accordance with the methodical instructions.[18-20] The plants were grown in the sod-podzolic heavy-textured soil. The experiment was repeated three times.

The experiences were laid in phytotron of Pryanishnikov All–Russian Scientific Research Institute of Agrochemistry. Temperature during the experiment rose to 20°C in daytime and 24–25°C at night. The light regime was provided with fluorescent lamps. Illumination was 12 klx and photoperiod lasted for 14 h.

TABLE 4.1 Various Variants of Soil and Cucumber Seedlings Treatment Scheme When Growing Cucumber from Seed.

Number of the variant	Various variants of soil treatment before the experiment and cucumber seedlings
1	Watering seedlings with tap water (control)
2	Watering seedlings with extract from crushed cones
3	Application of chicken manure in soil + watering seedlings with tap water
4	Application of chicken manure in soil + watering seedlings with extract from crushed cones

While laying out experiment in pots, we placed soil carefully blended with sand and, according to the scheme with chicken manure (in variants 3 and 4). In each 1-l flask 50 g of expanded clay aggregate as drainage and 700 g of a mixture of soil and sand (in a ratio of 3:1) was placed. The litter dose was 0.03 g nitrogen per pot, or, in terms, 100 kg/ha nitrogen (4 t/ha of dry litter physical mass).

The experiment was carried out on sod-podzolic heavy-textured soil. It was characterized by the following agrochemical indexes such as pH_{KCl} 5.9, the humus content 1.1%, total nitrogen 0.24%, and mobile elements: phosphorus (P_2O_5) 247 mg/kg and potassium (K_2O) 188 mg/kg.

The dry granulated fowl manure, according to chemical analysis, contained 39.6% of carbon, 79.2% of organic matter, 2.3% of total nitrogen, 3.1% of phosphorus (P_2O_5), and 2.7% of potassium (K_2O) at pH_{KCl} 6.3–7.1. Litter moisture was 3.9%, carbon–nitrogen ratio $(C:N) = 12$. The content of NPK in 1 t of manure was 81 kg.

The sowing of seeds of cucumber in soil were completed in early June, 2016 at the depth of about 2 cm. Shoots appeared after a week, after which the plants during a week initially were watered with tap water, and subsequently they were watered with 0.5% aqueous extract of crushed spruce and pine cones according to scheme of the experience.

In the beginning of the experiment, 120 g of spruce and pine cones were crushed and soaked in 12 l of water for 10–12 h. Then 48 l of water was added to the extract concentration of 0.2%, that is, 120 g cones for 60 l of water.

Accounting for the aboveground and root mass was performed in 20-day-old cucumber plants at the end of June.

The soil moisture in pots was maintained at 75–80% of field moisture capacity. For this watering with tap water or extracts of pine and spruce cones according to the experimental scheme was performed daily rate 30 ml per one pot at the initial stage and 60 ml in the second half of the experiment.

4.3   RESULTS AND DISCUSSION

As a result of the studies (Table 4.2), we found a significant positive impact of chicken manure on the accumulation of cucumber plant biomass under watering both a water tap (variant 3), and an extract of the cones of coniferous species (variant 4). In these two variants, the above-ground biomass was 3.43 and 3.76 g / pot, respectively, for most cucumbers, which was 65.7 and 81.6% higher than in case of irrigation with tap water.

Against the background of chicken manure effect of the cones extract compared with irrigation with tap water (the difference in the accumulation of aboveground biomass between variants 3 and 4) was 9.6%. Without manure effect of the extract compared with the control

TABLE 4.2 The Cucumber Biomass When Growing Cucumber Seedlings According to Application of Chicken Manure in Soil and Watering Seedlings with Extract from Wood Wastes.

Number of the variant of soil treatment (see Table 4.1)	Aboveground biomass			Roots			The ratio of aboveground biomass: root	Evapotranspiration, (g/24 h)
	g/pot	Increment		g/pot	Increment			
		g/pot	%		g/pot	%		
1	2.07	–	–	1.04	–	–	2:1	513
2	2.06	–	–	1.18	0.14	13.5	1.7:1	438
3	3.43	1.36	65.7	1.47	0.43	41.3	2.3:1	525
4	3.76	1.69	81.6	2.05	1.01	97.1	1.8:1	550
LSD_{05} (g/pot)	–	0.97	–	–	0.50	–	–	–

LSD: least significant difference

(comparison of biomass in the variants 1 and 2) was not manifested in the experiment.

It should be noted that in the variant 4, where we applied chicken manure into the soil and implemented irrigation with wastes wood extract, we observed more intensive root system whose biomass formed was about 2.05 g/pot, which was significantly higher compared with control, as well as in the variant 3, where the chicken manure was added when watering with tap water.

Moreover, it is important to add that application of chicken manure into the soil and watering seedlings of cucumber with extract from crushed cones of coniferous species versus the control without fertilizers decreased the risk of the following plant diseases such as powdery mildew, bacteriose, and askohitose.

Thus, the studies have shown that the most effective way for growing seedlings of cucumber on sod-podzolic heavy-textured soil was the combined use of chicken manure and watering extract of crushed cones of coniferous species. As a result of this combination of organic wastes cucumber plants showed significant increase in aboveground biomass compared with control and their root system was more intensively maturated.

Figure 4.1 shows that in the version 4, where manure was applied, and the extract of the crushed pine and spruce cones was irrigated, the plant had three developed true leaves, whereas in version 3, when irrigation with tap water, the third leave was not sufficiently developed, but in the two other variants of the experiment, only two true sheets were formed.

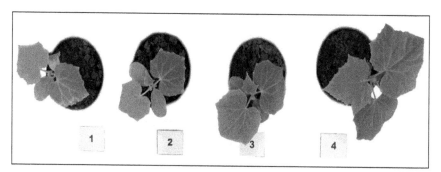

FIGURE 4.1 Effect of irrigation extract from crushed cones of coniferous species on the formation of cucumber biomass. (1): watering with tap water, (2): watering with extract of cones; (3): chicken manure + watering with tap water, and (4): chicken manure + watering extract of cones.

In options 1 and 2, watering was carried out with an aqueous solution of cones, and in version 3, litter was introduced, but watering was carried out only with tap water, only two real leaves were developed in cucumber plants. At the same time, in version 4, where bird droppings were added + watering with an aqueous solution of cones, the plant had three developed true leaves. In variant 4, the plants exhibited a higher evapotranspiration rate (see Fig. 4.1).

4.4   CONCLUSIONS

The experiment was carried out in phytotron of Pryanishnikov All–Russian Scientific Research Institute of Agrochemistry when growing plants of cucumber transplant seedlings. The combined use of chicken manure and extract of the coniferous species cones showed high efficiency.

When comparing this method with the control (watering only tap water), we obtained cucumber aboveground biomass increase at the level of 81.6% and root weight increasing biomass was 97.1%.

In general, we can conclude that on the sod-podzolic soil watering aqueous extract of the spruce and pine cones when applying chicken manure rate of 100 kg/ha N is an important agroecological techniques that allow, on the one hand, to create favorable conditions for plant growth and development, and on the other—to provide recycling wood wastes and protect the environment. It is important to consider that the applied technology is aimed primarily at recycling fowl manure at strengthening the agroecological effect by stimulating biological processes due to the effect of the extract of coniferous species cones.

KEYWORDS

- organic wastes
- chicken manure
- biomass of cucumber

REFERENCES

1. Owen, O. S.. *Protection of Natural Resources*, Bannikov, A. G., Ed.; Kolos: Moscow, 1977; p 416 (in Russian).
2. Mineev, V. G.; Debreceni, B.; Mazur, T. Biological Agriculture and Mineral Fertilizers. Kolos: Moscow, 1993; p 415 (in Russian).
3. Chernikov, V. A.; Alexakhin, R. M.; Golubev, A. V.; et al. *Agroecology;* Chernikov, V. A., Chekeres, A. I., Eds.; Kolos: Moscow, 2000; p 536 (in Russian).
4. Milaschenko, N. Z., Ed. The Strategy of Sewage Sludge Using and Composts Based on Them. Agroconsult: Moscow, 2002; p 140 (in Russian).
5. Derzhavin, L. M.; Afanasiev, R. A.; Merzlaya, G. E. Agrochemical Service Role in the Modernization of Agricultural Production and Food Security of the Russian Federation. Pryanishnikov All–Russian Scientific Research Institute of Agrochemistry: Moscow, 2016; p 116 (in Russian).
6. The Study of Soils in the European North. IV Sibirtsevskie Reading Collection. Tezis. Arkhangelsk Institute of Forest and Wood Chemistry: Arkhangelsk, 1990; p 176 (in Russian).
7. Merzlaya, G. E. The Use of Organic Wastes in Agriculture. *Russ. Chem. J. XLIX. Part 3* **2005**, 48–54. (in Russian).
8. Afanasiev, R. A.; Merzlaya, G. E. *Guidelines on the Effectiveness of Non-Conventional Organic and Organic–Mineral Fertilizers*; Russian Academy of Agricultural Sciences: Moscow, 1999; p 40 (in Russian).
9. Bekuzarova, S. A.; Alborov, I. D.; Khubaeva, G. P.; Lushchenko, G. V. Recycling Industrial Wastes in North Ossetia. Ecological Bulletin of the North Caucasus: Krasnodar, 9. 2, 2013; 16–18 (in Russian). **2013.**
10. Mineev, V. G. *Agrochemistry and the Biosphere;* Kolos: Moscow, 1984; p 245 (in Russian).
11. Shapoval, O. A.; Mozharova, I. L.; Korshunov, A. A.; Vakulenko, V. V. Effect of Plant Growth Regulators of Multipurpose Effect on the Growth, Development and Productivity of Crops. Prospects for the Use of New Forms of Fertilizers, Plant Protection Products and Growth Regulators in the Agro–Technologies of Agricultural Crops. *Proceedings of the participants of the 7th Conference "Anapa–2012".* Pryanishnikov All–Russian Scientific Research Institute of Agrochemistry: Moscow, 2012; 132–138 (in Russian).
12. Mukhin, V. D. *Manual of Amateur Grower*; Publishing and Printing Complex Moskovskaya Pravda: Moscow, 1995; p 96 (in Russian).
13. Tosunov, Y. K.; Barchukova, A. J.; Korshunov, A. A. Effect of the Drug Verva–El on the Growth, Yield and Quality of Winter Wheat. Prospects for the Use of New forms of Fertilizers, Plant Protection Products and Growth Regulators in the Agro–Technologies of Agricultural Crops. *Proceedings of the participants of the 8th conference "Anapa 2014".* Pryanishnikov All–Russian Scientific Research Institute of Agrochemistry: Moscow, 2014; 276–280 (in Russian).
14. Shapoval, O. A.; Mozharova, I. P. Classification Criteria of Innovative Fertilizer forms and Examples Assess Their Biological Activity (Based on the Registration Tests). Prospects for the Use of New forms of Fertilizers, Plant Protection Products and Growth Regulators in the Agro–Technologies of Agricultural Crops. *Proceedings*

of the participants of the 8th conference "Anapa 2014". Pryanishnikov All–Russian Scientific Research Institute of Agrochemistry: Moscow, 2014; 309–320 (in Russian).

15. Khurshkainen, T. V.; Kutchin, A. V. Technology for Obtaining of Biopreparations and Investigation of Their Effectiveness. In *Chemical and Technological of Plant Substances. Chemical and Biological Aspects;* Kutchin, A. V., Shishkina, L. N., Weisfeld, L. I., Eds.; [Reviewers and Advisory Board Members: G.E. Zaikov, I. N. Kurochkin, and A.N. Goloshchapov]. Apple Academic Press. 2017; 227–241.

16. Resources of Organic Fertilizers in Agriculture of Russia. In *Information and Analytical Reference Book;* Eskov, A. I., Ed.; Vladimir. Russian Research Institute of Organic Fertilizers and Peat. 2006; p 200 (in Russian).

17. Russian Academy of Agricultural Sciences. The use of Chicken Manure in Agriculture. Scientific and Methodological Guidance; Russian Academy of Agricultural Sciences: Moscow, 2011; p 272 (in Russian).

18. Dospehov, B. A. *Methodology of Field Experiment*; Kolos: Moscow, 1979; p 416 (in Russian).

19. Mineev, V. G., Ed. Workshop on Agricultural Chemistry. Moscow State University: Moscow, 2001; p 608 (in Russian) **2001.**

20. The Program and Methods of Research in the Geographical Network of Field Experiments on Integrated Application of Chemicals in Agriculture funds. Russian Academy of Agricultural Sciences. All–Russian Scientific Research Institute of Fertilizers: Moscow, 1990; p 187 (in Russian).

PART II
Arctic Berries: Ecology, Biochemistry, and Useful Properties

CHAPTER 5

BIOCHEMICAL COMPOSITION OF FRUITS OF WILD-GROWING BERRY PLANTS

EKATERINA A. LUGININA and TATIANA L. EGOSHINA[*]

Prof. B. M. Zhitkov Russian Research Institute of Game Management and Fur Farming, 79, Preobrazhenskaya St., Kirov, Russia 610000

[*]*Corresponding author. E-mail: etl@inbox.ru*

CONTENTS

ABSTRACT

Analyses of the chemical composition of Arctic raspberry, cloudberry, and cranberry revealed that wild berries can be used as a source of the most valuable carbohydrates (fructose, cellulose, and pectin substances) and organic acids.

Cloudberry fruits can also be a source of neutral lipids in the nutrition of the northern population. To achieve that, it seems necessary to promote the production of traditional foodstuffs as well as cloudberry oil, bagasse from juice, and alcoholic drinks production by oil or CO_2 extraction.

Cloudberry, Arctic bramble, and cranberry fruits contain oil- and water-soluble vitamins such as ascorbic acid, vitamins of B group (B_1, B_2, B_6, PP, and so forth.), and bioflavonoids.

Fruits of the studied wild-growing berries contain significant quantities of biologically active substances and vitamins, which are necessary for the organization of functional nutrition.

5.1 INTRODUCTION

In order to achieve the concept of healthy nutrition in Subarctic conditions and prevent diseases caused by unfavorable ecological conditions (low temperatures, hypoxia, and lighting regime), special attention should be given to the use of medicinal and wild-growing edible plants of the north as sources of vitamins and biologically active supplements.[1,2]

Wild-growing fruits have a low energy value: only 30–100 kcal per 100 g. The important energy component of fruits and berries is easily digestible carbohydrates. The most valuable biologically active substances in wild-growing fruits are[3–5] vitamins, macro- and microelements.[6,7] Due to the presence of these groups of substances, fruits improve digestion, cardiovascular activity, and emotional condition.[8] The average annual requirement of wild-growing fruits and berries per person is 7 kg according to Russian nutritionists, and is 16 kg according to Canadian nutritionists.[9]

The Yamal-Nenets Autonomous District has a significant species variability, high productivity and resources of raw materials of wild-growing berries. Cowberry, blueberry, bilberry, cloudberry, Arctic bramble, and other species of berry plants with significant resources are marked within the area. The productivity of cranberry in the region varies from 31 to 1000 kg/ha[10] (average 150 kg/ha)[12–14] bilberry—from 80 to 150 kg/ha (average 120 kg/ha), cowberry—from 50 to 1500 kg/ha (average from 200 to 250 kg/ha),[13,15] cloudberry—from 55.5 to 167.3 kg/ha[14] (average 85.5 kg/ha).[16,17] The productivity parameters of these species in Yamal-Nenets Autonomous District are slightly lower than in the southern taiga regions of Russia.[18–20]

Biological stock of cranberry in the Yamal-Nenets Autonomous District is 352 t, cowberry—10,439 t, bilberry—80,000 t,[13] and cloudberry—85.5 t.[16]

5.2 MATERIALS AND METHODOLOGY

The samples of fruits of wild-growing plants are cloudberry (*Rubus chamaemorus* L.), Arctic bramble (*Rubus arcticus* L.), cowberry (*Vaccinium vitis-idaea* L.), bilberry (*Vaccinium myrtillus* L.), blueberry (*Vaccinium uliginosum* L.), and cranberry (*Oxycoccus palustris* L.), which were collected from Purovskiy and Tazovskiy regions of the Yamal–Nenets Autonomous District in 2011–2014. For comparison, the results of chemical studies of fruit samples of wild-growing and cultivated plants such as cranberry tree (*Viburnum opulus* L.), bird cherry (*Padus avium* L.), wild ash (*Sorbus aucuparia* L.), raspberry (*Rubus idaeus* L), sea buckthorn (*Hippophae rhamnoides* L.), black currant (*Ribes nigrum* L.), cinnamon rose (*Rosa majalis* L.) collected in Kirov Region and schizandra (*Schisandra chinensis* (Turcz.) Baill), and Amur grape (*Vitis amurensis* Rupr.) collected in Khabarovskiy Krai, as well as literary data, were used.

The study of chemical composition included a wide range of parameters referring to different groups of active substances. Fresh samples of ripe fruits were tested for dry matter content by All-Union State Standard 8756.2-82,[21] ascorbic acid (vitamin C) by standard polyphenol method,[22] titratable acidity by volumetric method[22] with further identification of organic acids by high-performance liquid chromatography (HPLC). Air-dried samples were tested for the content of chemical elements such as phosphorus, sodium, calcium, iron, magnesium, zinc, manganese, lead, and chromium—by the method of atomic absorption analyses.[23] The content of soluble sugars was determined by the speedy semimicro-method,[24] pectin substances (water-soluble pectin and protopectin) was determined by carbazole method,[22] total anthocyanin pigments was determined by the method of T. Swain and W. Hillis,[25] true anthocyanins was determined by the method of L. O. Schneidman and V. S. Afanasyeva;[26]

total flavonols was determined by photoelectric colorimetry,[22] total cate-chols was determined by photometry using vanillin reagent,[27] phenol carbonic acids (in conversion to chlorogenic acid) was determined by the method of descending paper chromatography by carbazole method.[22,28] The total content of P-active substances was defined by L. I. Vigorov[29] colorimetry method modification. The mass fraction of neutral lipids was defined by gravimetric analyses. Fatty acids profile was studied with chromatography–mass–spectrometry. Spectrophotometric analyses were used to define vitamin K_1 content.[22] Vitamins B_1 and B_2 were tested with fluorometry.[30] Vitamins A, E, D, B_{12}, B_6, carotenoids, and niacin were detected with HPLC method.[31,32] The fiber content was estimated with the Kirschner–Ganek method.[33]

All analytical studies were carried out in three replications.

5.3 RESULTS AND DISCUSSION

The processing of chemical test results of the samples of wild-growing fruit plants revealed the following.

5.3.1 CARBOHYDRATES

Wild fruits are an efficient source of various carbohydrates, including sugars, polyols, pectin substances, fiber, and hemicelluloses.

Basic digestible carbohydrates in wild-growing plants are glucose, fructose, and saccharose. The total content of sugars in wild fruits varies from 2.1% in stone bramble to 20% in cinnamon rose fruits.[9] Simple sugars such as glucose and fructose prevail. In majority of fruits, these sugars are present in equal quantities. The concentration of saccharose in fruits of most of the studied species (excluding bird cherry), does not exceed 1%.

Sugar content in fruits of the studied species is significantly high and varies from 1.83 to 6.10% in European cranberry fruits, from 3.2 to 6.0% in cloudberry fruits, and from 5 to 8% in Arctic bramble fruits (Fig. 5.1). Fruits of cloudberry and cranberry mostly contain fructose, the sweetest and the most nutritionally valuable sugar, which is desirable in low-calorie diets and in children's and diabetic nutrition.

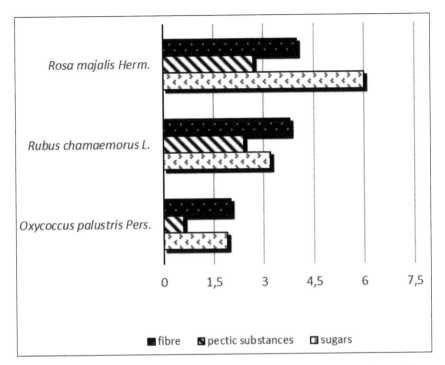

FIGURE 5.1 Carbohydrates concentration in fruits of *Oxycoccus palustris* Pers., *Rubus chamaemorus* L., and *Rosa majalis* Herm. (a) X-axis—concentration (%); (b) Y-axis—concentration

A significant seasonal variation of the soluble sugars content was marked. Other researchers detected a similar pattern. K. E. Vogulkin et al. stated in papers that [34] the seasonal soluble sugar content in cloudberry fruits collected in Belorussia differs two times (from 2.9 to 5.65%).

Pectin substances and fiber are carbohydrate polymers that are not digestible but their physiological role is significant. Pectin substances are involved in the formation of cells and extracellular matters. These are derivatives of galacturonic acid. Two types are distinguished, namely, soluble pectin and insoluble protopectin. The ratio of soluble pectin and protopectin in fruits and berries composition during the periods of growth, ripening, and storage, is changing. Correspondingly, the changes that are consistence become noticeable. Wild-growing berries include 0.2–1.8% of pectin substances with good jelling properties, which develop in a certain ratio of pectin substances, sugar, and acids.

The concentration of pectin substances in berries of European cranberry varies from 0.6 to 2.81%. Fruits of cloudberry contain twice as much as from 2.13 to 2.63% (see Fig. 5.1). Similar values of pectin substances concentration in cloudberry fruits were defined in samples collected from different areas of Russia and Northern European countries. For example, cloudberry fruits samples collected from Turukhansk region of Krasnoyarsk Krai contained 2.43% of pectin substances,[35] reaching almost maximum pectin concentration for wild-growing plants revealed in cinnamon rose fruits (about 2.7%). According to Zh. A. Rupasova et al.,[36] pectin substance concentration is a genetic determinant with a low variability level, which is proved by data of the study conducted by other researches.[35] Pectin content in fruits increases during warm sunny summers.[9]

Pectin is favorable for treating various gastric disorders. It does not form energy stock in the organism, regulates intestinal microflora, egests cholesterol, and more importantly is capable of forming insoluble complexes with toxic and radioactive metals, with microorganism's toxins and of clearing them from the organism. The properties of pectin substances are widely used in producing dietary low-calorie foodstuff and also in the products for employees working in conditions of lead, mercury, and other intoxications, in conditions of extreme north and alike. Therefore, the use of cranberry and cloudberry fruits in nutrition of the polar region population is important and reasonable from biochemical point of view.

The fiber is the most widespread polymer in plants formed from the numerous glucose moiety etaerio. The maximum concentration of fiber is customary for drupaceous fruits as well as for the berries in form of etaerio (raspberry, bramble, and so forth.).[9] The fiber concentration in cloudberry fruits (also etaerio) reaches 3.8%, which is slightly lower than maximum values for wild-growing plants (4.0% for cinnamon rose) (see Fig. 5.1). The fiber content in cranberry fruits is almost half (2.0%).

5.3.2 ORGANIC ACIDS

The most common acids of wild-growing fruits and berries are involatile: citric, malic, tartaric, oxalic, and amber. These acids are located in vacuole cell sap. The acid concentration in fruits and berries varies from 0.6 to 6.0%. The highest acidity is marked for schizandra, cranberry, sea buckthorn, and cranberry trees. The studied samples of cranberry and cloudberry contained 2.10–4.84% and 0.8% of organic acids, respectively.

Prevailing acids in organic acid composition of cranberry are citric, malic, glycolic, oxalic, benzoic, and salicylic; in cloudberry are citric, malic, acetic, propionic, benzoic, and hydroxycinnamic acid complex; in the Arctic bramble are malic and citric. Benzoic acid is a natural preserving agent, has antibacterial properties; hydroxycinnamic acids have systematic immune-enhancing effect. Phenolic acids (l-hydroxybenzoic, protocatechuic, o-pyrocatechuic, and gallus) were found in microquantities, excluding benzoic acid.

Acids are involved in forming the taste of wild-growing fruits, decreasing pH, favorably affecting digestion, promoting the development of certain microflora composition, and inhibiting decay processes in the gastrointestinal tract. Phenolic acids have antibacterial properties. During the oxidation of 1 g of citric acid in human organism, 2.5 kcal are released and 2.4 kcal in the case of malic acid.

Intensity of acidic taste of fruits and berries is determined by qualitative and quantitative composition of acids and also by the ratio of free and combined acids. Each acid has its own taste and perception threshold. Malic and citric acids have clear, soft, and non-astringent taste.

5.3.3   LIPIDS

Wild-growing fruits contain insignificant quantities of lipids (0.1–0.3%) that are located mostly in seeds. Lipids in fruits and berries are presented by oils, waxes, cutin, steroids, and other substances.

The concentration of neutral lipids in fruits of European cranberry and cloudberry fruits is 0.3% and 2.10–3.68%, respectively. This allows considering cloudberry fruits an important source of neutral lipids in northern diet. It is essential though to provide the possibility of producing traditional foodstuff from cloudberry along with cloudberry oil that can be produced from pulp after juice production or production of alcohol drinks with oil or CO_2 extraction.

Lipids play a significant role in metabolism as they are included in cell membranes, increase resistance to viral infections and catarrhal diseases. Lipids of wild-growing fruits have favorable fatty acid compositions with the dominance of desaturated fatty acids. For example, fatty acids composition of cloudberry neutral lipids is presented by seven acids with long chain C_{14}–C_{18} and even number of carbon atoms. The important acids by composition were linoleic (about 40% total fatty acids) and linolenic (about 35% total fatty acids) acids, which are higher polyunsaturated fatty acids (HPUFA omega-3) and play an important role in vegetal and animal organism. HPUFA are included in a lipid bilayer of cell membranes and regulate their microviscosity, penetrance, electric properties, decreasing the excitability of forming corresponding surrounding of membrane proteins and enzymes. As an anti-sclerotic factor, HPUFA promotes cholesterol metabolism in liver and its elimination from the organism and also play a role as an enzyme inhibitor (hydroxymethylglutaryl reductase) to control cholesterol biosynthesis.[8] The proportion of unsaturated fatty acids in cloudberry fruits is 90% of total neutral lipids.

Waxes, cutin, and ursolic acid form wax cover typical for many fruits (bilberry, blueberry, cranberry, and so forth) and have general tonic and anti-inflammatory action.

5.3.4   VITAMINS

Nutritional value of wild-growing fruits is determined by the presence of vitamins and vitamin-like substances. Fruits contain both water-soluble and liposoluble vitamins. Water-soluble vitamins are ascorbic acid, group B vitamins (B_1, B_2, B_6, PP, and so forth.), and bioflavonoids.

Vitamin C in its restored form is called as L-ascorbic acid and is common in wild-growing fruits. Many of these if consumed regularly can fulfill daily requirement of ascorbic acid. However, 70% of Russian population experience vitamin C deficiency. A sufficient intake of vitamin C is extremely important for populations of Northern regions of the country where the lack of sun and long winter period lead to its deficit.

Ascorbic acid is an active agent in many oxidation–reduction processes in fruits as well as in human body. The vitamin is essential for normal function of connective and bone tissues; it increases resistance to infectious diseases (overheating, cooling, and oxygen deficiency) and increases working capacity. Ascorbic acid influences hematopoiesis, carbohydrates, and cholesterol metabolism. Biological role of vitamin C is developed fully in the presence of P-active substances, which is typical for fruits.

The essential daily dose of ascorbic acid is 70–100 mg. The concentration of ascorbic acid in studied cranberry samples varied from 21 to 70.8 mg/100 g, increasing till fully ripe. Fruits of cloudberry and Arctic bramble are characterized by high concentration of vitamin C (cloudberry and Arctic bramble contain from 100 to 200 mg/100 g and 100 to

400 mg/100 g, respectively). This is almost the same amount as in black currant (from 47 to 374 mg/100 g lesser than in cinnamon rose (from 670 to 3800 mg%), but significantly higher than in Amur grape (from 7 to 15 mg/100 g).

A comparison of the received results of vitamin C content in cloudberry with literary data have shown that the concentration of ascorbic acid in cloudberry fruits from the area of investigation was higher than in fruits from Southern Karelia (12–22 mg/100 g),[37] Angara region, and Tomsk region (16.8–20.7 mg/100 g),[38] but slightly lower than in samples from the Komi Republic.[39]

Revealed data on vitamin C content in cranberry and cloudberry allow recommending a daily intake of 50–70 g of cloudberry or Arctic bramble, or 100 g of cranberry to compensate essential daily dose of vitamin C. An accumulation of vitamin C is closely connected with weather conditions and ripening stage of fruits and berries. Unripe fruits and berries have maximum vitamin C content, but it decreases during the storage, fruit sleeping, and processing. P-active substances administer economical and efficient spending of vitamin C.

The group of P-active substances includes anthocyanins, leucoanthocyanins, catechols, chlorogenic acids, and flavonols, differing by chemical composition but having similar activity in a human organism. P-active substances have hypotensive and sclerotic activity. The total content of P-active compounds is 0.02–0.6% in fruits and berries of light-red coloring. Intensively colored and black-fruited cultures such as bilberry, blackcurrant, cranberry, blueberry, honeysuckle, blackthorn, and Amur grape contain 1–1.5% of P-active substances.

The content of anthocyanins in cranberry fruits reached 600–890 mg/100 g. It is only lower than in bilberry (940–1900 mg/100 g), bird cherry (600–1490 mg/100 g), and cowberry (430–1280 mg/100 g). Anthocyanin content in cloudberry was only 62–90 mg/100 g. Leucoanthocyanins were not detected in cranberry fruits, but their content in cloudberry reached 91–175 mg/100 g.

The concentration of flavonols in cranberry fruits (210 mg/100 g) and cloudberry (110 mg/100 g) is relatively high. More flavonols were detected in cranberry fruits.

Each of the fruits of cranberry and cloudberry contains 0.15% P-active substances. The daily requirement of P-active compounds is 30–50 mg; therefore, a daily intake of 200 g of cranberry and cloudberry is sufficient.

Group B vitamins are included in enzyme prosthetic groups that provide energy metabolism and biosynthesis of purine bases; they regulate the metabolism of carbohydrates, proteins, and lipids. The highest requirements of these vitamins are in conditions of predominantly carbohydrates nutrition, which is typical in Russia. The lack of thiamine, riboflavin, niacin, and folacin causes vitamin deficiency disease having expressive symptoms. The most typical is an increase of intellectual fatigability. Currently, 20–40% of Russians experience deficit of group B vitamins.

The wild-growing plants do not typically have high content of vitamins B_1 and B_2 and cannot fulfill the human requirements. However, cinnamon rose, ashberry, and black currant are considered effective sources of thiamine as they contain 0.05–0.1 mg/100 g. Cranberry fruits contain 0.02 mg/100 g of vitamin B_1 and cloudberry fruits contain 0.025 mg/100 g.

The most wild-growing fruits contain vitamins of B_6-type pyridoxine, pyridoxal, and pyridoxamine in small quantities (0.1–3.9 mg/100 g). Physiological requirement of vitamin B_6 of an adult is 1.8–2.0 mg per day. The lack of B_6 leads to neurotic disorders, sleepiness, eye and skin inflammation, niacin requirement increases, use of amino acids, and proteins is damaged.

Niacin (vitamin PP) daily requirement is significant (14–28 mg). The lack of niacin in a diet causes irritation, depressed state, headaches, defective memory, skin roughness, gastrointestinal disorders, and an increase of blood cholesterol. Niacin is found in all wild-growing fruits but quantities of it are far from optimum. Slightly larger quantities of niacin are detected in bilberry, cinnamon rose, sea buckthorn, raspberry, cranberry, and cloudberry (0.15 mg/100 g).

Some wild-growing fruits are a valuable source of folic acid (vitamin B_9 and folacin). This vitamin is an antianemic factor; it stimulates the formation of red and white blood cells. The restored form of folic acid is included in enzymes, which participate in biosynthesis of some amino acids, purine, and pyrimidine bases. The lack of vitamin B_9 causes the increase of organism sensitivity to radiation and poliosis. Among wild-growing fruits and berries easily fulfilling the human requirement of folacin are ashberry, cinnamon rose, Arctic bramble, and cloudberry. It was defined that two to three times higher quantities of folic acid are accumulated in fruits and leaves during sunny summer compared to cold and rainy season.[9]

HPUFA content of vitamin D in wild-growing fruits is insignificant (0.1–0.2 µg/100 g). Vitamin A is not typical for the studied objects. The wild-growing fruits with bright-yellow flesh coloring are rich in carotenoids, including beta-carotene (provitamin A_1), which transforms to vitamin A in human most effectively. Accumulators of beta-carotene are sea buckthorn (8–10 mg/100 g), cinnamon rose (2.0—2.6 mg/100 g), and cloudberry (1.1—2.4 mg/100 g). Cloudberry excels carrot at beta-carotene concentration.[39]

Carotenoids prevent tissue hypoxia, participate in forming adrenal hormones and visual pigment.[40] Recommended daily intake of carotenoids is contained in 200 g of cloudberry.[41]

Vitamin E stabilizes and protects unsaturated lipids from excess oxidation, prevents sterility in animals and humans. Considering its chemical nature, vitamin E is a group of similar substances—tocopherols—that are present in three forms. Tocopherols are synthesized by plants exclusively. Sea buckthorn is particularly rich in tocopherols, from 2.9 to 18 mg/100 g in fruits, and ten times more in oil. Cinnamon rose also contains large quantities of tocopherols (1–8.8 mg/100 g), ashberry (0.6–5.1 mg/100 g), and cloudberry (0.6–0.9 mg/100 g). Tocopherol content in cranberry fruits is insignificant (0.02 mg/100 g).

5.3.5 MINERAL SUBSTANCES

The wild-growing fruits are a source of mineral substances playing an important role in metabolism. The total volume of mineral substances in wild plants is 0.2–0.54%. The qualitative composition and quantity of certain elements in wild-growing plants depend on biological properties, species specialization in element accumulation, and soil richness in available forms of elements. In some cases, the mineral composition of plants can simplify the identification of processed products and prove their naturality. During the biological evolution, the composition of elements essential for living organisms is gradually changing due to the change in chemical composition of the environment.

The role of 10 elements called metals of life, is indisputable: potassium, sodium, calcium, magnesium, iron, manganese, zinc, copper, cobalt, and molybdenum.

The wild plants rich in potassium, sodium, magnesium, and calcium produce alkaline substances and thus regulate alkaline–acid balance. Potassium and sodium regulate the water–salt metabolism and, possibly, memory mechanisms. The highest potassium content is typical for black currant, sea buckthorn, raspberry, *Vacciniaceae* berries, bird cherry, and cloudberry.

Fruits of cranberry and cloudberry in Yamal are characterized by relatively high content of sodium and calcium (Fig. 5.2). According to the studies of V. A. Rush and V. V. Lizunova,[42] samples of cloudberry fruits from Krasnoyarsk Krai contained high quantities of calcium.

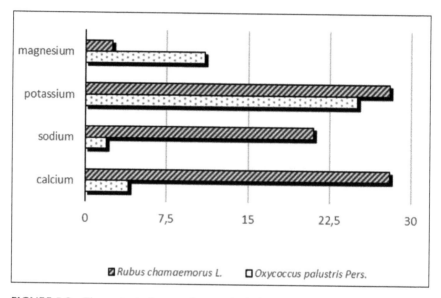

FIGURE 5.2   The content of macroelements in fruits of *Oxycoccus palustris* Pers. and *Rubus chamaemorus* L. (a) X-axis—content (air-dry weight); (b) Y-axis—macroelements (mg/kg).

Exclusive role in the plant world is played by magnesium, which is the base of chlorophyll molecule. The richest in magnesium among wild-growing fruits is cowberry. In taiga regions of North-Eastern Russia, magnesium concentration reaches 166.5 mg/kg, air-dry weight. It was marked that all *Vaccinium* and *Oxycoccus* species have the ability of magnesium accumulation.[43] The analyses of received data proved the pattern; especially, high content of magnesium was detected in unripe fruits. The daily requirement of magnesium for an adult is 400 mg, and deficit is rare.

Phosphorus is an important microelement in terms of magnesium and calcium utilization. Phosphates and phosphorus organic substances realize various functions in the organism such as plastic, maintenance of alkaline–acidic balance, synthesis of phospholipids, nucleotides and nucleic acids, enzyme catalyzes, and so forth. Human muscular and intellectual activity depends on phosphorus intake. Daily requirement of an adult is 1200 mg. Phosphorus deficit does not happen in humans. The most important issue is the maintenance of phosphorus–calcium ratio (1:1.5) and avoidance of excess phosphorus intake.

The content of phosphorus in wild-growing fruits and berries is low (8–126 mg/kg). The maximum values are marked for cloudberry (113.7–250 mg/kg) (Fig. 5.3).

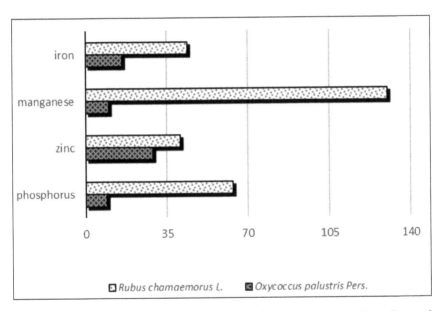

FIGURE 5.3   The content of some microelements in fruits of *Oxycoccus palustris* Pers. and *Rubus chamaemorus* L. (a) X-axis—content (air-dry weight); (b) Y-axis—microelements (mg/kg).

Iron is widely spread in nature. All species of wild-growing berries contain iron and can compensate its lack in diets that is detected in 80% of Russians. The well-known iron accumulators among fruit plants are cinnamon rose (24.4–115 mg/kg), bird cherry (10–87.2 mg/kg), cranberry tree (11–79.2 mg/kg), and raspberry (16–69 mg/kg). Bilberry fruits contain 50 mg/kg of iron.[44] Iron concentration in cloudberry and cranberry fruits is 50 mg/kg, and 4–28 mg/kg, respectively.

Daily iron requirement is 15 mg. Iron in blood has an important biological function—transport and activation of molecular oxygen. Two-third of all iron in the organism is located in hemoglobin. It is also found in myoglobin and enzymes, such as catalase, peroxidase, and cytochrome oxidase.

Iron from vegetable products is digested for just 10%. Thus, daily requirement of the element can be provided by 3 kg of cloudberry. Nevertheless, a majority of wild fruits remain effective sources of iron in diets. Especially, if to consider that iron is even worse digested at shortage of vitamin C and organic acids.[8] Manganese is contained in plants in higher quantities than in animal organisms. In plants, it promotes photosynthesis and ascorbic acid production. In human organism, it participates in the formation of bones, blood, influences insulin metabolism, and stimulates growth. Manganese daily requirement is 5 mg, minimum is 2–3 mg. Well-known manganese accumulators are *Vacciniaceae* berries (cowberry, bilberry, and cranberry),[44–46] which contain up to 63 mg/kg of manganese. Cloudberry and Arctic bramble contain 7–68 mg/kg manganese. The average assimilation of manganese is 50%; thus, 100 g of any above-listed berries easily fulfills the daily requirement.

Fruits of studied species are characterized by high zinc content. Cloudberry fruits contain about 40 mg/kg, air-dry weight. The maximum zinc concentration is detected in ashberry.[47]

5.4   CONCLUSION

The analyses of the chemical composition of fruits of northern wild-growing berry plants revealed the following.

Fruits of the studied species contain significant quantities of biologically active compounds, vitamins, micro- and macroelements necessary for the organization of functional nutrition.

The wild fruits can be considered as a source of the most valuable carbohydrates (fructose, fiber, and pectin substances) and organic acids. The content of sugars in fruits of the studied species varies from 2.36–6.10% in European cranberry to 5–8% in cloudberry; pectin substances: from 0.6 to 1.41% in cranberry to 2.13–2.63% in cloudberry; fiber: from 2.0% in cranberry to 3.8% in cloudberry and 4.0% in cinnamon rose. The studied cranberry samples contained 2.10–4.84% of basic organic acids and cloudberry—0.8%.

Fruits of cloudberry can be an important source of neutral lipids in northern diet as it contains 2.10–3.68% of neutral lipids.

Due to the high vitamin C concentration in cranberry, Arctic bramble, and cloudberry, it is possible to recommend a daily intake of 50–70 g of cloudberry and Arctic bramble fruits or 100 g of cranberry to fulfill daily requirement.

Cloudberry is an accumulator of beta-carotene (1.1–2.4 mg/100 g) and daily dose can be reached by consuming 50 g of fruits.

Samples of the studied species had high concentrations of microelements. *Vacciniaceae* berries contained up to 63 mg/kg of manganese, cloudberry, and Arctic bramble up to 68 mg/kg. Cloudberry fruits contained about 40 mg/kg of zinc.

KEYWORDS

- cranberry
- Arctic bramble
- cloudberry
- carbohydrates
- organic acids

REFERENCES

1. Boiko, E. R. *Physiological Biochemical Bases of Human Vital Activities in North;* Ural Division, Russian Academy of Sciences: Ekaterinburg, 2005; p 190. (In Russian).
2. Volodin, V.; Chadin, I.; Volodina, S. Role of Ethnobotanical Studies in Search of Biologically Active Substances of Adaptogenic Action. *Ann. Inst. Biol. Komi Sci. Cent. Ural Dep. Russ. Acad. Sci.* **2008,** *8,* 6–10. (In Russian).
3. Petrova, V. P. *Biochemistry of Wild Fruit and Berry Plants;* Kiyev, 1986; p 287. (In Russian).
4. Hakkinen, S. Flavonols and Phenolic Acids in Berries and Berry Products. Doctoral Dissertation, Kuopio, 2000, p 94.
5. Thiem, B. *Rubus Chamaemorus* L.—A Boreal Plant Rich in Biologically Active Metabolites: A Review. *Biol. Lett.* **2003,** *40*(1), 3–13. (In English).
6. Vigorov, L. I. *Garden of Medicinal Plants;* Nature: Sverdlovsk, 1979; p 176. (In Russian).

7. Gubina, M. D. *Wild fruits and berries. Wild berries, fungi, ferns of Siberia;* Novosi-birsk, 1991; pp 18–44. (In Russian).

8. Shabrov, A. V.; Dadali, A. V.; Makarov, V. G. *Biochemical Bases of Food Micro—Components;* Avvallon: Moscow, 2003; p 183. (In Russian).

9. Tsapalova, I. E.; Gubina, M. D.; Poznyakovskyi, V. M. *Expertise of Wild Fruits, Berries and Herbaceous Plants;* Siberian University Publishing House: Novosibirsk, 2002; p 180. (In Russian).

10. Kovrigina, T. A.; Musikhina, E. D.; Obotnin, S. I.; Tuzharov, E. S. Materials on Moni-toring of Cranberry (*Oxycoccus palustris* L.) Productivity in Conditions of Southern Tundra in Purovskiy Region of Yamal-Nenets Autonomous District in 2011–2013. *Proceedings of XI Allin 2011-Nenets Autonomous District in 2011–20rent Problems of Regional Ecology and Biodiagnistics of Live Systems.* K. V. Anisimov, Eds.; 2013, pp 456–459. (In Russian).

11. Egoshina, T. L. Resources of Berry Plants and Edible Fungi in Russia. Use and Protection of Natural Resources in Russia. **2003,** 7–8, 101–103. (In Russian).

12. Martynenko, V. A.; Gruzdev, B. I. *Vascular Plants of Komi Republic*; Institute of Biology of Komi Scientific Centre of the Ural Department of Russian Academy of Sciences NIA-Priroda: Moscow, 2008; p 136. (In Russian).

13. Egoshina, T. L.; Luginina, E. A. Vaccinium vitis-idaea and *Oxycoccus palustris* in Natural Populations and Culture in Taiga Zone of Russia. *Acta horticulturae et regiotecturae* **2007,** *10,* 57–61.

14. Kovrigina, T. A.; Musikhina, E. D.; Egoshina, T. L. Material on Monitoring of Productivity of Wild Berries in Conditions of Southern Tundras of Purovskyi Regain of Yamal–Nenets Autonomous District. *Proceedings of X All. rict. ict. ene of Russia. 11rict in 2011–20rentProblems of Regional Ecology and Biodiagnostics of Live Systems.* Kirov: Vesi.; 2012; pp 47–50. (In Russian).

15. Egoshina, T. L.; Luginina, E. A. Vaccinium vitis-idaea and *Oxycoccus palustris* in Natural Populations and Culture in Taiga Zone of Russia. *Acta horticulturae et regiotecturae* **2007,** *10,* 57–61.

16. Kositsin, V. N. *Cloudberry: Biology, Resources Potential, Introduction;* Russian National Research Institute for Silviculture and Forestry Mechanisation: Moscow, 2001; p 140. (In Russian).

17. Kovrigina, T. A.; Musikhina, E. D. Productivity Parameters of Cloudberry (*Rubus chamaemorus* L.) and Cranberry (*Oxycoccus palustris*) in Southern Tundras of Yamal–Nenets Autonomous District. *Proceedings of International Scientific–Prac-tical Conference. Current Problems of Nature Management, Game Management and Fur Farming.* Russian Research Institute of Game Management and Fur Farming: Kirov, 2012; pp 375–376. (In Russian).

18. Egoshina, T. L.; Luginina, E. A. Resources of Cowberry (Vaccinium Vitis-Idaea L.) and Cranberry (*Oxycoccus Palustris* Pers.) in Natural Populations of Taiga Zone of Russian and Prospects of Cultivation. *Ann. Tver State Univ. Ser. Biol. Ecol.* **2008,** *10,* 147–154. (In Russian).

19. Chirkova, N. Yu.; Egoshina, T. L.; Kolupayeva, K. G. Some Peculiarities of Phenology and Productivity of Vaccinium Vitis-Idaea (Ericaceae) in Southern Taiga Subzone of Kirov region. *Plant Resour.* **2009,** *1,* 12–21. (In Russian).

20. Kislitsina, A. V.; Egoshina, T. L. Basic Resources and Population Parameters of Vaccinium Myrtillus L. in Southern Taiga Forest Ecosystems in Kirov Region. *Ann. Privolzhkyi State Tech. Univ. Ser.: For. Ecol. Nat. Manage. Yoshkar-Ola* **2016,** *3*(31), 77–86. (In Russian).

21. Allumnion State Standard 8756.2ern Methods of Dry Matter Detection. Standard Publishing House: Moscow, 1982; p 5. (In Russian)

22. Ermakov, A. I.; Arasimovich, A. A.; Yarosh, N. P.; Peruanskiy, Yu. V.; Lukovnikova, G. A.; Ikonnikova, M. I. *Methods of Biochemical Study of Plants;* Agropromizdat: Leningrad, 1987; p 430. (In Russian).

23. Obukhov, A. I.; Plekhanova, I. O. *Atomic Absorption Method in Biological Soil Studies;* Moscow State University: Moscow, 1991; p 183. (In Russian).

24. Pleshkova, B. V. *Plant Biochemistry Course;* Kolos: Moscow, 1985; pp 110–112. (In Russian).

25. Swain, T.; Hillis, W. The Phenolic Constituents of Prunus Domestica. 1. The Quantitative Analysis of Phenolic Constituents. *J. Sci. Food Agric.* **1959,** *10*(1), 63–68.

26. Schneidman, L. O.; Afanasyeva, V. S. Method of anthocyanin substances detection. Abstracts of 9th Mendeleev Meeting for general and applied chemistry. No. 1965, 8, 7965. (In Russian).

27. Zaprometov, M. N. *Biochemistry of Catechols;* Nauka: Moscow, 1964; p 250. (In Russian).

28. Mzhavanadze, V. V. Quantitative Detection of Chlorogenic Acid in Leaves of Foxberry (*V. arctostaphylos* L.). Rep. Acad. Sci. Georgian SSR. **1971,** *63*(1), 205–210. (In Russian).

29. Vigorov, L. I. Wild Berries and Fruits as a Source of Biologically Active Substances. Productivity of Wild Growing Berries. Kirov; 1972; pp 30–32. (In Russian).

30. State Pharmacopeia of USSR. *Common Methods of Analyses. Medicinal Raw Material,* Xi ed.; Issue 2. Moscow, 1990; p 400. (In Russian).

31. *Guide To Methods of Quality Control and Safety of Biologically Active Food Additives,* P4.1.1672lly AInstitute of Nutrition of Russian Academy of Sciences: Moscow, 2003; Date of validity 2003.06.30. (In Russian).

32. *Guide to Methods of Quality Control and Safety of Biologically Active Food Additives,* P ologically Ministry of Health of Russia. Institute of Nutrition of Russian Academy of Sciences: Moscow, 2004; p 106. (In Russian).

33. Nechayev, A. V.; Trautenberg, S. E.; Kochetkova, A. A.; et al. *Food Chemistry;* GIORD: Saint–Petersburg, 2007; p 640. (In Russian).

34. Vogulkin, K. E.; Vogulkina, N. V.; Shandrikova, L. N. Seasonal Dynamics of Biochemical Composition of Cloudberry (*Rubus chamaemorus* L.) in North of Belorussia. *Proceedings of the Conference ian Academy ofActive Substances of Plants—Study and Use"e* Central Botanical Garden of the Academy of Sciences of Belorussia: Minsk, 2013; 82–83. (In Russian).

35. Sharoglazova, L. P.; Velichko, N. A. Development of Recipe of Nonelopment of Reci with use of Cloudberry. Bulletin of Krasnoyarsk State Agrarian University. *Tech. Sci.* **2016,** *2,* 88–92. (In Russian).

36. Rupasova, Zh. A.; Garanovich, I. M.; Shpitalnaya, T. V. Prognosis of Change in Biochemical Composition of Uncommon Fruit Cultures in Conditions of Belorussia. *Rep. Natl. Acad. Sci. Beloruss.* **2013,** *57*(1), 88–92. (In Russian).
37. Baranova, I. I.; Smirnova, L. M.; Ershova, G. F. Biologically Active Substances of Wild Growing Berries of Southern Karelia. Abstracts of All-Union Meeting "Study, collection and protection of forest wild growing berries in European part of USSR in connection with natural resources management of nonchernozem belt of USSR". Karelian branch of the Academy of Sciences of USSR: Petrozavodsk; 1980, 15–16. (In Russian).
38. Rush, V. A.; Lizunova, V. V. Chemical Compositions of Wild Berries in Siberia. Productivity of Wild Berries. Russian Research Institute of Game Management and Fur Farming: Kirov, 1972; pp 42–44. (In Russian).
39. Matistov, N. V.; Valuyskikh, O. E.; Shirshova, T. I. Chemical Composition and Concentration of Micronutrients in Cloudberry Fruits (*Rubus Chamaemorus* L.) in European North–East of Russia. *Bull. Komi Sci. Cent. Ural Dep. Russ. Acad. Sci.* **2012,** *1*(9), 41–45. (In Russian).
40. Stahl, W.; Sies, H. Bioactivity and Protective Effects of Natural Carotenoids. *Biochem. Biophys. Acta.* **2005,** 1740, 101–107.
41. Golovko, T. K.; Dymova, O. V.; Lashmanova, E. A.; Kuzivanova, O. A. Content and Composition of Yellow Pigments in Fruits of Cloudberry and Bilberry in Conditions of European Part of Russia. *Ann. Samara Sci. Cent. Russ. Acad. Sci.* **2011,** *13,* 813–816. (In Russian).
42. Rush, V. A.; Lizunova, V. V. *Macro- and Micro-Elements of Wild Berries of Siberia. Productivity of Wild Berries;* Russian Research Institute of Game Management and Fur Farming: Kirov, **1972;** pp 44–47. (In Russian).
43. Egoshina, T. L.; Orlov, P. P.; Shulyatyeva, N. A. Peculiarities of Microelements Composition of Some Medicinal and Berry Plants. *Agrar. Sci. Eur. North-East* **2000,** *2,* 53–55. (In Russian).
44. Egoshina, T. L.; Orlov, P. P.; Shulyatyeva, N. A. Element Composition of Some Medicinal and Berry Plants of Ericaceae. *Proceedings of the Second International Conference. Biodiversity and Bioresources of Ural and Bordering Territories.* Gazprompechat: Orenburg, 2001; pp 97–98. (In Russian).
45. Shikhova, L. N.; Egoshina, T. L. Heavy Metals in Soils and Plants of North–Eastern Part of Russia. Scientific Research Institute of Agriculture of the North–East: Kirov, 2004; p 263. (In Russian).
46. Luginina, E. A.; Egoshina, T. L. The Peculiarities of Heavy Metals Accumulation by Wild Medicinal and Fruit Plants. *Ann. Warsaw Univ. Life Sci.–SGGW Agric.* **2013,** *61,* 97–103.
47. Egoshina, T.L.; Shikhova, L. N.; Lisitsyn, E.M.; Voskresenskaya, O. L. Anthropogenic Influence on Heavy Metals Content in Natural Flora. *Int. J. Ecol. Develop.* **2016,** *31*(4), 1–13.

Species names of wild-growing berry plants studied	
Common names	**Scientific names**
Amur grape	*Vitis amurensis* Rupr.
Arctic bramble or Arctic raspberry	*Rubus arcticus* L.
Ashberry	*Sorbus torminalis* L.
Cloudberry	*Rubus chamaemorus* L.
Bilberry	*Vaccinium myrtillus* L.
Bird cherry	*Padus avium* L.
Black currant	*Ribes nigrum* L.
Bog berry	*Vaccinium uliginosum* L.
Cinnamon rose	*Rosa majalis* Herm.
Cowberry	*Vaccinium vitis-idaea* L.
Cranberry	*Oxycoccus palustris* L.
Cranberry	*Viburnum opulus* L.
Raspberry	*Rubus idaeus* L.
Schizandra	*Schisandra chinensis* (Turcz.) Baill
Sea buckthorn	*Hippophae rhamnoides* L.

CHAPTER 6

CULTIVATION OF ARCTIC RASPBERRY ON MESOTROPHIC BOGS: PHYTOCENOTIC STUDY IN VARIOUS ECOLOGICAL CONDITIONS

YULIA V. GUDOVSKIKH, TATIANA L. EGOSHINA*,
ANASTASYA V. KISLITSYNA, and EKATERINA A. LUGININA

Prof. B. M. Zhitkov Research Institute of Game Management and Fur Farming, 79, Preobrazhenskaya St., Kirov, Russia 610000

Corresponding author. E-mail: etl@inbox.ru

CONTENTS

ABSTRACT

The study investigates the possibility of cultivation of Arctic raspberry (*Rubus arcticus* L.) on cutover peatlands in conditions of Kirov region (Volga-Vyatka region). Morphometric and productive parameters of the species are determined. Plants of Arctic raspberry on the experimental plot grow and bear fruit successfully under regional conditions; and are also characterized by high winter hardness. Arctic raspberry is a promising species for cultivation in southern taiga subzone of European part of Russia.

6.1   INTRODUCTION

The study of complex resource characteristics of Arctic raspberry *Rubus arcticus* L. in conditions of cultivation is significant because of its value as an important source of edible and medicinal raw material. Familiarization of the studies of *R. arcticus* is an infinitesimal. There are no available published data on ecological-phytocoenotic and resources investigations of *R. arcticus* L. populations, and there are few papers about the species cultivation in Russia and abroad.[1–3]

Arctic raspberry (*R. arcticus* L.) is perennial half-shrub of the family Rosaceae, facultative weed forming.[1] Arctic raspberry is pollinated by insects and is ornitho and zoochore.[4] It is also an optional burr plant which ground shoots are polycyclic with lignified lower perennial part. It has compound leaves with stipules.[5]

Flowers are actinometric, pentamerous, and pentacyclic with double perianth[5] and monoclinous; it is 1–3.7 cm in diameter, have 7–9 bright pink and red petals and 5–7 sepals; blossoms from June to August. The fruit is fleshy and complex drupe. The ripe fruits are raspberry red with strong pleasant aroma.[1]

Arctic raspberry ontogeny is presented by the following stages such as germs, old reproductive, sub-senile, senile, productive (young and middle-aged), juvenile, immature, and virginile plants (Table 6.1). Dying plants were not marked. It is impossible to define the total duration of ontogeny because of lack of observation in natural populations.

The aboveground shoots of Arctic raspberry reach 25 cm and consist of five to nine internodes. Leaves are ternate with two stipules. Folioles are obovate with sharp tip. Long cord-like roots of mature plants grow horizontally 10–25 cm deep. Species phenology is poorly studied. The most complete data is found for the Polar Urals.[6]

TABLE 6.1 Ontogeny of Arctic Raspberry (*Rubus arcticus* L.)[5] (Modified and Enlarged by the Authors).

Figure	Characteristics of plants at different stages of ontogeny
	Germs (p). Rosellate shoot with two assimilating oval seed lobes, glabrous with reticulated venation, and—two to three simple orbicular leaves with five large dents and long stake.
	Stipules are oval and not separated from the basal sheath; leaves and leaf stakes are tomentous.
	Tap-root system has noticeable major root with—one to two branch roots which sometimes branch out up to the third order. Height of the germ is 3 cm and length of underground part is 1–1.5 cm.
	Juvenile plants (j). Plants are small and rise 2–3 cm above the ground. The main shoot is rosellate. The lower leaves are small and simple with five dents. Leaf tip is blunt, leaf base is heart-shaped, venation is reticulated, and stakes length is medium (leaf type I). Upper leaves are larger with—seven to nine dents, stakes are long. Dry seed-lobe leaves can remain on a plant. Two stipules non-completely accrete to the basal sheath, the tip is free. Leaves and stakes are tomentous. The average number of leaves is 2.0–3.6. Tap-root system is presented by a well noticeable major root with branch roots of orders I, II, and IV, length of the root exceeds the height of the upperpart of the plant.
	Immature plants 1 (im 1). The height of the aboveground part is 3–3.5 cm. The shoot is rosellate with leaves of three types such as type I: lower—simple orbicular with —five to nine dents blunt tip and heart-shaped base, type II: middle—simple trifid dentate with blunt tip, and type III: upper—simple asymmetrical, one foliole is separated, blunt tip, heart-shaped base.
	Leaves and stakes are tomentous, tomentous stipules are even more separated from the basal sheath. Tap-root system is more furcated (up to III–IV orders), adventive buds are being formed on the roots.

TABLE 6.1 *(Continued)*

Figure	Characteristics of plants at different stages of ontogeny
	Immature plants 2 (im 2). Plant height is 4–5 cm, rosellate shoot with 1–3 leaves of types II and III and —two to four mature leaves of type IV. The major root starts to bend horizontally, numerous lateral secondary roots, and growing root shoots may appear.
	Virginile plants (v). Shoot type change—main shoot is vegetative, lengthened, and poorly furcated with —three to six tricompound leaves. Internodes can bend during extension. Stipules are completely separated from the leaf stake. Pubescence is less significant than the one of juvenile and immature plants. The average number of assimilated leaves is 3.1–6.2; root shoots—1.3–1.7; vegetative shoots—1.0–1.4; leaf area—4.7–7.5 cm². The major root reaches to a significant length. Horizontal secondary cord-like roots abundantly furcate and form —one to two root shoots each. Polycentric system develops. Old and new roots differ in color, new are light brown, old are dark brown.
	Young generative plants (g1). A developed polycentric system of shoots of orders I and III. The main shoot is generative, —one to two vegetative shoots are lengthened with several internodes. Internodes are shorter in the lower parts of the shoots. The growth is sympodial, bud scales and three to four large tricompound leaves with duplicodentate folioles (type IV) remain in the lower part of the shoot. Well-developed root system and partial tuft rhizome system include root shoots and short communication rhizomes. Bed continues to form. Dying-off process is not yet shown: small amount of dead leaves and few dead shoots.

TABLE 6.1 *(Continued)*

Figure	Characteristics of plants at different stages of ontogeny
	Middle-aged generative plants (g2). Polycentric shoot system including —one to two generative and two to three vegetative shoots is being developed. Four to five leaves are tricompound. Leaf area is 7.1–14.25 cm². Up to three dead leaves remain. About 0.9–4.0 root shoots; 0.6–1.0 dead shoots appear. Root system is well-developed, furcated, secondary roots are thin, non-lignified, appear in bunches. Rhizome is dark and knotty. The tuft is formed. The processes of formation and dying-off are balanced.
	Old generative plants (g3). Polycentric system of generative and vegetative shoots of orders II–III is getting simpler. The number of assimilating leaves decreases to three to four per shoot, root shoots from—one to two. Increased number of dead leaves (up to 2.5) and dead shoots (up to 2.0–3.0). Dying-off process begins to dominate. Root system includes more old dark roots.
	Sub-senile plants (ss). Plants are strong—one to three vegetative shoots and single root shoots remain. Processes of ageing continue to increase: number of assimilating leaves decreases from one to two, dead leaves—two, number of dead shoots reaches three to five. Immature and transitional leaves appear (types II and III). The edge of mature leaf (type IV) becomes dentate, not duplicodentate. Root system remains in the conditions of the previous state.
	Senile plants (s). Ageing features are distinctively seen. One to two vegetative shoots, four to five dead shoots. Number of assimilating leaves decreases and number of dead leaves rises. Juvenile and immature leaves of types II and III appear, the edge of type IV leaves is dentate. Leaf area decreases. Root shoots are rare. Both their development and presence of immature leaves can be seen as a sort of rejuvenation. The major root and rhizomes are strong, lignified, of larger diameter compared to the previous states, dark brown.
—	Dead plants not found.

Generative cycle starts with budding which is normally marked around June 11 in the region. The mass flowering is marked in the end of the second decade of June, and fruits are ripe in the end of July. The length of summer vegetation from the beginning of growth reaches the average 72.2 ± 1.6 days. Leaves start to fade in the first decade of August. Leaves vegetation is short and does not exceed 55 days.

The published studies on the peculiarities of seasonal development of Arctic raspberry are few. In the conditions of forest tundra of the Polar Urals budding starts 18–20 days after growth begins and it lasts for 13 days.[7] Arctic raspberry needs day/degrees from 210 to 250°C for mass flowering, flowering continues for 6–8 days. 46–49 days pass from the beginning of flowering to the beginning of fruit ripening, and it takes 730–780°C day/degrees. Arctic raspberry gets ripe 1 month later in the Polar Urals than in Upper Kolyma, because necessary accumulated temperatures can only be reached in the second part of August.[6] In conditions of Kirov region (Volga-Vyatka region of Russia), Arctic raspberry blossoms from the end of June till the middle of July, and fructification is solitary and marked in the beginning-middle of August.

Arctic raspberry *(R. arcticus* L.) is a Eurasian-North American Arctic-boreal species.[5]

The species is widely distributed in Northern Europe, North America, in the northern parts of Mongolia and Japan, and Korean peninsula.[5] In Russia, it is found in northern and northeastern part (from Arctic tundras to the subzone of southern taiga inclusively)[8] in the Urals, Siberia, and the Far East.[4,9–11]

Arctic raspberry is a mesophyte, according to some researchers,[5] the others attribute it to hygrophytes[11] and hemicryptophyte.[5]

It is often found as disseminated individuals or in the small groups. It grows on the marshy forest edges, on wet clearings, burnt areas, flood meadows, in sphagnum pine forests, sedge-herbaceous and sedge-sphagnum forests, on peat hillocks, forest–tundra sparse forests, and tundra.[1,12] In conditions of the Polar Urals, the species is found within taiga and tundra zones, in forests, on swamps, and in tundras.[11]

The Arctic raspberry grows everywhere in Yakutia, found in marshy and wet sparse forests, dwarf birch thickets, willow forests, tundra, and mountain meadows.[13] In conditions of upper Kolyma, the species is distributed on floodplains and riverbanks, in willow forests, dwarf birch–willow thickets, alder forests, meadows, and disturbed habitats.[6]

In Kirov region, the species is rarely found in mesotrophic herbaceous bogs and even rarer on sedge bogs overgrown with spruce and birch, marshy spruce and spruce–birch forests, on forest edges, in bushes around floodplains of forest streams, and rivers.

In the Udmurt Republic, Arctic raspberry is marked in the marshy forests, more often in the northern part and rarely in the southern such as Golyany village, Zavialovskyj district. It is the southern range border.[10]

Arctic raspberry is an important food, forage, and medicinal plant. Its fruits are used in food and liquor industries, consumed as fresh and processed products, used in national cuisine of northern populations, including Yakuts,[6,7] used for making fruit drinks, juices, and jams, and are widely used in folk medicine.

The fruits of Arctic raspberry are known for analgesic, hypolipidemic, antiemetic, antipyretic, and diuretic action and are used in treating respiratory infections, stenocardia, anemia, and podagra.[9–11,13–15]

The leaves and fruits are used for treating digestive system diseases including diarrhea and dyspepsia, leaves for rheumatic disease.[16] Fruits are used in folk medicine of northern population.[6,13,14,17] In Yakut folk medicine Arctic raspberry is used for the liver diseases and rheumatism. The experiment shows that water–alcohol extract of fruits has radioprotective action.[15]

The aboveground part of the plant contains flavonoids. The leaves contain vitamin C and fruits—carbohydrates (glucose and fructose), organic acids (pyruvic, oxalacetic, and so forth), essential oil (containing carbohydrates and derivatives), and antocyans (cyanidine, and so forth).[16]

The gourmet fruits of Arctic raspberry are the most expensive among exported wild-growing berries and demand special attention (according to Russian Federal Customs Service).[18]

Arctic raspberry is also an ornamental and technical dye plant.[16] It is nectariferous and bee-bread plant.[19] There is no data on Arctic raspberry fruit stock in Russian Federation.[1]

Arctic raspberry in natural populations does not fructify annually, the yield is low. Therefore, the hybrids of Swedish Arctic raspberry with North American subspecies *Rubus stellarcticus* G. Larsson are usually used in cultivation. This allowed collecting the hybrid seeds (*R. arcticus* L. ssp. × *R. stellarcticus* G. Larsson) which were used for forming Aura and Astra cultivars in Finland, and 'Anna', 'Linda', 'Beata', and 'Sofia'—in Sweden.[20] Currently, the most widely spread are 'Anna', 'Linda', 'Beata', 'Sofia', 'Valentina', 'Aura', and 'Astra'.[1]

The studies of Arctic raspberry cultivation began in 1960s in Finland and Sweden.[2,3] The cultivation experiments are being held in Estonian Agricultural University (Tartu) since 1995, specialist's work with planting stock from Finland and Sweden and local clones.[3] Later, specialists from the Central Forest Experimental Station (Kostroma)[1,21] and Botanical Garden of Petrozavodsk State University, who proved the possibility of use of the Finnish cultivars 'Pimaa' and 'Maspi' in the region, joined the studies.[22,23] Currently, the study of Arctic raspberry introduction is also held in Forest Institute of National Academy of Sciences in Belarus and Central Siberian Botanical Garden of Russian Academy of Sciences.[20]

Since 2005, the researchers from Central Forest Experimental Station have been studying the possibility of use of the Arctic raspberry for biological remediation of cutover peatlands, along with cranberry, cowberry, blueberry, and other berries. Cultivar planting stock of Arctic raspberry provided by Dr. Kadri Karp (Estonia) in 1998 was used for the experiment.[1]

6.2   MATERIALS AND METHODOLOGY

The experimental ground for Arctic raspberry cultivation in Kirov region (Volga-Vyatka region of Russia) was organized in 2012 in Orichevskyi district (southern taiga subzone) on partly cutover mesotrophic bog, which was withdrawn from agricultural use. The work on the area development started in June 2011 and included clearing, smoothing, rotary cultivation, keeping black fallow, use of herbicide (Raundup), and tilling.

The experimental ground is suitable for the cultivation of forest berries: 100% luminance; thickness of residual peat layer from 0.5 to 1 m; peat pH—4.8. The groundwater level during the vegetation period is at the level of 30–60 cm from the soil surface.

About 100 nursery transplants of each cultivar: 'Beata', 'Sofia', and 'Anna', were planted in the end of April. Containerized plants formed from denned cuttings in spring 2011 were placed in trays (5 cm slots) and received from Central Forest Experimental Station (Kostroma). Trays were immediately temporary planted on the experimental ground. Plants remained in the trays till bedding out which was accomplished in August 2012.

The planting pattern for permanent planting was four-rowed. The plants were placed in rows 40 cm apart, distance between the plants was 30 cm. Technological passage was 1 m. The complex mineral fertilizer (NPK) was used during the planting in the amount of 60 kg/ha of each active ingredient.

The rooting and viability during the first two winters were 100%. Further tendency included fertilization (twice per season, NPK—30 kg/ha of each active ingredient), regular weeding, mellowing (three to four times a season), and watering during the dry periods. Cutting and fragmentation were done annually. Total number of plants on the experimental ground reached at 15,000 in 2014.

The regular phenological observation was organized on the experimental ground. The number of shoots, yield volume, and major leaf morphometric parameters were marked on the experimental ground.

Morphometric parameters of the second leaf in the lower part of the shoot were defined such as length and width of each foliole, number of veins and the angle between the lower lateral and central veins on each side of the foliole (Figure 6.1).

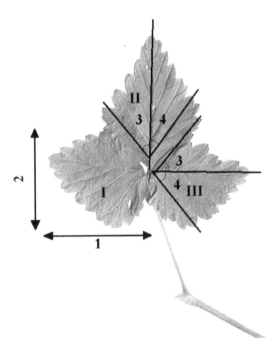

FIGURE 6.1 Definition of leaf morphometric parameters: 1—foliole length; 2—foliole width; 3—angle between central and lower vein (left side); 4—angle between central and lower vein (right side); I—1 foliole; II—2 foliole (central); III—3 foliole.

The statistical data processing was accomplished according to common methods with use of STATISTICA and EXCELL software. For each arithmetical average, we defined error value (M±m) and variation coefficient

(*CV*). The following levels of parameters variability were taken: $CV > 20\%$ high, $CV = 11–20\%$ mean, $CV < 10\%$ low.

6.3   RESULTS AND DISCUSSION

In the end of June 2012, 60% of last year shoots of Arctic raspberry remaining in trays formed fruits, one to two per shoot. In August 2013, all plants fructified, the average level of productivity was 680 ± 2.1 kg/ha with no significant difference within studied cultivars. Thus, the productivity of 'Beata' reached 780 ± 8.4 kg/ha, 'Sofia' 380 ± 2.3 kg/ha, and 'Anna' 710 ± 7.3 kg/ha. All plants fructified in 2014. The level of productivity was slightly lower than in 2013 being 220 ± 1.9 kg/ha for 'Beata' and 'Sofia' was 205 ± 2.4 kg/ha, and 'Anna' was 245 ± 2.8 kg/ha. In 2015, the productivity of cultivars reached 'Beata' 3384 ± 11.88 kg/ha, 'Sofia' 4088 ± 27.88 kg/ha, and 'Anna' 1044 ± 1.39 kg/ha.

It is worth mentioning that during the whole period of cultivation of Arctic raspberry, its productivity in southern taiga conditions of Kirov region was high and even exceeded the values reached by other researchers from different regions.[1–3]

In 2015, we studied morphometric parameters of shoots (shoot height and number of leaves), morphometry of fruit (height and diameter), number of generative and vegetative shoots, average number of fruits per shoot, and their weight, number of drupes per fruit. The results are given in the Table 6.2.

The largest number of shoots is marked for 'Beata', the least for 'Anna'. The total number of shoots per square meter varied from 188 ('Anna') to 1056 ('Beata'). The maximum number of vegetative shoots was marked for 'Beata' (732) and the minimal was for 'Sofia' (544). The largest value of average generative shoot height parameter is found for 'Anna' cultivar (14 cm), the smallest was for 'Sofia' (9.8 cm). The average height of vegetative shoot is the highest for 'Beata' (10 cm), the lowest was for 'Sofia' (7.1 cm). The 'Sofia' cultivar is characterized by the largest number of fruits (416), for 'Anna' cultivar is by the smallest (200).

Fruit diameter of the studied cultivars varied from 14.74 ± 0.52 mm ('Sofia') to 15.39 ± 0.51 mm ('Beata'). Fruit height average is 11.88 ± 0.34 mm, number of drupes is 19.36 ± 2.15. The results of measurements of fruit morphometric parameters (diameter and height) and number of drupes are presented in the Table 6.3.

TABLE 6.2 Morphometric Parameters and Productivity of Arctic Raspberry of Cultivars 'Beata', 'Sofia', and 'Anna' in 2015 year.

Cultivar	Total number of shoots per m²	Number of vegetative shoots, pcs	Average height of generative shoot (cm)	Average height of vegetative shoot (cm)	Number of fruits	Productivity (g/m²)
Beata	1056.00 0 19.78	732.00 3 13.61	11.10 1 0.81	10.00 0 0.33	416 1 19.15	3384.00 311.88
Sofia	308.00 014.84	544.00 4 8.33	9.80 . 0.95	7.40 . 1.44	480 832.42	4088.00 027.88
Anna	188.00 83.79	660.00 610.60	14.00 4 2.15	8.10 . 0.49	2000 6.12	1044.00 01.39

TABLE 6.3 Morphometric Parameters of Fruits of Arctic Raspberry Cultivars.

Cultivar	Fruit diameter	Fruit height	Number of drupes
Beata	15.39±0.51 16.71	12.19±0.38 16.71	21.47±1.9 34.27
Sofia	14.74±0.52 16.71	12.47±0.35 13.85	20.73±2.54 47.52
Anna	14.86±0.31 16.71	10.97±0.3 13.74	15.87±2.02 49.39

Note: In numerator—average±standard error; in denominator—variation coefficient (*CV*).

The largest indicator of the value of average number of drupes in a fruit is marked for 'Beata' and minimum for 'Anna' (see Table 6.3). The average number of leaves on a shoot varied from 3.2±0.2 ('Anna') to 4.0±0.3 ('Sofia'). The maximum number of leaves for 'Anna' cultivar reached 6, 'Beata'—6, and 'Sofia'—8. The average shoot height varied from 128.1±5.5 mm ('Anna') to 133.2±4.3 mm ('Sofia'); maximum shoot height—from 84.2 mm ('Anna') to 181.0 mm ('Sofia'), minimum—from 84.2 mm ('Anna') to 97.6 mm ('Sofia'). Shoot height and number of leaves per shoot of the studied Arctic raspberry cultivars are shown in the Table 6.4.

TABLE 6.4 Shoot Height and the Number of Leaves per Shoot of Arctic Raspberry of the Studied Cultivars.

Cultivar	Parameter	Number of leaves per shoot	Shoot height, mm
Anna	Average±standard error	3.2±0.2	128.1±5.5
	Maximum value	6	180.3
	Minimal value	2	84.2
Beata	Average±standard error	3.5±0.2	132.8±5.8
	Maximum value	6	189.6
	Minimal value	1	89.8
Sofia	Average±standard error	4.0±0.3	133.2±4.3
	Maximum value	8	181.0
	Minimal value	2	97.6

Morphometric parameters of model shoots of Arctic raspberry are presented in the Table 6.5.

TABLE 6.5 Morphometric Parameters of Model Shoots of Arctic Raspberry.

Parameter		'Anna'			'Beata'			'Sofia'		
		One foliole	Two foliole (central)	Three foliole	One foliole	Two foliole (central)	Three foliole	One foliole	Two foliole (central)	Three foliole
Foliole width, mm		22.6±1.0	23.4±0.9	23.4±1.0	19.8±0.7	22.1±0.8	19.8±0.8	21.6±0.8	24.7±1.0	21.4±1.0
Foliole length, mm		29.7±1.1	34.0±1.1	29.1±1.4	27.1±0.8	33.6±1.2	27.1±0.9	28.8±1.2	36.1±1.6	28.8±1.0
Number of veins of foliole	left	6.0±0.2	6.1±0.1	5.6±0.1	6.0±0.2	6.1±0.1	5.7±0.1	6.1±0.2	6.3±0.1	6.0±0.2
	right	5.9±0.2	6.0±0.2	6.2±0.2	5.9±0.2	5.9±0.2	6.1±0.2	5.9±0.1	6.0±0.1	5.8±0.1
Angle between veins	left	41.1±1.3	38.6±1.2	40.8±1.3	41.5±1.7	41.6±1.5	37.4±1.5	43.7±1.3	44.8±1.3	40.9±1.3
	right	41.9±1.4	40.1±1.4	39.2±1.1	38.1±0.9	39.5±1.9	41.1±1.3	40.9±1.7	46.8±1.5	44.7±1.7

The width of central foliole varies from 22.1 ± 0.8 mm ('Beata') to 24.7 ± 1.0 mm ('Sofia'). The length of foliole is in from 33.6 ± 1.2 mm ('Beata') to 36.1 ± 1.6 mm ('Sofia'). The number of veins on the left half of a foliole is the smallest for 'Anna' cultivar (5.6 ± 0.1), the highest is for 'Sofia' (6.3 ± 0.1); the same parameter on the right half of follicle is minimal for 'Sofia' (5.8 ± 0.1) and maximum for 'Anna' (6.2 ± 0.2). The angle between the veins of the left half of foliole varies from 38.6 ± 1.2 ('Anna') to 44.8 ± 1.3 ('Sofia'), the same parameter for the right half is in the range from 38.1 ± 0.9 ('Beata') to 46.8 ± 1.5 ('Sofia').

6.4   CONCLUSIONS

Arctic raspberry is a promising species for cultivation in conditions of Kirov region (Volga-Vyatka region of Russia).

The experience of introduction of Arctic raspberry in southern taiga subzone of Kirov region allowed determining the following:

1. Plants of Arctic raspberry successfully grow and fructify on the experimental ground in the conditions of the region, they give abundant vegetative shoots and are winter-hardy;
2. Short-term spring and autumn waterlogging does not have negative effect on the state of plants;
3. Productivity of the species on cutover mesotrophic peatlands is high and reached 3384 ± 5.8 kg/ha ('Beata'), 4088 ± 8.7 kg/ha ('Sofia'), and 1044 ± 3.0 kg/ha ('Anna') in 2015.

During the whole period of Arctic raspberry cultivation its productivity in southern taiga conditions of Kirov region was high and exceeded other known results received in different regions.

KEYWORDS

- southern taiga
- ontogeny of *Rubus arcticus L.*
- peatlands
- introduction

REFERENCES

1. Tyak, G. V.; Tyak, A. V. Growth and Fructification of Arctic Raspberry on Cut-over Peat Lands, Modern Problems of Botany, Microbiology and Nature Management in Western Siberia and Bordering Areas, *Proceedings of Russian Scientific Conference Dedicated to 10-years Anniversary of Department of Botany and Microbiology in Surgut State University,* Surgut, 2015, 183–185 (in Russian).
2. Kokko, H.; Hämäläinen, J.; K.mälainen, S. Cultivation of Arctic Bramble in Finland is Seriously Disturbed by Downy Mildew. Forestry Studies XXX. *International Conference on Wild Berry Culture: An Exchange of Western and Eastern Experiences,* Tartu, 1998; 82–86.
3. Karp, K.; Starast, M. Domestication of Estonian Natural Arctic Bramble, Forestry Studies XXX. *International Conference on Wild Berry Culture: An Exchange Experiences,* Tartu, 1998; 70–75.
4. Gubanov, I. A.; Kiselyova, K. V.; Novikov, V. S.; Tikhomirov, V. N. *Rubus arcticus* L., *Illustrated Plants Guide for Middle Russia;* Scientific Publishing Association KMK, Institute of Technological Studies: Moscow, 2003; Vol. 3, p 402 (in Russian).
5. Ontogenetic Atlas of Medicinal Plants. *Study Guide;* Mari State University: Yoshkar-Ola, 1997; p 240 (in Russian).
6. Rozhdestvenskyi, Y. F. *Rhythm of Seasonal Development of Some Plants in the Polar Urals;* Infomation Materials, Sverdlovsk, 1981; p 44 (in Russian).
7. Sinelnikova, N. V.; Pakhomov, M. N. *Seasonal Life of Upper Kolyma Nature;* Institute of Technological Studies. Scientific Publishing Association KMK: Moscow, 2015; p 329 (in Russian).
8. Vascular Flora of Taymyr Peninsuls and Neighboring Territories. P. 1. KMK. Moscow. p 457. (in Russian).
9. Siberian Flora In *Rosaceae;* Polozhyi, A. V.; Malysheva, L. I., Ed.; Nauka Publishers: Novosibirsk; 1988; Vol. 8, pp 29–30 (in Russian).
10. Baranova, O. G.; Puzyrev, A. N. *Flora Review of the Udmurt Republic;* (*Vascular plants*), Monograph, Institute of Computer Studies: Izhevsk; 2012; p 212 (in Russian).
11. Martynenko, V. A.; Gruzdev, B. I. Vascular Plants of Komi Republic, Institute of Biology of Komi Scientific Centre of the Ural Department of Russian Academy of Sciences, Syktyvkar, 2008; p 136 (in Russian).
12. Egoshina T. L. Floristic Studies in the North of Kirov Region. Flora of the Northern Part of USSR. Arkhangelsk; 1987; pp 55–56 (in Russian).
13. Makarov, A. A. *Medicinal Plants of Yakutia and Usage Prospects;* Publishing House of the Siberian Department of Russian Academy of Sciences: Novosibirsk, 2002; p 264 (in Russian).
14. Egorov, A. D. *Vitamin C and Carotene in Plants of Yakutia;* Publishing House of USSR Academy of Sciences: Moscow, 1954; p 248 (in Russian).
15. Lapinskyi, A. G.; Gorbachyov, V. V. Anti-Residual Activity of Extracts of Some Wild Plants of Northern Areas of Sea of Okhotsk. *Chem. Pharm. J.* **2006,** *40*(6), 27–29 (in Russian).
16. Budantsev, A. L.; Lesiovskaya, E. E. *Wild Useful Plants of Russia;* Publishing House of Saint-Petersburg Chemical Pharmaceutical Academy: Saint Petersburg, 2001; p 663 (in Russian).

17. Makarov, A. A.; Pryamkova, N. A. *To the Studies of Tannin Content of Plants of Central Yakutia;* Publishing House of Yakutia State University: Yakutsk, 1977; pp 8–39 (in Russian).
18. Kurlovitch, E. L. Prospects of Use of Edible and Medicinal Forest Resources of Russian Federation. *Proceedings of Russian Conference with International Attendance, State of Far Eastern Forests and Current Problems of Forest Management,* Far Eastern Research Institute of Forest Economy: Khabarovsk, 2009; pp 138–140 (in Russian).
19. Plant Resources of USSR. Flowering Plants, Their Chemical Composition, Use. Familia Hydrangeaceae—Haloragaceae. Leningrad: Nauka, 1987; p 326 (in Russian).
20. Gorbunov, A. B.; Moiseeva, N. V.; Simagin, V. S.; Snakina, T. I.; Boyarskikh, I. G.; Fotev, Yu, V.; Kudryavtseva, G. A.; Belousova, V. P. Introduction and Selection of Edible Plants in Central Siberian Botanical Garden of Siberian Department of Russian Academy of Sciences. *Bull. Vavilov Soc. Genet. Selectionists* **2005,** *9*(3), 394–406 (in Russian).
21. Tyak, G. V. Use of Mineral Furtilizers in Growing Raspberry on a Cut-over Peat Land. Theoretical and Practical Aspects of Rational Use and Restoration of Non-Wood Forest Products, Gomel, 2008; pp 305–308 (in Russian).
22. Kirilkina, T. I. Development of Agricultural Technics for Arctic Raspberry (*Rubus arcticus* L.) for the Agriculture of Karelia. *Proceeding of International Conference Dedicated to the 50-years Anniversary of the Department of Agronomy and Soil Science of Petrozavodsk State University, Crop Farming in the European North: State and Prospects,* Petrozavodsk 2004; pp 58–60 (in Russian).
23. Kirilkina, T. I. Development of Agricultural Technics for Arctic Raspberry (*Rubus arcticus* L.) for the Agriculture of Karelia. Proceeding of International Conference Dedicated to the 125-years Anniversary of Tver University Botanical Garden, Life in Harmony: Botanical Gardens and Society, Tver. Tver University Publishing House, 2004; p 102 (In Russian).
24. Yatsina, A. A.; Kontsevaya, I. I. Reproduction and Introduction of Arctic Raspberry (*Rubus arcticus* L.) in Belarus. *Horticulture* **2004,** *15*, 207–211 (in Russian).

CHAPTER 7

CHARACTERISTICS OF BLUEBERRY RESOURCES IN NATURAL POPULATIONS AND THE PROSPECTS OF CULTIVATION IN THE TAIGA ZONE

NATALYA V. KAPUSTINA, EKATERINA A. LUGININA, and TATIANA L. EGOSHINA*

Prof. B. M. Zhitkov Russian Research Institute of Game Management and Fur Farming, 79, Preobrazhenskaya St., Kirov, Russia 610000

Corresponding author. E-mail: etl@inbox.ru

CONTENTS

ABSTRACT

The chapter presents the results on phytocoenotic confinement, productivity, and resources of *Vaccinium uliginosum* L. in taiga regions of Russia. It also marks irregular character of distribution of bilberry fruits relative density within the territory of Russia, and as on fructification fluctuation, blueberry areas decrease, and increase of anthropogenic press. The possibility of *Vaccinium angustifolium* Ait. cultivation on cutover peatlands in southern taiga zone of Kirov region (Volga-Vyatka region of Russia) is demonstrated. Winter hardness and productivity parameters of the species were defined.

7.1 INTRODUCTION

Modern socioeconomic conditions reveal a sharp increase of the anthropogenic press on natural populations of economically important species of the wild growing plants. This leads to understand the necessity of wild berries resources and their rational use based on data on productivity, patterns of special distribution, resources, and usage peculiarities. Nevertheless, phytocoenotic confinement, yield, and stocks of these species in Russia remain under insufficient study. One of the ways of decreasing anthropogenic press of natural populations of berry plants and supplying people and food industries with valuable raw material is organization of industrial plantations. It demands revealing the most resistant species and cultivars for taiga conditions. The most popular wild growing berries in Russia are genus *Vaccinium* L. species, including the only blueberry species (*Vaccinium uliginosum* L.) presented in natural populations in taiga zone of the country.

The bog blueberry (*V. uliginosum* L.) is a circumpolar Holarctic species. The range of bog blueberry includes two continents: North America and Eurasia. The northern border of the range is in the European part (Norway). Within the Russian, west bog blueberry is spread on the Kola Peninsula and Arctic Ocean Islands. The range also includes the whole tundra region from Siberia to Kamchatka. The blueberry is widespread in Japan, Kuril Islands, Arctic parts of North America, and Greenland.[1]

The bog blueberry range reaches north Turanian and East Asian floristic regions.[2] Within its range, the species is decided into two subspecies

V. uliginosum ssp., inhabiting taiga zone, and *V. microphyllum* (Lange) Tolm. spread in northern taiga, forest tundra, tundra, and mountain tundra.[3]

The bog blueberry (*V. uliginosum* L.) is a deciduous low shrub of 0.05–1.6 m height, with alternating leaf arrangement. Leaves are obovate or elliptic, 10–30 mm long and 5–20 mm wide, with exserted veins. Flowers are regular, bisexual, droop, binate perianth, and quadripartite or quinquefid cups. Corolla is urceolate, white or pink, 3–6 mm long, sympetalous with 4–5 ovate or broad-ovate dents. It blossoms in June–July and fructifies in August–September.[4,5]

The bog blueberry fruits are rich in biologically active compounds, which allow using them as a raw material for food and pharmaceutical industries.[6]

The several species of blueberry are being cultivated in different northern regions such as highbush blueberry (*Vaccinium corymbosum* L.), smallflower blueberry *(Vaccinium virgatum* Ait.)*, and so forth. The most often *V. corymbosum* L. is cultivated. This species and its hybrid cultivars are being cultivated for more than 100 years. The highbush blueberry (*V. corymbosum* L.) is successfully cultivated in Belarus[6–8] and Baltics.[9–11] In 1986–1993, there were attempts of cultivation of *V. corymbosum* L. and its hybrid varieties in Kirov region, but in regional conditions, shoots were annually frostbitten to the level of snow cover. The possible reason for that is deficient length of vegetation period and low accumulated positive temperatures. Kurlovich and Bosak[6] determined that mid-ripening cultivars of blueberry demanded 2000–2250°C of total heat, late ripening is in diapason 2200–2500°C.

Some researchers suppose that in taiga subzone of Russia, lowbush blueberry (*Vaccinium angustifolium* Ait.) is the most adapted one to various conditions and has the highest economic value among introduced short-growing blueberries.[1,12–14]

The lowbush blueberry (*V. angustifolium* Ait.) is a short-growing plant[15] which naturally inhabits the northwestern USA—Maine and Oregon; and southeastern and eastern Canada—British Columbia, Quebec, and Ontario.[7] The blueberry rises to 1900 m above sea level. The species forms thick bushes on forest clearances, fire-sites, abandoned agricultural lands on dry sandy soils and open rocky hills.[13] *V. angustifolium* gradually pushes out highbush blueberry in areas where cyclical burning occurs. It often forms the thick bushes.[15,16] In Canada and the USA, the area of wild growing *V. angustifolium* Ait. bushes reaches 50,000 ha, and exploitation yield of berries is about 60,000 t.[17]

The lowbush blueberry (*V. angustifolium*) in natural populations is a 0.1–0.6 m deciduous shrub with tomentose shoots. Leaves are elliptic with scalloped edge, 15–41 mm long, and 6–20 mm wide. Inflorescence, truss in a leaf base or terminal, includes up to 15 white or pink cylindrical flowers of 4–6 mm length. The small-size berries (5–10 mm in diameter) have a bright light-blue to black color and contain 10–65 small seeds. Fruits are tasty and contain biologically active compounds, have medicinal and preventive properties, and are of high demand on American market.[13,18] During the recent years, there is an increased interest to cultivation of wild berries, especially *Vaccinium* species which is viewed as a way of widening the assortment of fruit cultures for northern horticulture and forming stable base for food industries. However, there is a lack of studies dealing with species introduction in the country. Experiments of *Vaccinium* species cultivation in taiga regions of the country carried out in Kostroma Region in Central Forest Experimental Station of Russian Scientific Institute of Forestry and Mechanization selected the best varieties of plants and introduced them to culture.[12,19,20] Some species of wild berries cultivated in Kirov region are cowberry (*Vaccinium vitis-idaea* L.), European cranberry (*Oxycoccus palustris* Pers. sin. *Vaccinium oxycoccus* L.), and American cranberry [*Oxycoccus macrocarpus* (Ait.). Pers. sin. *Vaccinum macrocarpon* Aiton].[21,22] The research on introduction of *V. uliginosum* L. to culture began in 1966 in Central Botanical Garden of Russian Academy of Sciences (Moscow), in 1969 in Central Siberian Botanical Garden of Siberian Department of Russian Academy of Sciences (Novosibirsk), in 1986 in Kostroma Forest Experimental Station of Russian Scientific Institute of Forestry and Mechanization, and in 1975 in Biological Institute of Karelia Scientific Centre of Russian Academy of Sciences. Introduction of *V. angustifolium* began in 1970 in Central Botanical Garden of Russian Academy of Sciences[23] and in 1981 in Central Siberian Botanical Garden of Siberian Department of Russian Academy of Sciences.[13] Results of the study show prospects of the cultivation of both *V. uliginosum* and *V. angustifolium* in Kostroma region[20] and bog blueberry in Novosibirsk region.[13]

The purpose of the work was to determine the parameters of productivity of *V. uliginosum* in natural populations in Russia and study ecological and biological peculiarities of lowbush blueberry *V. angustifolium* while introducing and estimating the prospects of the species in conditions of southern taiga subzone of Russia.

7.2 MATERIAL AND METHODOLOGY

The research was carried out in taiga regions of Russia located in different federal districts. A significant part of the work including blueberry introduction was accomplished in Kirov region which was chosen as a model region because of its natural conditions typical for taiga zone of Russia and presence of significant quantity of bog blueberry natural populations.

The productivity and resources of *V. uliginosum* L. were determined by using the traditional methods of resources investigation.[24,25] To study the natural *V. uliginosum* L. populations, we used the method of constant sample plots and key areas with further extrapolation of data on same type phytocoenoses,[26] yield scoring method,[24] questionnaires, polling, and structured marketing observations.

Archives of the Department of Ecology and Resources of Russian Research Institute of Game Management and Fur Farming were also used in the study.

An investigation of ecological and biological parameters of *V. angustifolium* in conditions of introduction was pursued by using the common methods.[27] Phenological observations of plants development were carried out according to I. N. Beideman[28] methodic approach.

7.3 RESULTS AND DISCUSSION

V. uliginosum is found within taiga zone in coniferous long moss, pleurocarpous moss, low shrub-sphagnum forest types, and low shrub tundras.

The blueberry habitats in Kirov region include low-density (crown density 0.2–0.4) mature sedge-sphagnum pine forests and middle-aged low-density sedge-sphagnum birch forests; The northeast of the region consists meso-oligotrophic sedge-sphagnum bogs. Abundance of blueberry is higher in low-density sedge-sphagnum pine forests.

The stand of pine and birch sphagnum forest is usually formed by more or less depauperated pine and birch trees, whose growth is slow and reproduction is weak. Herbaceous-shrub layer is poorly developed and presented by *V. uliginosum* L., *Ledum palustre* L., *Chamaedaphne calyculata* (L.) Moench, *Andromeda polifolia* L., *V. oxycoccus*, *on Oxycoccus palustris* Pers., *O. microcarpus* (Turcz. ex Rupr.), *Eriophorum vaginatum* L., *Salix myrtilloides* L., *Menyanthes trifoliata* L., species of genus *Carex*, *Comarum palustre* L., and on hillocks, *Vaccinium myrtillus* L. and

Eriophorum polystachyon L. are found. The solid moss cover is presented by *Sphagnum wultianum* Girg, S. *girgensohnii* Russ., and *S. sguarcosum* Grome; rarely separate plants of *Thuidum tamariscinum* (Hedw.) Schimp. are found between hillocks and genus *Polytrichum* species, the most widespread of which is *P. commune* L. is found on hillocks. In these forest types, blueberry is scattered or found in the large groups on hillocks and elevated areas. Fructifying area in middle taiga subzone is about 10% and 5 to 10% of low-density stands area is present in southern taiga and coniferous-broadleaved forests.

The area of blueberry habitats in Kirov region is 53,000 ha. A significant part of it (63%) is located in northern areas of the region in middle taiga subzone (Table 7.1).

TABLE 7.1   Fructifying Area and Stocks of *V. uliginosum* L. Berries in Different Subzones of Kirov Region.

Subzone	Habitats area, thousand ha	Fructifying area, t/ha	Biological stock (BS), t	Exploitation stock (ES), t
Middle taiga	33.0	3.32	598.1	187.0
Southern taiga	19.7	1.42	152.9	81.8
Coniferous-broadleaved forests	0.3	0.040	2.8	2.1
Total	53.0	4.78	758.3	270.9

More than one-third (37%) of blueberry habitats are located in southern taiga subzone. Less than 1% (261 ha) of blueberry habitats are marked in coniferous-broadleaved forests subzone. Fructifying area is 10% of total blueberry habitats area in middle taiga, 7.2% in southern taiga, and 14.2% in coniferous-broadleaved forests subzone; its largest value defined for middle taiga subzone is 3320 ha (69% of total fructifying area in the region). Among all districts of the region, Verkhnekamsk is found to have maximum fructifying area (2504 ha).

Accessibility of habitats has increased during the last years and varies from 70 to 100% in different districts.

V. uliginosum biological stock (BS) varies from 275.5 to 1082.0 t, and average 758.3 t. These values are much lower than in other regions of Russia that are richer in blueberry resources. For example, Krasnoyarsk regions has 370,000 t of blueberry BS, Khabarovsk Krai 190,000 t (thousand metric tons), and Irkutsk region 180,000 t. In average fruitful years, Kirov region only reaches 0.6% of total blueberry BS in Russia.

The maximum values of blueberry BS in Kirov region are marked in northern districts, located in middle taiga subzone, and they provide 80% of *V. uliginosum* BS in the region. The BS value defined in Verkhnekamsk district is 83%. The BS of more than 20 t was defined in six districts: Verkhnekamskiy (500.8 t), Omutninskiy (47.0 t), Luzskiy (40.5 t), Kotelnichskiy (33.0 t), Nagorskiy (26.5 t), and Belokholunitskiy (26.0 t).

A relative density of blueberry fruits BS in the region is low and reaches 6.2 t/thousand km². It is almost 40 times less than in Irkutsk region, Khabarovsk Krai, Khanty-Mansiysk Autonomous District, and Tomsk region (234.4–220.9 t/thousand km²). The maximum BS density among Kirov region districts is marked in Verkhnekamskiy (4.8 kg/thousand ha) and Omutninskiy (1.0 kg/thousand ha).

At average level of fructification which is common for blueberry areas of the region, exploitation stock (ES; trade yield) is about 35% of BS and reaches 270.9 t. The value is 2–10 times lower than in bordering territories—Vologda (400 t), Nizhegorodskaya (400 t), Perm (1300 t), Arkhangelsk (3000 t) regions, and Komi Republic (3000 t).

The ES is about 69% concentrated in northern districts of the region, including 55% in Verkhnekamskiy district. The blueberry ES (150 t) marked in southern taiga subzone is about 30%. A significant ES is defined in Omutninskiy (24 t), Belokholunitskiy (18 t), and Kotelnichskiy (16 t) districts. In these districts, limited commercial cropping of blueberry is possible.

Organized collection of *V. uliginosum* never occurred in the region. The blueberry was only collected by population for personal consuming. Usually, blueberry collection was combined with collection of cranberry and cowberry as an associate species, but during the last 15 years, blueberry collection by population became common. Habitat development varies from 50 to 70% in northern and central districts to 100% in southern.

The maximum *V. uliginosum* productivity in European part of Russia reaches 505 kg/ha, [29] Siberia 620 kg/ha,[24,30,31] and the Far East (530–1300) 168.0 kg/ha.[32,33] *V. uliginosum* productivity varies depending on geographic position and phytocoenosis type. Thus, in Kirov region, blueberry productivity parameters vary in different phytocoenoses types and subzones (Table 7.2).

TABLE 7.2　*V. uliginosum* L. Berry Productivity in Different Plant Subzones in Kirov Region (kg/ha).

Subzone	Productivity*
Middle taiga	30–260; 128.3±18.0
Southern taiga	20–200; 107.1±16.0
Coniferous-broadleaved forests	40–110; 62.5±9.1

Note: *max–min; average

The blueberry coenopopulations in low-density sedge-sphagnum bogs in the northern districts of the region (Verkhnekamsk) are characterized by the highest productivity. The long-term average annual productivity in these areas reaches 200±28 kg/ha, varying from 60±8.4 kg/ha in low fruitful years to 260±32.0 kg/ha in highly fruitful years (see Table 7.2).

In middle taiga subzone, the long-term average annual productivity of blueberry is 128.3±18.0 kg/ha, varying from 30±3.8 to 260±32.0 kg/ha; in southern taiga 107.1±16.0 kg/ha, varying from 20±3.1 kg/ha to 200±19.6 kg/ha; and in coniferous-broadleaved forests subzone 62.5±9.1 kg/ha, varying from 40±5.3 kg/ha to 110±13.8 kg/ha.

The blueberry productivity in the region is gradually decreasing southward. The blueberry collection is only profitable in middle and south taiga subzones.

The bog blueberry productivity in Kirov region is also lower than in the most bordering areas. Thus, average productivity in the northeast of European part of Russia in highly fruitful years is 300 kg/ha, in medium fruitful years 200 kg/ha, and in the northwest 350 and 210 kg/ha, correspondingly.[34] Puchnina[35] mentioned blueberry productivity varying from 117.1±3.1 to 946.0±168.0 kg/ha in blueberry-green moss birch forest in Pinezhskiy State Reserve (Arkhangelsk region). In middle taiga with swampy pine forests in northwestern part of Perm region, the long-term annual average blueberry productivity reaches 250 kg/ha.[36] Literary data on blueberry productivity in northern taiga subzone deal with swampy pine forests and bogs. Antonova[37] estimated the blueberry productivity on pine-cotton low-grass shrub bog in Pechyoro-Ilych Reserve (Komi Republic) as 44–84 kg/ha. Barykina[38] mentions that blueberry productivity in sphagnum pine forests in Kandalakshskiy Reserve (Murmansk region) varies from 42 to 160 kg/ha. According to Pushkina,[39] the maximum blueberry productivity reached 550 kg/ha in 15–20 years old fire-sites in birch-pine blueberry-shrub forests in Laplandskiy Reserve

(Murmansk region). Grimashkevich[40] also marked higher average productivity on constant sample plots in pine forests of sedge-moss and long moss pine types (102 ± 16.2 to 676 ± 56.2 kg/ha), and bilberry and ledum (134 ± 15.8 to 842 ± 23.8 kg/ha) than in sphagnum pine forests (68 ± 8.3 to 311 ± 27.4 kg/ha). According to Grimashkevich, differences in productivity can be explained by variations in groundwater level which is optimum 40–90 cm. The groundwater level in sphagnum and sedge-sphagnum pine forests usually reaches 15–30 cm and conditions favorable for fructification are only marked during the dry years. Rusakov[41] gives the values of average blueberry productivity for southern part of Novgorod region (border of southern taiga and coniferous-broadleaved forests) that are close to the values given for coniferous-broadleaved forests subzone of Kirov region—72–75 kg/ha (2–31 kg/ha minimum and 159–319 kg/ha maximum).

The blueberry productivity increases eastward and maximum levels are marked in conditions of medium level of groundwater. Thus, according to Nekratova,[30] in Western Siberia (north of Tomsk region), blueberry productivity on low shrub-green moss clearance in 10–15 years old reaches 328.4 ± 44.1 kg/ha, on low shrub-green moss-sphagnum high-moor bogs 233.1 ± 30.8 kg/ha, on transitional low shrub-sphagnum bog 114.3 ± 14.5 kg/ha, in central part of Western Siberia 150–500 kg/ha, in Eastern Siberia 150–450 kg/ha, and in south of the Far East 200–650 kg/ha.[34]

BS of blueberry fruits in Russia in average fruitful year reaches 1,261,900 t and ES 354,700 t (Table 7.3).

TABLE 7.3 Productivity and Stocks of Blueberry Fruits in Some Federal Districts of Russia.

Federal district	Productivity, kg/ha		BS, thousand t	ES, thousand t
	limits	average		
Northwestern	2–505	100	38.6	9.4
Central	50–150	100	1.3	0.4
Volga	50–150	100	4.8	2.2
Ural	100–946	300	245.1	122.0
Siberian	114–620	300	647.5	155.8
Far Eastern	150–1300	500	324.6	64.9
Total	—	—	1261.9	354.7

Note: BS—biological stock, ES—exploitation stock.

More than half of blueberry fruits resources is concentrated in Sirerian (51% BS and 44% ES), and about a quarter in Far Eastern (26% BS and 18% ES) federal districts. Among separate territories, the highest stocks are marked in Krasnoyarsk Krai (BS—370,000 t, ES—74,000 t), Khabarovsk Krai (BS—190,000 t, ES—38,000 t), Irkutsk region (BS—180,000 t, ES—40,000 t), Khanty-Mansiysk Autonomous District (BS—120,000 t, ES—60,000 t), and Yamal-Nenets Autonomous District (BS—100,000 t, ES—50,000 t). The minimum stocks of blueberry fruits were marked in Central (Yaroslavl, Smolensk, and Moscow regions) and Volga (Tatarstan Republic) Federal Districts. Relative BS density dispersal within Russia is irregular. The maximum density of raw material stocks (more than 200 t/thousand km^2) is common in Irkutsk region, Khabarovsk Krai, Khanty-Mansiysk Autonomous District, and Tomsk region.

The level of berry habitats development differs between species and regions: from 10% in the most remote areas in Siberia and Far East to 90% in European part of the country, in urban surroundings, along the roads, which leads to degradation of highly productive phytocoenoses, fading of the most productive varieties, and loss of genetic material. The analyses of bog blueberry fruits stocks dynamics in taiga zone of Russia during 1691–2004 showed that habitats decreased 40% because of area transformation.

Unstable fructification of bog blueberry in natural populations, sharp decrease of raw material stocks, and the presence of significant areas of dried peatland, which occupy 60,000 ha in Kirov region,[42] makes the works on the species cultivation important. In the harsh climatic conditions in taiga zone of Russia where the assortment of horticultural crops is small, cultivation of berry shrubs can be a stable base for national horticulture.

The investigation of *V. angustifolium* introduction was carried out on the experimental plot in Orichevsk district of Kirov region (southern taiga subzone). The study area is situated in the moderate continental climate zone with long snowy and cold winter and moderately warm summer. The average annual temperature is 1.5°C.[43] In winter, air temperature can reach at −54°C. Duration of frostless period is about 102 days. Accumulated positive temperatures reach at 2107°C. Some climatic parameters of the study area are presented in Tables 7.4 and 7.5.

The plot is located on partly meliorated mesotrophic bog where the upper peat layer was cut off during the peat extraction. The experimental plot is relatively favorable for blueberry cultivation: 100% lighting, thickness of remaining peat layer 0.5–1 m, peat acidity 4.8, and groundwater level during the vegetation period 30–60 cm.

TABLE 7.4 Precipitation Occurrence (mm) of Vegetation Periods During the Study in 2011–2016 Years.

Year/month	May	June	July	August	September	October	Total annual precipitation
Long-term annual	49.0	61.0	71.0	70.0	62.0	59.0	574.0
2011	44.3	85.8	90.0	17.8	82.3	64.2	680.9
2012	32.7	102.8	101.9	61.9	86.1	115.8	727.1
2013	41.7	44.5	67.6	37.7	97.3	85.7	671.1
2014	11.3	109.0	25.5	53.6	22.5	87.0	540.4
2015	27.0	70.0	100.0	103.0	26.0	77.0	711.0
2016	30.0	24.0	117.0	45.0	99.0	31.0	620.0

TABLE 7.5 Average Monthly Air Temperature (°C) of Vegetation Periods During the Study in 2011–2016 Years.

Year/month	May	June	July	August	September	October	Average annual temperature
2011	12.8	16.7	21.2	16.1	10.8	4.7	3.2
2012	12.9	17.3	19.3	16.6	10.6	4.3	3.3
2013	12.5	19.0	19.7	18.0	10.3	2.8	4.2
2014	14.9	15.3	16.8	17.9	10.5	−1.1	3.4
2015	14.9	18.7	15.6	13.9	12.9	0.9	3.5
2016	13.9	16.9	20.8	20.9	10.0	2.1	4.1
Long-term annual	9.8	15.5	17.8	15.4	9.0	1.5	1.5

The plot development started in June 2011 and included area clearance from tree vegetation and shrubs, abrasion, rotary tillage, black fallow, herbigation (roundup), and plowing.

In the beginning of September 2011, 928 containerized plants of lowbush blueberry were delivered to the experimental plot. The plants were provided with Central Forest Experimental Station of Russian Scientific Institute of Forestry and Mechanization (Kostroma). The plants were grown from seeds selected on the station in spring 2011. The size of plants varied from 5 to 18 cm. During the period of September 5- 29, 2011, 128 lowbush blueberry plants were set. The rest was heeled in trenches and planted in July 1–7, 2012 on the experimental plot.

The planting pattern of constant growing is two-row. The plants were set in rows 1 m apart; distance between the plants is 0.4 m.

Technological pass between rows is 1.5 m. During the setting, complex fertilizers were applied for each plant in the amount of 2.5 g of each active substance. Further care included application of mineral fertilizers (once a season) at a dose of NPK 60, regular weeding, interrow mellowing, and watering during the dry periods. Each spring since 2013, the plants come to life.

The lowbush blueberry plants were successfully reset, rooting reached 100%.

The lowbush blueberry showed high frost resistance in regional conditions. Frostbitten shoots and dead plants were not marked. Viability of plants during the winter 2011–2012 and subsequent years was 100%. However, it is worth mentioning that the years of investigation were relatively warm, average annual temperature was significantly higher than long-term annual average (see Table 7.5), and minimum winter temperatures were not lower than −40°C.

To the end of the second vegetation period (October, 10, 2012), plants had 4–18 sprouts. Individuals with 5–15 sprouts prevailed (Fig. 7.1). The average number of sprouts was 9.3±0.3 per individual.

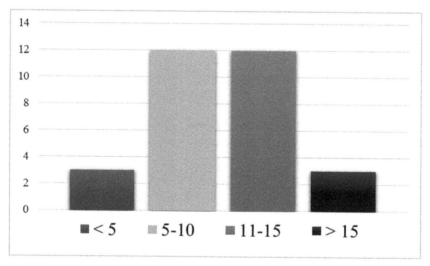

FIGURE 7.1 The number of sprouts of 2 years old blueberry plants per individual (October 10, 2012). (a) X-axis—number of sprouts and (b) Y-axis—number of individuals.

Sprout length varied from 9 to 43 cm, and average 20.3 ± 1.8 cm; sprout of 10–30 cm long prevailed (Fig. 7.2).

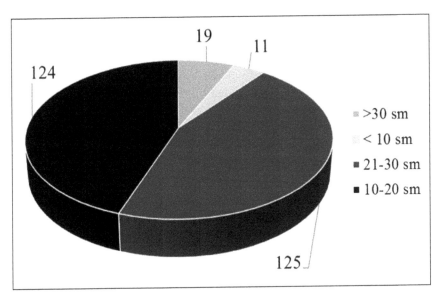

FIGURE 7.2   Length of sprouts of 2 years old blueberry plants (cm; October 10, 2012).

During further observation, there was no plant height growth marked. In autumn 2014, the plant height varied from 0.27 to 0.80 m and average 0.43 ± 0.8 m. In the end of vegetation season 2015, variation reached 34.7–85.2 cm and average 58.2 ± 2.1 cm. In October 2016, the height of 5 years old plants varied from 35.2 to 87.6 cm and average 63.4 ± 5.1 cm. This peculiarity of crown forming is explained by emergence of a large number of plagiotropic formation sprouts on the second-year plants. The feature was also marked by other researchers.[44]

Plagiotropic sprouts of 3–4 years old plants form secondary roots. Forming of new partial plants with plagiotropic growth out of rhizome resting buds was marked on the bush periphery.

Thus, it was defined that in conditions of southern taiga in the Volga-Vyatka area, *V. angustifolium* life-form is a deciduous low shrub, whose height varies from 35.2 to 87.6 cm on the fifth year and average 63.4 ± 5.1 cm. Its aboveground vegetative sphere mostly grows horizontally during the ontogenesis as a result of emergence of new formation sprouts and partial bushes out of rhizome resting buds. This characteristic of lowbush blueberry ontogenesis allows assuming the possibility of forming of solid bushes out of the maternal and partial plants on the experimental plot. Forming of partial bushes and secondary roots on plagiotropic formation sprouts enlarges possibilities of lowbush blueberry vegetative reproduction with use of rhizome cuttings and partial bushes.[14,45]

Flowering and fructification were marked on the second-year of plant vegetation for about 10% of individuals. Flowering continued from June 8 till June 24, 2012. The fruits got ripe to July 21, 2012. In consequent years, the blueberry flowering was marked in the period from May 28 till June 20, and fruit ripening from July 20till August 5.

In the end of July and August 2013, 83% of plants were fructifying. The level of fructification reached 83.2 ± 5.1 g per plant. In the beginning of August 2014, all plants were fructifying, fructification level varied from 18.2 to 106.0 g per plant and average 57.5 ± 6.3 g per plant. In 2015, all plants were fructifying, average fructification level was 230.0 ± 12.4 g per plant. The productivity of separate plants exceeded 350 g. In 2016, average productivity was 330 ± 5.1 g per plant.

The berry weight during the study period varied from 0.9 to 5.3 g and average 2.3 ± 0.2; the value was relatively stable and variation coefficient was 6.00–8.67%.

No plant pests or diseases were marked during the study.

Winter hardiness in plants was high. All plants set in 2011 remained. They had no frostbite marks. It might be explained by short period of plants vegetation and correspondingly complete wood maturation. Phenological data show that autumn leaf coloring of some lowbush blueberry plants appears on July 30 (Fig. 7.3) even in conditions of humid and warm summer (like in 2012). In colder and drier vegetation seasons (like 2014), autumn leaf coloring began on July 27.

The resistance of flowers and flower buds to spring frosts is high: no dead flowers or flowers buds were marked during the short decrease of air temperature at −5°C. Successful forming of fruits was observed.

The blueberry successfully develops in regional conditions during the hot summer years and shows no damages while short periods of elevated temperatures (up to +36°C in shade—July 2012). The species is resistant to groundwater level rise and short (2–4 days) flooding. In summer 2012, during May and June, groundwater level raised up to 30–40 cm six times and in August, three times up to 40 cm from the soil surface. In September, groundwater rose three times up to 40 cm and 3 times up to 30 cm, and once up to 20 cm. Root system, a lower part of sprouts of some plants, was under the water for 2–4 days. The plants were not damaged.

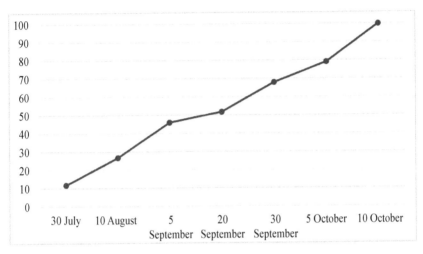

FIGURE 7.3 Appearance of autumn leaf coloring of lowbush blueberry in 2012. (a) X-axis—dates and (b) Y-axis—plants having autumn leaf coloring (%).

Thus, the results of conducted experiment of lowbush blueberry cultivation in southern taiga subzone of Kirov region make it possible to suggest possibility of successful introduction of the species in that region.

7.4 CONCLUSION

1. The most productive blueberry habitats in Kirov region are low-density (0.2–0.4) mature sedge-sphagnum pine forests of middle taiga subzone, where long-term average blueberry productivity reaches 128.3 ± 18.0 kg/ha.
2. The average value of blueberry BS in Kirov region is 758.3 t and ES 270.9 t. The BS value of 80% and ES value of 69% within the region is concentrated in middle taiga subzone in Verkhnekamsk district.
3. The bog blueberry productivity in Kirov region is lower than in most bordering areas and within the range. The maximum productivity is marked in the south of the Far East of Russia.

4. The blueberry BS in Russia in average fruitful year is 1,261,900 t and ES 354,700 t. More than a half of resources is concentrated in Siberian Federal District (51% of BS and 44% of ES) and about a quarter in the Far East Federal District (26% of BS and 18% of ES).

5. In conditions of southern taiga within the Volga-Vyatka region, *V. angustifolium* life-form is a deciduous low shrub, and 35.2–87.6 cm high on the fifth year and average—63.4±5.1 cm; average productivity is 330±5.1 g per plant.

6. High winter hardiness, resistance to unfavorable environmental conditions, possibility of successful vegetative reproduction, and absence of pests and diseases complexes were determined for lowbush blueberry in southern taiga subzone of Russia.

7. Results of lowbush blueberry cultivation in southern taiga subzone of Kirov region allow assuming the possibility of successful introduction of the species in the region.

KEYWORDS

- blueberry
- peatlands
- cranberry
- cowberry

REFERENCES

1. Konobeyeva, A. B. *Vacciniaceae in Central-Chernozem Region (Vaccination in the Central Black Earth Region): Scientific Edition;* Michurinsk State Agrarian University Publishing House: Michurinsk, 2007; p 230 (in Russian).
2. Plant Life (Flowering Plants). Prosvescheniye: Moscow, 1980; p 430 (in Russian).
3. Gorbunov, A. B.; Snakina, T. I. *Introduction of Blueberry in Western Siberia. Vegetation of Northern Asia: Problems of Study and Biodiversity Conservation;* Central Siberian Botanical Garden of Siberian Department of Russian Academy of Sciences: Novosibirsk, 2013; pp 31–33 (in Russian)

4. Koropachinskiy, I. Y.; Vostovskaya, T. N. *Ligneous Plants of Asian Russia;* Siberian Department of Russian Academy of Sciences Publishing House: Novosibirsk, "Geo" branch, 2002; p 707 (in Russian).

5. Snakina, T. I. Bog Blueberry (*V. uliginosum* L.) Introduction in Western Siberia. Ph.D. Thesis, Central Botanical Garden of Siberian Department of Russian Academy of Sciences: Novosibirsk, 2007; p 212 (in Russian).

6. Kurlovich, T. V.; Bosak, V. N. *Highbush Blueberry in Belarus;* Belorussian Nauka: Minsk, 1998; p 175 (in Russian).

7. Pavlovskiy, N. B. Systematic Position and Classification of Section Cyanococcus Blueberry Cultivars. *Horticulture* **2013,** *25,* 533–542 (in Russian).

8. Pavlovskiy, N. B. Results of Highbush Blueberry Introduction in Belarus. Collected Articles of International Scientific Conference. State and Prospects of Non-Wood Forest Resources use. Kostroma, September 10–11, 2013. Russian Research Institute of Forestry and Mechanization: Pushkino, 2014; pp 116–121 (in Russian).

9. Butkus, V.; Pliszka, K. The Highbush Blueberry—a New Cultivated Species. *Acta. Hortic.* **1993,** *346,* 81–86.

10. Paal, T.; Starast, M.; Karp, K. Influence of Different Fertilizers and Fertilizing Frequency on the Development of *V. angustifolium* Ait. Seedlings. *Bot. Lith.* **2004,** *10*(2), 135–140.

11. Paal, T.; Starast, M.; Karp, K. Response of Lowbush Blueberry Seedlings to Different Fertilizers. *Vaccinium spp.* and Less Known Small Fruits: Cultivation and Health Benefits: Abstracts of International Conference, Sept 30–Oct 5, 2007; Slovak Agricultural University: Nitra, 2007; p 43.

12. Makeev, V. A.; Makeeva, G. Y. Influence of Mineral Fertilizers on Growth and Fructification of Lowbush Blueberry on Cutover Peatland of Upper Type. *Proceedings of International Scientific Practical Conference, Devoted to 75 Years Anniversary of Andrey Bolotov. Modern Cultivars and Technologies for Intensive Horticulture.* Russian Research Institute of Fruit Plants Selection: Oryol, 2013; pp 147–149 (in Russian).

13. Gorbunov, A. B.; Simagin, V. S.; Fotev, Y. V. Introduction of Non-Traditional Fruit, Berry and Allotment Plants in Western Siberia. Academic Publishing House "Geo": Novosibirsk, 2013; p 290 (in Russian)

14. Tyak, G. V.; Kurlovich, L. E.; Tyak, A. V. Biological Remediation of Cutover Peatlands by Organization of Forest Berries Plantations. *Bull. Kazan State Agrar. Univ.* **2016,** *2*(40), 43–46 (in Russian).

15. Shoemaker, J. S. *Culture of Berry Plants and Grape;* International Literature Publishing House: Moscow, 1958; p 562 (in Russian).

16. Kurlovich, T. V. *Cranberry, Blueberry, Cowberry;* "Niola-Press," "UNION-public" Publishing House: Moscow, 2007; p 200 (in Russian).

17. Gallant, T. L.; Percival, D. C.; Kemp, J. R.; Olson, A. R. Intra- and Interclonal Pollination Affects Fruit Set and Berry Weight of the Lowbush Blueberry (*V. angustifolium* Ait.) in Eastern Canada//Problems of Rational Utilization and Reproduction of Berry Plants in Boreal Forests on the Eve of the XXI Century. Glubokoye: Gomel, 2000; pp 135–140.

18. Morozov, O. V.; Reshetnikov, V. N.; Yakovlev, A. P.; Morozova, T. A. Ecological and Biological Peculiarities of Lowbush Blueberry (*V. angustifolium* Ait.) During

Introduction in Conditions of Southern Parts of Belarus. *Proceedings of International Conference Devoted to 75 Years Anniversary of Central Botanical Garden of National Academy of Sciences of Belarus. Theoretical and Applied Aspects of Plant Introduction as a Prospect of Scientific and Economical Development,* Publishing House: Minsk, "Edit VV»0.1; pp 238–341 (in Russian).

19. Tyak, A. V. Growing of Lowbush Blueberry Seedlings on Cutover Peatland. *Proceedings of X International Symposium. New and Non-Traditional Plants and Prospects of use. Vol. 1. Puschino.* June, 17–21 2013. Russian Peoples' Friendship University: Moscow, 2013; pp 37–40 (in Russian).

20. Tyak, G. V.; Tyak, A. V. Selection of economically important varieties of lowbush blueberry V. angustifolium Ait. for cultivation on cutover peatland. *Proceeding of International conference devoted to 170-years anniversary of Russian Research Institute of Fruit Cultures Selection.* Oryol: Russian Research Institute of Fruit Cultures Selection; pp 209–211 (in Russian) **2015.**

21. Egoshina, T. Prospects of cranberry cultivation in Kirov Region. Culture of Vaccinium berries: results and prospects: *Proceedings of International Scientific Conference,* August 15–19, 2005. Central Botanical Garden of National Academy of Sciences of Republica Belarus: Minsk, 2005; pp 11–14 (in Russian).

22. Egoshina, T. L.; Luginina, E. A. *Vaccinium Vitis-Idaea* L. and *Oxycoccus palustris* Pers. in Natural Populations and in Culture of Taiga Zone of Russia. *Vaccinium spp.* and Less Known Small Fruits: Abstracts of International Conference, Sept 30–Oct 5 2007, Slovak Agricultural University: Nitra, 2007; pp 5–61 (in Russian).

23. Bukharin, P. D.; Danilova, I. The Experience of Introduction of Highbush Blueberry, American Cranberry and Honeysuckle in the Department of Cultural Plats in Central Botanical Garden of the Academy of Sciences of USSR. *Proceedings of the Conference Ways of Increasing Efficiency of use and Reproduction of Forest Food, Forage and Medicinal Resources for USSR Food Program,* Penza Agricultural Academy: Penza, 1983; pp 352–354 (in Russian).

24. Egoshina, T. L.; Shikhova, L. N.; Safonov, V. G.; et al. Modern State of Non-wood Plants Resources of Russia. Russian Research Institute of Game Management and Fur Farming: Kirov, 2003; p 263 (in Russian).

25. Egoshina, T. L. Non-Wood Plant Resources of Russia. Scientific Information Agency: Moscow, "Priroda." 2005; p 80 (in Russian).

26. Maznaya, E. A. *Estimation of Plant Resources having Economical Importance;* Methods of Study of Forest Societies: Saint Petersburg, 2002; pp 95–102 (in Russian).

27. Program and Methods of Study of Fruit, Berry and Nut Plants Cultivars. Russian Research Institute of Fruit Cultures Selection: Oryol, 1999; p 606 (in Russian).

28. Beideman, I. N. Methods of Phenological Studies of Plants and Plant Societies. Nauka: Novosibirsk, p 154 (in Russian).

29. Yudina, V. F.; Maksimova, T. A. Productivity of *V. uliginosum* L. Fruits on a Reserved Bog "Nenazvannoye" (Southern Karelia). *Rastitelnyye resursy (Plant Resour.)* **1995,** *31*(4), 33–36 (in Russian).

30. Nekratova, N. A.; Nekratov, N. F.; Mikhailova, S. I. Resources of Medicinal and Berry Plants in Northern Districts of Tomsk Region. *Rastitelnyye resursy* **1986,** *22*(3), 297–310 (in Russian).

31. Plotnikov, D. A.; Schmidt, A. S. Stocks of Wild Medicinal, Berry and Fruit Plants in Parabelskiy District of Tomsk Region. *Rastitelnyye resursy* **1988,** *24*(2), 177–182 (in Russian).

32. Skryabina, A. A. Productivity of Wild Berries and Their Use. Russian Research Institute of Game Management and Fur Farming: Kirov, 1972; pp 125–128 (in Russian).

33. Raus, L. K. State and Use of Wild Berries in Kamchatka. Productivity of Wild Berries and Their Use. Russian Research Institute of Game Management and Fur Farming: Kirov, 1972; pp 131–133 (in Russian).

34. Skryabina, A. A.; Kolupayeva, K. G. Methodic Recommendations on Counting and Prognosis of Wild Fruits and Berries Resources in Consumer's Cooperatives Establishments. Russian Research Institute of Game Management and Fur Framing: Kirov, 1986; p 46 (in Russian).

35. Puchnina, L. V. (*V. uliginosum* L.) Productivity in Pinezhskiy State Reserve. Murmansk Regional Publishing House: Murmansk, 1990; Vol. 26(2), pp 179–191 (in Russian).

36. Gonnov, V. V.; Egoshina, T. L.; Lopatina, N. A. *Natural Resources of Komi-Permyatskyi Autonomous District;* Komi-Permyatsk Publishing House: Kudymkar, 2005; p 192 (in Russian).

37. Antonova, N. I. *Productivity of Wild Growing Berry Plants of Yakshinskiy Plot. Annals of Pechyoro-Ilychskiy State Reserve;* Komi Publishing House: Syktyvkar, 1976; Vol. 3, pp 20–40 (in Russian).

38. Barykina, V. V. Yield of Some Berries on the Islands of Kandalakshskiy Creek. Annals of Kandalakshakiy State Reserve. Murmansk Regional Publishing House: Murmansk, 1969; Vol. 7, pp 178–189 (in Russian).

39. Pushkina, N. M. Natural Reproduction of Vegetation on Forest Fire-Sites. Annals of Lanlandskiy State Reserve. Murmansk Regional Publishing House: Murmansk, 1960; Vol. 4, pp 5–126 (in Russian).

40. Grimashkevich, V. V. Fructification of *V. uliginosum* L. in Polesye. *Rastitelnye resursy* **1987,** *23*(3), 323–333 (in Russian).

41. Rusakov, O. S. Wild Berries Productivity in Southern Part of Novgorod Region. *Rastitelnyye resursy* **1969,** *5,* 337–341 (in Russian).

42. Ulanov, A. N.; Zhuravlyova, E. L. Bogs. Encyclopedia of Vyatka Land. Kirov Regional Publishing House: Kirov, 1997; Vol. 7, pp 222–233 (in Russian).

43. Smirnova, V. V.; Frenkel, M. O. *Climate of Kirov;* Frenkel, M. O., Shvera, A., Ed.; Gidrometeoizdat: Leningrad, 1982; p 215 (in Russian).

44. Yakovlev, A. P.; Morozov, O. V. Development of Vegetative Sphere of Lowbusk Blueberry during Introduction in Conditions of Belarus. *For. Econ. Inf.* **2008,** *12,* 40–44 (in Russian).

45. Estabrooks, E. N. The use of *V. angustifolium* Clones for Improved Fruit Quality and Yield. Wild Berry Culture: An Exchange of Western and Eastern Experiences; Estonian Agricultural Academy: Tartu, 1998; pp 46–49.

Species names of wild-growing berry plants studied	
Common names	**Scientific names**
American cranberry	Vaccinium macrocarpon Aiton
Blueberry	*Vaccinium uliginosum* L.
Bilberry bog	*Vaccinium myrtillus* L.
Bog blueberry rabbit-eye, low bush	*Vaccinium virgatum* Ait.
Blueberry low bush	*Vaccinium angustifolium* Ait.
Blueberry highbush (tall)	*Vaccinium corymbosum* L.
Cowberry	*Vaccinium vitis-idaea* L.
European cranberry	*Vaccinum oxicoccus* L.

PART III
Breeding and Biochemistry of Decorative Plants

CHAPTER 8

TOLERANCE IMPROVEMENT OF INDOOR PLANTS

ANATOLY I. OPALKO[1,2], LARISSA I. WEISFELD[3,*],
SARRA A. BEKUZAROVA[4], ALEXEY E. BURAKOV[3],
OLGA A. OPALKO[1], and FYODOR A. TATARINOV[5]

[1]*National Dendrological Park "Sofiyivka," National Academy of Sciences of Ukraine, 12-a Kyivska Str., Uman, Cherkassy Region 20300, Ukraine*

[2]*Uman National University of Horticulture, Instytutska St., Uman, Cherkassy Region 20305, Ukraine, E-mail: opalko_a@ukr.net*

[3,*]*Emanuel Institute of Biochemical Physics, Russian Academy of Sciences, 4, Kosygin St., Moscow, Russia 119334, Tel: +79162278685*

[4]*Gorsky State Agrarian University, 37, Kirov St., Vladikavkaz, Republic of North Ossetia Alania, Russia 362040, E-mail: bekos37@mail.ru*

[5]*Severtsov Institute of Ecology and Evolution, Russian Academy of Sciences, 33, Leninsky Prosp., Moscow, Russia 119071, E-mail: f.tatarinov@gmail.com*

Corresponding author. E-mail: liv11@yandex.ru

CONTENTS

Happiness is being with nature,
seeing her, and conversing with her.

—Leo Tolstoy (1852).[1]

ABSTRACT

The chapter analyzes the historical aspects of ornamental horticulture evolution, starting from the ancient Egyptians, Greeks, and Romans until the medieval period of "dark ages" and subsequent Renaissance. It notes the vital importance of the development of safe for human health ways to improve the resistance of indoor plants to pathogens and pests. The results of the study of indoor plants for their treatment effectiveness in the winter gardens of the Baltschug Kempinski hotel (Moscow, Russia) and in the garden of Marble Statue in the House of Scientists (Uman, Cherkasy Region, Ukraine) have been obtained. Various concentrations of para-aminobenzoic acid (PABA) in combination with a biological preparation named Kornevin were applied. In all variants of the experiment, the increase in the plants growth and the decrease in their morbidity and damageability by insects relative to control were observed. The best was the variant that included consecutive treatment of leaves by PABA water solution in the concentration of 0.03% and 3–4 days later by Kornevin in the dosage of 1.0 g/l of water with the addition of 5 ml of concentrated fertilizer containing nitrogen (N) 7%, phosphorus (P) 3%, potassium (K) 6%, and microelements ($2MgO + 2S + 15B + 0.3Fe + 0.3Mn + 0.05Cu + 0.2Zn + 0.008Mo$).

8.1 INTRODUCTION

Human happiness weakly depends on the consumption level. Moreover, when the consumption becomes the aim and the sense of life, and the wishes to newly wedded couples and persons whose anniversaries are celebrated to have a lot of gold, diamonds, and banal money start to dominate the congratulatory postcards and wassails, then the moral dies and the joy disappears.... Consequently, modern consumerism becomes more and more often an analogue of superfluousness which is a kind of pernicious dependence similar to the dependence on alcohol or drugs. In all these cases as well as in other similar cases, the overdose is mortally dangerous.

However, every human independently of his social status, place of residence, profession, and a lot of other external characters seeming to be significant remains a child of the nature. Human can suddenly feel happiness when seeing the sun-like flower of the first dandelion or when inhaling the odor of lily of the valley in spring. Lilies, gladioluses and many other summer plants please the eye and improve the mood as well. Brightly blazing haws and hips on a background of green pines and spruces as well as the scarlet and gold of autumn foliage evoke particular feelings. The Christmas tree in winter always supplements with different flowers on the windowsill, cultivation of which gives their owners unmatched joy. Inhabitants of urbanized megalopolises, where all kinds of life except human's one are ignored, experience permanent attraction on an unconscious level to the contact with nature and get pleasure from plants in the urban streets, parks, and public gardens as well as in a few small islands of nature, preserved by enthusiasts somewhere in urban conditions. The growing interest in the cultivation of ornamental plants, which increased in the post-Soviet era during recent years as well as the perception of enormous value of the beauty of the environment, are the expression of original human intensions to surround oneself by beauty by means of planting different trees, shrubs, and flowers, thus filling life with positive emotions, joy, and creative energy.

Green urbanism, which is now observed in Western Europe, proves that alternative urban future is possible and practically feasible if green urbanism principles will turn from a strange utopia (from the point of view of a cheerless man in the street) into natural components of his mode of life.[2]

Decorative plants improve the architectural appearance of cities, impart them color diversity and create a spatial silhouette. The richness of ornamental properties of most of the plants and their capacity to change in time create unlimited possibilities for the formation of the appearance of settlements in green territories.[3]

In order to determine the advisability of massive multiplication of a particular plant species (form, cultivar), its cultivation and use in landscape design, as well as greenery for balconies, flats, lobbies, and offices, it is necessary to take into account not only ornamental properties but also ecological and biological peculiarities. Both ornamental properties and ecological and biological peculiarities are determined by the inheritance of plant organism in the interaction with environmental factors,[4,5] first of all by the probability of survival in conditions not always optimal for the cultivated plant and by the ability to preserve in such conditions its ornamental qualities, which is the aim of the cultivation of this plant.

The criteria of the ornamental quality of landscape design element,[3] individual plant species and/or cultivar, as well as of an individual plant (sometimes cultivated using a special technology, e.g. "bonsai"),[6] were different in different historical periods, different climatic zones of our planet, and different social and ethnic entities.

For an enthusiastic ecologist, every plant is beautiful in its own manner. It changes with changing seasons and becomes remarkable. Of course, there are very bright plants, but even an ordinary dandelion (*Taraxacum officinale* Webb) is also indescribably beautiful, although it is missing in the list of ornamental plants (Fig. 8.1).

FIGURE 8.1   **(See color insert.)** The flowering of dandelion (*Taraxacum officinale* Webb.)

Plants uncommon for a particular region are often, surprisingly, considered as ornamental. Because of their exoticism, many tropical plants look attractive for inhabitants of the temperate zone and are cultivated in greenhouses, flats, lobbies of big companies and hotels, in the offices. However, many of them are considered as ordinary in their tropical motherland, nobody cultivates them here for the decoration of flats, offices, and so forth.

The cultivation of ornamental plants has a rather long history. However in contrast with cereals and other food plants, it is rather hard to reliably separate the historical antecedents of ornamental gardening from pomology, forestry, and vegetable production. Some actual ornamental plants were earlier utilized as food and/or forage cultures. It is known that the remarkable plant such as peach (*Prunus persica* (L.) Batsch), whose juicy fruits can satisfy the most refined taste, formerly was used as important hog feed.[7] Now along with fruit peach cultivars, ornamental cultivars of peach as well as of many other species of fruiters are cultivated. Ornamental cultivars of cherries (*Prunus* spp.), apple trees (*Malus* spp.), pear trees (*Pyrus* spp.), and other fruit plants are successfully used in the planting of greenery in cities.[8]

Several western cultures took the biblical Garden of Eden as the ideal standard, the model for imitation when creating their gardens. Biblical and other religious texts are full of gardening metaphors, such as the placement of Adam and Eve in the Garden of Eden close to the tree of knowledge, description of the olive branch as a symbol of peace, as well as the cultivation of vineyard by Noah as his first action after the Deluge.[7,9]

Beside the religious texts in the history of horticulture, there are documented records about gardens as far back as the ancient Egyptians and Sumerians over 5000 years ago. In spite of the fact that the main aim of the ancient gardens cultivation was food supply, Egyptians were one of the earliest civilizations which started to cultivate plants for their aesthetic values.[7]

More than 200 species of flowering and aromatic plants were identified from the remainings found in Egyptian royal tombs. Judging from the drawings of ornamental and utilitarian plants growing in pots and tubs, it is possible to suppose that ancient Egyptians used container gardening along with plant cultivation in gardens. Horticulture also developed in the pre-Columbian North and South America. For example, Aztec gardens were holy places full of ornamental, aromatic, and medicinal plants symbolizing different myths and gods, not food plants. Food cultures for upper classes were cultivated by people from lower classes in special gardens and vegetable gardens. The great Aztec chief Montezuma forbade at all to cultivate edible plants in his gardens, saying that they should be cultivated only in the gardens of lower classes and brought to his table as tribute. In contrast, flowers were cultivated as they were very important in religious ceremonies. Flowers were presented by gods to chiefs as the attestation of their right to rule. Spanish documents show that Aztecs had botanical gardens full of wild plant species, organized in a peculiar botanical system.[10]

In ancient Greece, gardens and parks were also a focus of attention. According to historians, a quarter of the city of Alexandria found in 332 before Christ (BC) by Alexander of Macedon was occupied by parks.[10] The city was projected by the architect Dinocrates of Rhodes according to grid street plan typical for Greek empire. However, the abundance of ornamental plantations and parks can be probably explained not only by Greek traditions but also by the influence of ancient Egyptian culture.

Ancient Greek stage of ornamental gardening can be considered as intermediate between Sumerian-Egyptian and Roman.[7] Gardening was an art of ancient Greece formed under the influence of early horticultural innovations and conceptions, such as the model of Persian garden (dating to 4000 BC). The idea of isolation and escape from the world into nature was introduced through Persian gardens.[10,11] The Persian word for garden was transmitted into Greek and then into Latin, and the word "paradise" started to mean paradise on the earth.[10]

Romans adopted horticultural knowledge from Egyptians and Greeks, successfully applied it, and provided further development. In particular, they improved the grafting and budding methods proposed by Greeks that are used even now and constructed garden knifes, garden ladders, and many other instruments for gardening. In ancient Rome, the most astonishing fantasies of well-off citizens were embodied in villas-gardens with ponds, sculptures, flowers, and clipped shrubs aimed for visual enjoyment and nice leisure. The ideology of Roman society was based on the aspiration for the ideal human, whose soul and body would be in harmony. A peculiar cult of physical pleasures appeared.[7] In this time, a garden was perceived as a space populated by deities and the worship of deities was spread to the nature and its man-made copies—gardens. But in the ancient times the ambiguous relationships between humans and the nature worsened because of the tension between the increase in resource exploitation and the need of nature protection.[12]

After the fall of Roman Empire, its provinces were conquered by barbarian tribes and by the end of fifth century on the ruins of a formerly powerful empire, several barbarian states were formed. In these states, the ruined Roman cities came to decline, and the construction of shrines and public buildings ceased. The barbarians did not see any value in the images of humans and deities and they easily destroyed beautiful paintings and melted bronze sculptures into ingots. The single kind of art that barbarians did not destroy was the decoration of utensils and weapons by ornaments.[13] Agriculture in the Roman Empire during the period of its rising was very prominent; that is why it is evident that the barbarians, who plundered Rome, could not adopt the antique agro-technique right away. The barbarians brought with them extensive forms of agriculture (two-field system, often even more primitive fallow system, when arable plots were left for few years without tillage in order to restore soil fertility) and the achievements of Roman agriculture were forgotten. Crop growth dramatically fell down and a massive famine swept over Western European countries in the second part of the first millennium. Hungry men did not care about landscaping beauties and high art. That is why the period of 5–14 centuries—the millennium between the fall of Rome at the end of the fourth century and the start of renaissance in Italy (14th century), usually called the epoque of Middle Ages, is characterized as the period of "stifling dogmatism" or "dark ages" (chiefly of fifth to eighth centuries). The landscape art, which is the most vulnerable of all kinds of art and which more than others requires for its existence a peaceful situation for permanent treatment and renovation, was practically eliminated, and individual plants were preserved in small gardens only in the monasteries and castles, that is, on the territories relatively protected from destruction. These gardens mainly had utilitarian character. Later, with the development of monasteries, which became plant repositories and places of European gardening technologies improvement, the search of the manuscript herbals of the ancient Greeks and Romans and their translation started. Some monks, who cultivated herbs and spices in monastery gardens, became specialists in the application of knowledge from herbals for the cure of different diseases. Between 11th and the end of 13th century, cities, which appeared already in the dark ages, started to grow, the agriculture and horticulture started to extend, and learning and knowledge were revived. However, only in 15th century, the monastic gardens started to be decorated with arbors, trellis, pergolas, and hedges, to be supplied with benches from sod formed as ledges along the fence, and small water fountains and flowers appeared in the gardens. Many of these gardens were already aimed for relaxation. In the late Middle Age, the separation of gardening from agriculture and forestry was officially recognized.[10]

However, the mentioned "dark ages" were only concerned with wreckage of the Western Roman Empire. India and China did not know desolation and barbarization. Sometimes, barbarians cut their way through their borders but the mass of local tradition-bearers assimilated them rather quickly.[14] Consequently, not only did the western world become leader of world history and culture in the medieval time but also semi-oriental Byzantium and Eastern China. China and Japan were centers of horticultural development and innovations, insulated from other development centers for many centuries, at least until 10th century BC. Indirectly, this date is supported by archeological findings of Chinese silk in Egypt of 10th century BC. Landscape art in China in the second century BC was already rather developed; documented descriptions of complex gardens with artificial hills, complex water pools and irregular rocks, as well as other elements of landscape design able to evoke idealized scenes from the nature are evidence of it. Pools, ponds, islands, small bridges, waterfalls, hillocks, trees, as well as stones and rocks were common elements of formal Chinese and later Japanese garden. It is known that some new fruits and vegetables were introduced from Southeast Asia and India to China in 6th century AD and in the 12th century, namely simple Chinese-built greenhouses, where oiled paper was used as transparent cover, appeared. In such greenhouses, Chinese successfully cultivated heat-loving flowers and vegetables.[10]

Early ornamental gardens in China as well as in other ancient civilizations were mainly created attached to emperor's palaces and later to the palaces of rich men. Such adoption intensified from 900 BC along the famous Silk Road which was caravan trade way. The trip of Marco Polo as well as the activities of Arabic merchants contributed to the spreading of Chinese gardening technologies and plants in Europe. The design of Japanese ornamental gardens was formed under direct influence of Chinese models. However, Japanese gardens, which appeared after the third century, were more abstract than the Chinese ones. The first Japanese treatise about landscape art "Sakuteiki or Records on Garden Making"[15], appeared at the beginning of the 13th century.[10]

Ancient Russian gardening started in the 11th century under Theodosius of Kiev in Kiev Pechersk monastery. In the hagiography of Abraham of Smolensk, a kitchen garden is also mentioned. The sources of 14th century mentioned gardens in Moscow, but with very poor assortment. The Italian traveler Joseph Barbaro, who visited Moscow in 1439, wrote that there were almost no fruits in the city, excluding "small amount of apples, nuts and hazelnuts."[16] Flowers were the main attribute of one of the ancient Slavic feasts—Green Yuletide or White Monday, which was celebrated in June. However, the tradition to decorate living quarters appeared with the advance of Christianity, according to Byzantine model.

Further trends in the landscape art development with different regional peculiarities are rather extensively described and illustrated in many fundamental monographs,[17–27] which we recommend to the interested reader. Among them, we can pick out a very recent "kindle edition".[27] However, we presume to discuss only some aspects of the use of indoor plants, whose cultivation improvement is the topic of our study.

Until recent times, window-sills in apartments were the paramount place of indoor plants placement. The actual diversity of indoor plants gives a wide range of options for the arrangement of the interior of not only a modern flat but also a hotel lobby, various office rooms, broad stair flights and resting places, and the like. Many factors like the background color, incoming light, neighboring plants, and surrounding interiors can represent the best qualities of any plant absolutely differently. Special indoor gardens allow ignorance of the inevitability of vegetation seasons and bring natural beauty into our indoor conditions, the whole year. In an indoor garden, indoor plants are represented not as simple accessories, but as the display of the mode of life of a nature-loving gardener.[28]

Indoor gardens are particularly important for the people with reduced mobility. A small peperomia (Peperomia Ruiz and Pavón) or aspidistra (*Aspidistra lurida* Ker Gawl.) on the window-sills can provide a daily dose of nature to a handicapped individual.[29] The design of any room properly decorated with indoor plants obtains living traits and no one in such a room feels himself deprived of contact with nature.

The use of houseplants, which clean the air in the room, also presents a considerable interest.[30,31] The point is that houseplants not only release oxygen into the air of the room where they are growing, although this function is leading, but also decrease the dustiness of rooms and neutralize toxic volatiles in the air, making them harmless for human lungs and skin. Moreover, because of indoor plants, the air cleaned from dust and pathogenic bacteria is enriched with moisture, phyto-organic volatiles with abiotic effect relative to a considerable amount of harmful microorganisms (bacteria, fungi, viruses, protozoa), and other useful natural components extracted by green leaves.[31,32]

Phyto-design as scientific determination of the introduction of indoor plants in the interiors of rooms as a part of the general design and implementing aesthetic, ecological, sanitary, and other functions appeared in the late seventies of the previous century and received theoretical ground in the works of Grodzinsky and his followers.[33,34] Plants of a winter garden with high biomass also have, according to the definition of Grodzinsky, a psychological and aesthetic effect.[33] As a practical application of phyto-design, it was proposed to introduce medicinal, aromatic, tropical, and subtropical plants with certain known healing properties in the interiors. The creation of possibilities for phyto-recreations with the application of a selected assortment of plant species for the prophylaxis and treatment of different diseases, including infectious ones, is now picked out as a particular branch—*medical phyto-design.*[31,32,34]

From the psychological point of view, houseplants decorating window-sills, and especially their daily care, have a sedative effect on shattered nerves, regulate thinking, and contribute to the stimulation of brain activity and generation of new creative ideas.[35] Houseplants beneficially influence humans through different receptors: they please eyes by the beauty of shapes, their smell is pleasing, and the calm green color of leaves has a sedative effect on the nervous system; they help to manage bad mood and even strong stress. Amaryllis (*Amaryllis belladonna* L.), hibiscus (*Hibiscus rosa-sinensis* L.) with crimson flowers, poinsettia (*Euphorbia pulcherrima* Willd. ex Klotzsch) and other plants with flowers and leaves of red tints rise the vital tonus and increase the people activity. They are useful for very busy and hardworking people on a daily basis, at least a little time to deal with flowers and ornamental plants, to admire them, to look after the plants, to photograph them should be given. It is recommended to choose, according to the possibility, a separate room for relaxation, to transform it into a small jungle and to spend some time there for the restoration of forces after a hardworking day or a working week.

During recent decades, along with the increase of flourishing ornamental plants popularity, the interest to the secret language of flowers, almost forgotten earlier, has gradually started to reappear. First were attempts to use spectacular asymmetric compositions more or less saturated with *ikebana* elements. The traditional structure of Japanese flower composition is based on an asymmetric scalene triangle marked by three main points, usually rods, symbolizing heaven, earth, and humans. The strong symbolic and philosophical implication presented in *ikebana*[36] was not always understandable for Europeans and was often modified according to ethnic traditions of different communities.[37]

Ikebana has more than 600 years long history of flower arrangement in Japan and it is based on asymmetry, whereas European compositions tend to be symmetrical. The differences in these two styles of flower arrangement are based on the corresponding different philosophical and religious backgrounds.[38] At that they are neither static nor unchangeable forms of art. Hence, in our dynamic century tending to eclecticism, these two styles began to strongly influence each other and be nonverbally combined, thus combining different religious traditions.[39]

The traditions of flower language also go deep into the centuries. This symbolic language was created by the women of the Orient, who, being within the harsh frame of Islam with covered faces and without normal communication, transmitted their feelings and all nuances of the mood into the flowers.[40] In the Victorian epoque, the language of flowers was also used for the secret expression of emotions. Flower dictionaries were published in France and Great Britain, and European young men, girls, and even married women calculated with enthusiasm the number of petals trying to guess the "language" of each bouquet given. Nowadays, apparently in the age of openness bordering on shamelessness, the fashion for symbolic language of flowers has returned and the dictionaries of flower language are being published again in many countries of the world.[41–44]

Well-groomed flowering ornamental plants at the entrance and inside the office have now become an important component of the image of a prospering company. A modern office is characterized not only by technical equipment and ergonomic furniture but also mandatorily by a cozy and psychologically comfortable setting. A carefully planned phyto-design of the office can enliven the strict interiors of private offices and create favorable atmosphere for work. Before planning the planting of greenery in the office, it is necessary to carefully analyze the peculiarities of the microclimate of each zone, first of all humidity and illumination, paying a special attention to the "risky zones," such as dark corners, draughty places, the space under the air conditioner, windowsills above hot radiators, places aimed for visitors waiting in the reception which can damage plants, and so forth. According to the results of such analysis, it is possible to correctly choose plants which will be attractive during the year and will not require too much efforts for their protection from plant pests and pathogens[45], and which will quickly restore mechanical damages.

It is very important to take into account drought resistance of plants because high-quality watering is not always possible during weekends and holidays.[29] It is even more important to select plants that combine attractiveness with the resistance to pathogens and plant pests. This is because in the offices, flats, lobbies of hotels, and other rooms, where people are located, it is forbidden to use pesticides. Hence, the task to elaborate ecologically safe ways to increase indoor plants resistance to pathogens and plant pests is very important. The last reason became the basis of our study aimed to elaborate the methods increasing resistance of indoor plants.

8.2 MATERIAL AND METHODOLOGY

The study was performed in two different ecological and geographic zones situated about 1000 km from each other from northeast (Moscow) to southwest (Uman). These regions are the following:

- Central part of the East European plain, country between Oka and Volga rivers at the turn of Smolensk–Moscow height (on the west), Moskva–Oka plain (on the east), and Meshchyora Lowlands (on the southeast), (winter garden of the Balchug Kempinski hotel, Moscow, Russia)
- Uman'-Man'kivka region of Central Pre-Dnipro Upland of the East European plain (the garden of marmoreal sculptures of the House of Scientists, Uman, Cherkasy region, Ukraine).

The Moscow city is situated at two shores of Moscow River in its mean current in the zone of mixed forests. The climate of Moscow is moderately continental. The period with mean daily temperature below 0°C lasts for 120–135 days, starting in mid-November and finishing in late March. January is the coldest month with an average monthly temperature −10 to −11°C. In past few years, frost has reached −45°C. In winter (especially in December and February), thaws evoked by Atlantic and (more rarely) Mediterranean cyclones are frequent; they are usually rather short—up to 4 days.[46]

Uman town is situated in Central Ukraine in the southwest of Cherkassy region. It is situated in the basin of South Bug river at the junction of two small rivers Kamyanka and Umanka.[47] The town is situated in the Atlantic continental climatic region which determines its climate as moderately continental, relatively warm with a positive mean annual air temperature of $+7.0$ to $+7.7°C$. January is the coldest month with a monthly mean temperature of -5.6 to $-6.1°C$. Absolute minimum of air temperature reaches $-34°C$ (January 9, 1987). During winter, there are frequent thaws which can often last for more than 5 days; in some years, the number of days with thaw reaches exclude 75. Frost-free period with daily mean air temperature above $+5°C$ continues in average for 205–210 days.[48]

In the winter garden of Balchug Kempinski hotel (Moscow, Russia) and in the garden of marmoreal sculptures of the House of Scientists (Uman, Cherkassy region, Ukraine), the growth and development of different species, cultivars, and forms of the following genera were studied: *Aglaonema* Schott, *Coleus* Lour., *Cordyline* Comm. ex R. Br., *Dieffenbachia* Schott, *Dracaena* Vand. ex L., *Epipremnum* Schott, *Euphorbia* L., *Ficus* L., *Hibiscus* L., *Monstera* Adans., *Schefflera* J. R. Forst. and G. Forst., *Schlumbergera* Lem., *Spathiphyllum* Schott, *Syngonium* Schott, *Tetrastigma* (Miq.) Planch., *Yucca* L., and others. In total, above 600 indoor plants were studied.

The variants of the experiment included:

• Control (without treatment by stimulators, but with sparging of leaves with distilled water and bringing in the soil with water of 5 ml of concentrated fertilizer containing nitrogen (N) 7%, phosphorus (P) 3%, potassium (K) 6%, and microelements ($2MgO + 2S + 15B + 0.3Fe + 0.3Mn + 0.05Cu + 0.2Zn + 0.008Mo$);
• Four variants of treatment (sparging of leaves) with the solution of para-aminobenzoic acid (PABA) in the concentrations of 0.01, 0.03, 0.06, and 0.12%;
• Three variants of Kornevin preparation solution in the soil in the doses of 0.5, 1.0, and 1.5 g/l of water with the addition of concentrated fertilizer containing nitrogen (N) 7%, phosphorus (P) 3%, potassium (K) 6%, and microelements ($2MgO + 2S + 15B + 0.3Fe + 0.3Mn + 0.05Cu + 0.2Zn + 0.008Mo$ as well as zircon preparation);
• Twelve variants of joint treatment (leaves sparging) with PABA solution in the above mentioned concentrations with further (3 days later) addition of Kornevin preparation in the soil in the aforesaid doses.

In order to prepare the sprinkling mixture, first the working PABA solution was prepared in the concentration of 200 g per 10 liters of water (2% solution), at the temperature range of 60–80°C. The hot water was mixed with cold one in order to reach the required concentration and after the solution cooled down to room temperature, the aforesaid components were added. The amount and concentration of concentrated fertilizer containing macro- and microelements in control and all other variants were identical. The treatment was performed from April to July with the interval of 2 weeks.

8.3 RESULTS AND DISCUSSUON

In order to demonstrate the results of optimal variants of plants treatment by immunostimulators, we selected four indoor plants from a total of 647 studied plants popular in the countries with temperate climate (Table 8.1).

The plants of genus *Dracaena* Vand. are placed in the family Asparagaceae Juss. subfamily Nolinoideae Spreng. These species originate mainly in Africa, although some are common in Madagascar, Mediterranean regions, tropical and subtropical Asia, Central America, Micronesia, North Australia, and Pacific islands. They have valued application in horticulture, medicine, the fiber industry, and in ceremonies of different cultures.[49] The *Dracaena fragrans* can tolerate lots of neglect and fight back to full health easily after periods of neglect and its leaves stay attractive. However, here are some potential problems that should be taken into account: brown leaf tips, soft leaves, and brown spots on leaves can be triggered by dry air, cold drafts, or lack of water. Among insects, mealybugs and aphids can harm them.[50] The *D. fragrans* 'Massangeana' wide leaves are long and arc gracefully from the stalk. A dramatic yellow variegation runs down the center of its leaves (Fig. 8.2)

The color of the leaves is green; they have well-expressed strips and veins, which can have various tints, usually from light green to bright yellow. *D. fragrans* 'Massangeana' is a small palm-like tree; because of bright yellow stripes on the leaves, it can brighten and make any office or living room attractive.

Monstera spp. [family Araceae Juss. subfamily Monsteroideae (Schott) Engl.] are the epiphytic plants with aerial roots; thus in the wild, this plant

TABLE 8.1 Some Indoor Plants Used in the Experiments.

Indoor plant	Family	Origin	Uses
Dracaena fragrans (L.) Ker Gawl. 'Massangeana' (corn plant or cornstalk)	Asparagaceae Juss.	Native throughout tropical Africa	Ornamental, hedge, boundary marking; popular as a houseplant
Monstera deliciosa Liebm. (ceriman or fruit salad plant)	Araceae Juss.	Native Central and South America (Mexico, Guatemala, Costa Rica, and Panama)	Cultivated as foliage ornamental
Spathiphyllum wallisii Regel 'Quatro' (peace lily or spathe flower)	Araceae Juss.	Native tropical America	Cultivated as ornamental
Tetrastigma voinierianum (Baltet) Gagnep. (chestnut vine or lizard plant)	Vitaceae Juss.	Endemic to Vietnam	Interior ornamental plant leaves

FIGURE 8.2 **(See color insert.)** *Dracaena fragrans* 'Massangeana'.

grows by climbing shrubs or trees. *Monstera* is an evergreen liana with carved leaves. They are not parasites: they do not steal nutrients or directly harm the plant they grow on. This means that growing the plants of genus *Monstera* Adams. on a pole perfectly creates the environment the plant needs to raise the heavy stem upright and provides a pleasing appearance. The *Monstera deliciosa* Liebm. (Fig. 8.3) grows in tropical rainforests of Central America. In mountainous forests, it can be found up to 1000 m above sea level.

FIGURE 8.3   (See color insert.) *Monstera deliciosa* Liebm.

This well-known climbing indoor plant can reach the length of up to 3 m in rooms. The cultivar *M. deliciosa* 'Borsigiana' with smaller leaves than the original species, up to 30 cm in diameter, and with thinner stems (Fig. 8.4) was selected in the population, obtained as a result of splitting under seminal reproduction of *M. deliciosa*. Some researchers as early as in late 19th century proposed to classify this cultivar as a separate species *Monstera borsigiana* Engler (*M. borsigiana* K. Koch) or the variety *M. deliciosa* Liebm. var. *borsigiana* (Engler) Engler and Krause.[51] However, a status as cultivar for this plant within the frame of *M. deliciosa* Liebm. (*M. deliciosa* 'Borsigiana') can be probably accepted. *M. deliciosa* 'Borsigiana' is suitable for the cultivation in living rooms as well as in other quarters, in particular in small ones.[52,53]

The plants of *M. deliciosa* can start looking very untidy once it matures in size and age. Among the causes of decreasing decorativeness, it is possible to notice the yellowing of leaves because of soil overwatering and the deficit of nutrients in the substrate; due to moisture deficiency in the soil and especially in the air, leaf tips and edges turning brown is observed, and under insufficient illumination, water deficit, or insufficient fertilization, leaves do not form slits or holes.[53]

FIGURE 8.4   **(See color insert.)** *Monstera deliciosa* 'Borsigiana'.

The plants of genus *Spathiphyllum* Schott in Schott and Endl. belong to the family Araceae Juss. subfamily Monsteroideae (Schott) Engl.[54] The species of *Spathiphyllum* originate mainly from the tropical regions of North and South America, some species originate from Southeast Asia. *Spathiphyllum* species is commonly called as peace lily, snowflower, spathe flower or white anthurium. It is a very common houseplant which is known for its brilliant white flowers and acceptance of dark places.[55–57] The peace lily has a minimalistic look and the charming white flower (spathe) that makes it a great choice of plant for the home or office.[54]

The inflorescences of *Spathiphyllum* represent ornamental ears covered with bisexual flowers. The color of ears in some varieties are fine-pink or cream. A recently opened flower of *Spathiphyllum wallisii* Regel is surrounded by white covering, which later becomes green. The shape of the covering often looks like a small flag fluttering in the wind; hence, *S. wallisii* is also called white flag (Fig. 8.5).

FIGURE 8.5   (See color insert.) *Spathiphyllum wallisii* 'Quatro'.

Under good cultivation conditions, the flowering of *Spathiphyllum* continues during a few weeks from April until July. Sometimes it flowers in autumn. Among problems of *Spathiphyllum* spp. cultivation, it is necessary to mention insufficient resistance to soil salinization, damage by spider mites, scales, and mealybugs; overwatering can stimulate the development of *Cylindrocladium* root rot (*Cylindrocladium scoparium* Morgan). Most of the *Spathiphyllum* spp. tissues contain calcium oxalates which can be toxic for humans, cats, and dogs.

The plants of genus *Tetrastigma* (Miq.) Planch. belonging to the family Vitaceae Juss. subfamily Vitoideae Eaton. *Tetrastigma* spp. are spread in tropical and subtropical regions of Asia from India to China, in the whole Southeast Asia reaching Australia, and going easterly until Fiji.[58] *Tetrastigma* spp. is a very spectacular perennial evergreen woody liana with a strong stem; it has big (up to 35 cm), palmate-composed leaves with a notched edge and is attached to the support by means of tendrils. The species *Tetrastigma voinierianum* (Baltet) Gagnep. is cultivated in the countries with temperate climate as houseplant under the name of Chestnut vine or Lizard plant (Fig. 8.6).

FIGURE 8.6   **(See color insert.)** *Tetrastigma voinierianum* (Baltet) Gagnep.

This species endemic to Vietnam is spread in Indo-China (Vietnam and Laos)[59], where under the canopy of woody plants it can reach up to 50-m length. Thick and rapidly growing stems of *T. voinierianum* attach by tendrils to any support and carry to the light big, alternate, compound-palmate leaves, composed from 3–5 oval or rhombic leaflets, dark-green from above and with rusty thin coating from the bottom, with notched edges. *T. voinierianum* is extremely enduring and shows high degree of bactericide activity against pathogenic staphylococcus. Single plant of *T. voinierianum* covered by magnificent foliage and tendrils can create in the room the effect of jungles. *T. voinierianum* develops in its full beauty in the big, spacious rooms like lobbies of theaters, concert halls, restaurants, hotels, big offices. The most intensive development of *T. voinierianum* occurs in halls with water pools, where the plant can develop a magnificent marquee because of elevated humidity if support of suitable shape and durability is provided.

Among few potential problems, it is necessary to notice that in a room with dry air, *Tetrastigma. voinierianum* can be attacked by arachnoidal mite. Watering with hard chlorinated water and/or under deficit of nutrients in the soil can make leaves yellowish and fall.

The plants described above as well as plants of other species under study were arranged in the winter garden of Baltschug Kempinski hotel (Moscow) on the balconies of rectangular shape. The upper balcony situated on the eighth floor at the elevation of 25 m above ground floor overhangs until the middle of the atrium width. The atrium represents an inner court of the Baltschug hotel glassed-in from the top, where plants were arranged on the balconies in five levels: the width of the smallest balcony was 0.90 m and the length was 2.0 m, the biggest balcony of the uppermost level had the size of 2.0×2.0 m². The air temperature in the atrium in summer reached +36°C and in winter, it did not fall below +13°C. Plants were enumerated from left to right according to the distribution of root necks.

In the city of Uman, plants were placed in the garden of the Marble Statuary at the ground floor of the House of Scientists, which belongs to the National Dendrological Park "Sofiyivka" of National Academy of Sciences (NAS) of Ukraine. The hall is built on three levels with the vertical difference of about 1 m. Large windows look southeast. Seasonal air temperature variations in this room ranged from +12.0 to +29°C.

The Marble Statuary garden of the House of Scientists as well as the winter garden of Baltschug Kempinski hotel are open for public access and are daily visited by dozens of people. Some days, the amount of visitors exceeds 200. Because of such amount of visitors, the application of protection measures as well as quarantine measures recommended for closed greenhouses,[60] are impossible in these rooms. Consequently, we applied nontoxic biologically active preparations: PABA, biostimulant preparation of Kornevin for plants, and active substance indole-3-butyric acid (IBA) in the concentration 5 g/kg.

PABA as a chemical compound has been known since 1863, but its high biological activity in low concentrations was first discovered in 1939 by a well-known geneticist Rapoport on *Drosophila*.[61] Rapoport showed that the positive PABA effect on living systems is based on a previously unknown phenomenon and its interaction with ferments. This interaction results in the restoration of the ferments activity, decreased in some cases at the genetic level (e.g. because of an excess of recessive genes) or because of damaging environmental factors.[62] PABA is classified as a nontoxic vitamin-like compound of group B, also known as vitamin H_1 or vitamin B_{10}. In subsequent studies, the ranges of PABA suitable for different objects were determined. It was proved that PABA is a promoter of phenotypic activity and increases immunity; it has viricidal and antimicrobial action, and showed biocidic functions. There are data available about the PABA effect decreasing harmful mutagens action, on all characters determining the yield structure, and increasing adaptive plant properties including the resistance to a series of diseases.[63]

PABA (4-aminino-2-hydrooxybezoic acid, molecular mass 137.1) is an organic compound with the formula $H_2NC_6H_4CO_2H$. PABA, a white-gray crystalline substance, is only slightly soluble in water (Fig. 8.7).

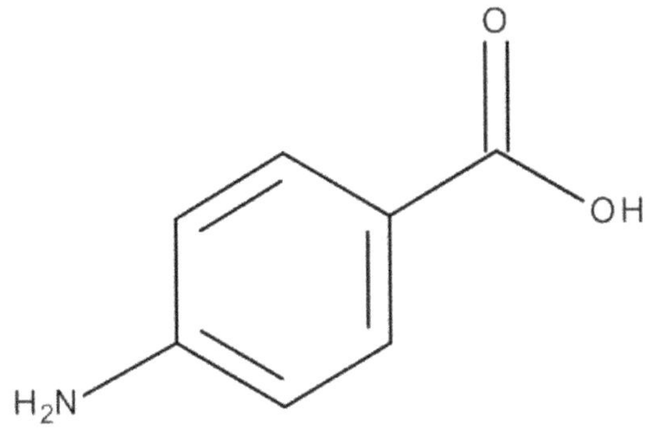

FIGURE 8.7   Structural formula of PABA.

Because of the presence of IBA, Kornevin preparation, when pene-trating into the plant, stimulates the appearance of callus ("living" cells, formed on the surface of wound) and roots.[64] IBA is a substance that is closely related in structure and function to a natural growth regulator found in plants (Fig. 8.8). IBA is used on many crops and ornamentals to promote growth and development of roots, flowers, and fruits, and to increase crop yields. Growers find it more effective and efficient than its natural counter-part because plants cannot break it down as quickly. No harm to humans or the environment is expected to result from the use of IBA. When entering the soil, IBA is transformed because of the natural synthesis into phytohor-mone heteroauxin, which in essence, stimulates root formation and conse-quently, by means of root system development, also stimulates growth and development of plants. Consequently, Kornevin effects are slower than pure heteroauxin, but its effect is more prolonged.[65]

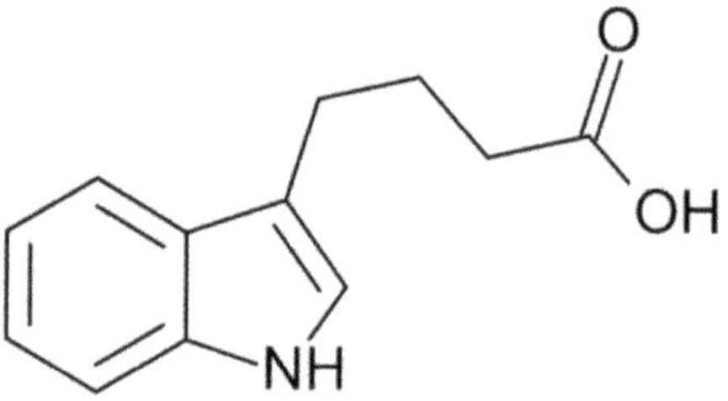

FIGURE 8.8   Structural formula of indole-3-butyric acid (IBA).

Leaves of plants in the winter garden were sparged using a garden sprayer. After 3–4 days, as PABA penetrated into the plants through leaves, the root system was fertilized by Kornevin in the above mentioned concen-trations with the addition of the prepared Zircon, which is an immunostim-ulator with functional and anti-stress effect. This preparation is prepared from the medicinal plant *Echinacea purpurea* (L.) Moench (eastern purple coneflower or purple coneflower) and represents a solution of hydroxycin-namic acids in ethanol in the concentration of 0.02%.[66] PABA was applied the whole year to all plants at once, once per 2 weeks. Under the joint application of PABA and Kornevin, the effect of stimulation consider-ably increased,[67] which was shown on four sample species, on which the working hypothesis about the stimulation of indoor plants growth because of their treatment by PABA and Kornevin was proved the most strongly (Table 8.2). As an example, the table shows only the results of the best

TABLE 8.2 Mean Seasonal Growth of Indoor Plants in the Best Variants of Stimulation in the Winter Garden of Baltschug Kempinski Hotel.

Indoor plant	Plant length in variants[*] of treatment, cm				Effect of growth stimulation after the treatment by PABA+watering with Kornevin	
	Sterile distilled water	Treatment of foliage with PABA 0.03%	Watering with Kornevin	Treatment of foliage with PABA 0.03%+ watering with Kornevin 1 g/l of water	cm	%
D. fragrans (L.) Ker Gawl. 'Massangeana' (corn plant)	45	52	54	70	25	56
M. deliciosa Lieb. (fruit salad plant)	90	108	122	150	60	66
S. wallisii Regel (spathe flower)	50	62	58	70	20	40
T. voinierianum (Baltet) Gagnep.(chestnut vine)	50	72	85	100	50	50

[*]NPK and microelements.

variants of stimulation for the four genotypes of different indoor plants species, but it is necessary to notice that all tested plants of other species in the best variants also considerably exceeded control both by the indicators of annual growth as well as by general resistance to diseases.

Joint plant treatment by PABA and Kornevin gave the highest growth of sample indoor plants, which amounted from 20 to 60 cm in dependence on plant species, that is, from 40 to 66%. Each one of the plant species has its own development peculiarities, which explains the difference in the growth rate.

On an average, under the joint treatment by PABA and Kornevin, the annual growth of foliage biomass increased by 49% and the mean shoot length increased by 51%. Some plants formed additional shoots.

Taking into account that the increase in size of ornamental houseplants does not always improve their ornamentality, the data about the increase of resistance to biotic and abiotic stress of plants treated by PABA, Kornevin, and Zircon confirm the effectiveness of immunostimulation.

Under the treatment of leaves of ornamental plants with 0.03% water solution of PABA with subsequent watering of substrate in containers by Kornevin in the dose of 0.1 g/l of water with the addition of 5 ml of concentrated fertilizer 3–4 days later, no decrease of ornamentality was observed, neither in winter nor in the summer season in both experimental sites. However in control variants, when *Dracaena* spp. was cultivated without PABA and Kornevin in winter period, the decrease of growth and drying of young leaf tips was observed. The serodiagnosis and biological control of the mentioned diseases showed the absence of mycotic, bacterial, and mycoplasmal infections. Consequently, the decrease of growth and drying of leaf tips was attributed to physiological disturbances evoked by the decrease of air temperature.

Similar symptoms were observed in the control variants of *Monstera* spp., *Spathiphyllum* spp., *Tetrastigma* spp., and other species of ornamental houseplants. All of them were attributed to physiological disturbances except *Spathiphyllum wallisii* 'Quatro'. In the container with the untreated plant of this cultivar in the garden of the Marble Statuary in the House of Scientists, mealy bugs were detected in the second part of October. After triple treatment by Zircon with the interval of 7–10 days by mid-December, the plants were purified from this pest without the application of pesticides. It can be supposed that the joint effect of hydroxycinnamic acids and ethanol, which are part of Zircon, provided the effect of immunostimulation.

8.4   CONCLUSION

The studies of the possibility of the use of indoor plants in the interior of classic winter gardens conducted in two locations distant from each other, namely in the winter garden of Baltschug Kempinski hotel (Moscow, Russia) and in the winter garden of the Marble Statuary in the House of Scientists (Uman, Ukraine), in order to create a favorable atmosphere for the staff and visitors, confirmed the expediency to introduce different species of ornamental houseplants. The effectiveness of ecologically safe ways of resistance of indoor plants to pathogens and pests without the use of pesticides was shown. In all variants of the experiment, the increase of growth and decrease of morbidity and damage by insects relative to control was observed. The best result was obtained in the variant with the treatment of leaves by water solution of PABA in the concentration of 0.03% and the subsequent (3–4 days later) treatment by Kornevin in the dose of 1.0 g/l of water with the addition of 5 ml of concentrated fertilizer containing nitrogen (N) 7%, phosphorus (P) 3%, potassium (K) 6%, and microelements ($2MgO + 2S + 15B + 0.3Fe + 0.3Mn + 0.05Cu + 0.2Zn + 0.008Mo$).

ACKNOWLEDGMENT

We are deeply grateful to the director of National Dendrological Park "Sofiyivka" of NAS of Ukraine, member-correspondent of the NAS of Ukraine, and doctor of biological sciences Ivan Kosenko, whose kind interest and encouragement has brought us to this podium.

KEYWORDS

- green urbanism
- house plants
- immunostimulation
- indoor gardens
- ornamental horticulture
- para-aminobenzoic acid

- **kornevin**
- **Zircon**
- **microelements**

REFERENCES

1. Tolstoy, L. N. *The Cossacks. A Tale of 1852 [Translated by Louise and Aylmer Maude];* The Floating Press: Portland, 2009; p 346.
2. Beatley, T. *Green Urbanism Learning from European Cities;* Island press: Washington, Covelo, 2012; p 512.
3. Kolesnikov, A. I. *Decorative Dendrology;* Forestry Industry Press: Moscow, 1974; p 615 (in Russian).
4. Dirr, M. A. *Manual of Woody Landscape Plants: Their Identification, Ornamental Characteristics, Culture, Propagation and Uses;* Stripes Publishing: London, 2009; p 1325.
5. Seneta, W.; Dolatowski, J. *Dendrologia. Warszaw;* Scientific Publishing. 2012; p 544 (In Polish).
6. Tomlinson, H. *The Complete Book of Bonsai;* Dorling Kindersley: London, 2004; p 224.
7. Ingels, J. E. *Ornamental Horticulture: Science, Operations, and Management,* 4th ed.; Cengage Learning: Delmar, 2010; p 712.
8. Gasanov, Z. M.; Iskandarova, T. G.; Bilmanli, A. I. Environmental Conditions and Use of Fruit Plants in Landscaping of Gyandzy Town. 2014, Contemporary Horticulture 2. http://journal.vniispk.ru/pdf/2014/2/28.pdf (accessed Aug 29, 2016) (in Russian).
9. Janick, J. The Origins of Horticultural Technology and Science. *Acta Hortic.* **2007,** *759,* 41–60.
10. Baeyer, E. *The Development and History of Horticulture;* Gardening: Ruston, 2010; p 25.
11. Farahani, L. M.; Motamed, B.; Jamei, E. Persian Gardens: Meanings, Symbolism, and Design. *Landscape Online* **2016,** *46,* 1–19.
12. Thommen, L. *An Environmental History of Ancient Greece and Rome;* Cambridge University Press: Cambridge, New York, 2012; p 186.
13. Janick, J. Horticulture and Art. Horticulture: Plants for People and Places [Eds.: Geoffrey R. Dixon and David E. Aldous]. Trilogy. Heidelberg: Springer Science and Business Media. 3. Social. *Horticulture* **2014,** *36,* 1197–1223.
14. Pomerantz, G. S.; Mirkin, Z. A. The Sunsets and Dawns Civilizations. *Soc. Sci. Present* **2012,** *2,* 155–160 (in Russian).
15. Takei, J.; Keane, M. P. *Sakuteiki: Visions of the Japanese Garden: A Modern Translation of Japan's Gardening Classic (Tuttle Classics);* Tuttle Publishing: Clarendon, 2001; p 256.

16. Aristov, N. J. *Industry of Ancient Russia;* St. Petersburg, 1866; p 335 (in Russian).

17. Cherny, B. *Russian Medieval Gardens: The Classification of Experience [Handwritten Monuments of Ancient Rus];* Bukva: Moscow, 2014; p 5063 (in Russian).

18. Haynes, G. *Landscape and Garden Design: Lessons from History;* Whittles Publishing: Dunbeath, Caithness, 2013; p 201.

19. Kosenko, I. S.; Pylypiuk, V. V. *Sofiyivka. National Dendrological Park: Photo Albums;* Palyvoda A.V.: Kyiv, 2016; p 275 (in Ukrainian).

20. Likhachev, D. S. *Poetry Gardens: To the Semantics of Landscaping Style Garden as Text;* Soglasie: Moscow, 1998; p 356 (in Russian).

21. Nehuzhenko, N. A. *Basics of Landscape Design and Landscape Architecture,* 2nd ed.; Corrected and Add. Publishing House "Piter": St. Petersburg, 2011; p 192 (in Russian).

22. Perry, R. C. *Landscape Plants for California Gardens: An Illustrated Reference of Plants for California Landscapes;* Land Design Publishing: Claremont, 2010; p 652.

23. Petrik, V. V. *The History of Landscape Art: The Text of the Lectures;* Northern (Arctic) Federal University: Arkhangelsk. 2010; p 243. (in Russian).

24. Rubtsov, L. I. *Design of Gardens and Parks;* Stroyizdat: Moscow, 1979; p 184 (in Russian).

25. Sullivan, C.; Elizabeth, B. *Illustrated History of Landscape Design;* John Wiley and Sons: USA, 2010; p 275.

26. Tepe, E.; Markert, P. *The Edible Landscape: Creating a Beautiful and Bountiful Garden with Vegetables, Fruits and Flowers;* Keefe, M., Ed.; Voyageur Press: Minneapolis, 2013; p 160.

27. Rogers, E. B.; Hiss, T. *Green Metropolis: The Extraordinary Landscapes of New York City as Nature, History, and Design;* Alfred A. Knopf: New York, 2016; p 240.

28. Martin, T. *Once Upon a Windowsill: A History of Indoor Plants;* Timber Press: Portland, 2009; p 312.

29. Martin, T. *The Indestructible Houseplant: 200 Beautiful Plants that Everyone Can Grow* [Photo.: Kindra Clineff]. Timber Press: Portland, 2015; p 289.

30. ElAziz, A. N. G.; Mahgoub, M. H.; Mazhar, A. M. M.; et al. Potentiality of Ornamental Plants and Woody Trees as Phytoremidators of Pollutants in the Air: A Review. *Int. J. ChemTech Res.* **2015,** 8(6), 468–482.

31. Jumeno, D.; Matsumoto, H. The Effects of Indoor Foliage Plants on Perceived Air Quality, Mood, Attention, and Productivity. *J. Civ. Eng. Arch. Res.* **2016,** 3(4), 1359–1370.

32. Tkachenko, K. G.; Kazarinova, N. V. Medical Phytodesign—Using Plants into Interior and Prophylactics Infectious Diseases. *Belgorod State University Scientific Bulletin: Nat. Sci.* **2008,** 3(43), 6, 53–59 (in Russian).

33. Grodzinsky, A. M. Phytodesign: Problems and Prospects UNESCO News. *Newsletter* **1979;** 9, 1–8 (in Russian).

34. Tarakanova, K. V.; Baklyskaya, L. E. Fitodesign in the Interior and Its Impact on Human Activity. The New Ideas of New Century. *Proceedings of the Fourteenth International Scientific Conference.* Pacific National University Press: Khabarovsk, 2014, 2, 235–239 (in Russian).

35. Ferrantea, A.; Trivellini, A.; Scuderi, D. et al. Post-Production Physiology and Handling of Ornamental Potted Plants. *Postharvest Biol. Technol.* **2015,** 100, 99–108.

36. Kubo, K.; Schrempp, E. Keiko's Ikebana: A Contemporary Approach to the Traditional Japanese Art of Flower Arranging [Photographer: Erich Schrempp]. Tuttle Publishing: North Clarendon, 2006; p 128.

37. Sultanova, G. Ikebana in Russian. Litres: Moscow; p 169 (in Russian). 2015.

38. Johnson, N. B. Religion, Spirit, and the Idea of Garden. *Relig. Stud. Rev.* **2010**, *36*(1), 1–14.

39. Moriyama, M.; Moriyama, M. A Comparison Between Asymmetric Japanese Ikebana and Symmetric Western Flower Arrangement. *Forma: Proceedings of the 2nd International Katachi U Symetry Symposium, Tsukuba* 1, **1999**, *14*(4), 355–361.

40. Gray, S. The Secret Language of Flowers [Illustr.: Sarah Perkins]. Ryland Peters and Small: London, 2011; p 128.

41. Contenson, E. *The Language of Flowers;* Archives and Culture: Paris, 2009; p 70 (in French)

42. Cruz, P. S. *The Language of Flowers Dictionary.* Xlibris Corporation: Bloomington, 2015; p 50.

43. Diffenbaugh, V. *The Language of Flowers* [Novel]. Ripol Klassik: Moscow, 2014; p 317 (in Russian).

44. Kirkby, M.; Diffenbaugh, V. *A Victorian Flower Dictionary: The Language of Flowers Companion.* The Random House Publishing Group: New York, 2011; p 192.

45. Zachos, E. *Growing Healthy Houseplants: Choose the Right Plant, Water Wisely and Control Pests;* Guare, S., Madigan, C. Eds.; North Adams: Storey Publishing, 2014; p 129.

46. Gorbanyov, V. A. Geographical Zoning of Russia. *The Herald of MGIMO-University* **2014**, *4*(37), 187–196 (in Russian).

47. Gerasymenko, N. M.; Davydchuk, V. S.; Marynych, O. M.; et al. Physiographic Zoning: Landscapes and Physiographic Zoning. National Atlas of Ukraine: Texts and Maps' Legends [Chairman of Editorial Board: Borys Yev. Paton]. Kartographia: Kyiv, 2007; Extra Vol.; 636–637.

48. Lipinsky, V. M.; Dyachuk, V. A.; Babichenko V. M.; et al. Climate of Ukraine. Lipinsky, V. M., Dyachuk, V. A., Babichenko, V. M., Eds.; Rayevsky Publishing House: Kyiv, 2003; p 345 (in Ukrainian).

49. Lu, P.-L.; Morden, C. W. Phylogenetic Relationships Among Dracaenoid Genera (Asparagaceae: Nolinoideae) Inferred from Chloroplast DNA Loci. *Syst. Bot.* **2014**, *39*(1), 90–104.

50. Bos, J. J.; Graven, P.; Hetterscheid, W. L. A.; van de Wege, J. J. Wild and Cultivated Dracaena Fragrans. *Edinburgh J. Bot.* **1992**, *49*(3), 311–331.

51. Madison, M. A Revision of Monstera (Araceae). Contributions from The Gray Herbarium of Harvard University, 1977; 207, 1–101.

52. Martin, T. J. A Mexican Migrant: The Naturalization of Monstera Deliciosa (Fruit Salad Plant) in New Zealand. *Auckland Bot. Soc. J.* **2002**, *57*(2), 151–154.

53. Muir, C. D. How Did the Swiss Cheese Plant Get Its Holes? *Am. Nat.* **2013**, *181*(2), 273–281.

54. Nauheimer, L.; Metzler, D.; Renner, S. S. Global History of the Ancient Monocot Family Araceae Inferred with Models Accounting for Past Continental Positions and Previous Ranges Based on Fossils. *New Phytol.* **2012**, *1959*(4), 938–950.

55. Cabrera, L. I.; Salazar, G. A.; Chase, M. W.; et al. Phylogenetic Relationships of Aroids and Duckweeds (Araceae) Inferred from Coding and Noncoding Plastid DNA. *Am. J. Bot.* **2008,** *95*(9), 1153–1165.

56. Kakoei, F.; Salehi, H. Effects of Different Pot Mixtures on Spathiphyllum (Spathiphyllum Wallisii Regel. Growth and Development. *J. Cent. Eur. Agric.* **2013,** *14*(2), 140–148.

57. Tam, S. M.; Boyce, P. C.; Upson, T. M.; et al. Intergeneric and American Journal of Botany Phylogeny of Subfamily Monsteroideae (Araceae) Revealed by Chloroplast trnL-F Sequences. *Am. J. Bot.* **2004,** *91*(3), 490–498.

58. Soejima, A.; Wen, J. Phylogenetic Analysis of the Grape Family (Vitaceae) Based on Three Chloroplast Markers. *Am. J. Bot.* **2006,** *93*(2), 278–287.

59. Dobbins, D. R.; Fisher, J. B. Wound Responses in Girdled Stems of Lianas. *Bot. Gaz.* **1986,** *147*(3), 278–289.

60. Kuznetsova, N. P. Complex System of Protection Greenhouse Plants from Pests and Diseases in Siberian Botanical Gardens of Tomsk State University. *Tomsk State Univ. J. Biol.* **2008,** *2*(3), 43–46 (in Russian).

61. Rapoport, I. A. Phenogenetically Analysis of Independent and Dependent Differentiation. *Proceed. Inst. Cytol. Histol. Embryol.* **1948,** *1*(2), 31–32 (in Russian).

62. Rapoport, I. A. Action PABA in Connection with the Genetic Structure. Chemical Mutagens and Para-Aminobenzoic Acid in Increasing the Yield of Crops. Nauka: Moscow, 1989; pp 3–37 (in Russian).

63. Opalko, O. A.; Bekuzarova, S. A. Efficiency of Immune-Stimulation of Fruit and Small Fruit Crops with Para-Amino-Benzoic Acid. *Pomiculture and Small Fruits Culture in Russia: A Collection of Scientific Works* **2016,** *44,* 201–206. (In Russian).

64. Kefeli, V. I. Photomorphogenesis, Photosynthesis and Growth as the Basis of Plant Productivity. COMBINING Scientific and Technical Publishers. Pushchino Research Center of Russian Academy of Sciences: Pushchino, 1991; p 132 (in Russian).

65. Bondorina, I. A. The Impact of Physiologically Active Substances on the Regeneration Processes Woody Plants. Thesis … of DSc of Biological Sciences. The Tsytsin Main Moscow Botanical Garden of Russian Academy of Sciences: Moscow, 2012; p 41 (in Russian).

66. Malevannaya, N. N. The Drug "Zircon"—An Immunomodulator of New Type. Proceedings of the III Moscow International Congress "Biotechnology: State and Prospects for Development" (Moscow, 14.03–18.03.2005), Moscow, 2005; Vol. 1, pp 273–274 (in Russian).

67. Burakov, A. E.; Bekuzarova, S. A.; Weisfeld, L. I.; et al. A Method of Caring for Decorative Plants. Patent of Russian Federation No 2463779. Published on Oct 20, 2012 (in Russian). 2012.

CHAPTER 9

EVALUATION OF SALT-RESISTANT FLOWERING PLANTS IN SIMULATED CONDITIONS

NINA A. BOME[1,*], MARINA V. SEMENOVA[1],
KONSTANTIN P. KOROLEV[1], NINA V. BISEROVA[2], and
ALEXANDER YA. BOME[3]

[1]*Department of Botany, Biotechnology and Landscape Architecture,
Institute of Biology, Tyumen State University, 10, Semakov St.,
Tyumen, Russia, 625003*

[2]*Municipal Autonomous Comprehensive Educational Institution
Secondary Comprehensive School No 9, 15, Shishkov St., Tyumen,
Russia, 625031, E-mail: nwnag@mail.ru*

[3]*Federal Research Center, N.I. Vavilov All-Russian Institute of
Plant Genetic Resources, 42–44, Bol'shaya Morskaya St., Saint
Petersburg, Russia, 190000, E-mail: office@vir.nv.ru*

*Corresponding author. E-mail: bomena@mail.ru

CONTENTS

ABSTRACT

Seventeen cultivars of annual flower plants from the collection of Tyumen State University were studied from the point of view of resistance to salinization. The cultivars belong to five species: California poppy (*Eschscholzia californica* Cham. family Papaveraceae), common flax (*Linum usitatissimum* L., family Linaceae), and three species from family Asteraceae: Tagetes erect (*Tagetes erecta* L.), small flowering tagetes (*Tagetes patula* L.), Chinese Callisaurus (*Callistephus chinensis* (L.) Nees.). These species are used in the floral compositions in Tyumen and other settlements of Tyumen region. The study was performed in controlled laboratory condition on the provocative background by using NaCl solution in three concentrations: 0.15, 1.05, and 1.95%. Cultivars differed considerably according to their reaction to salinization stress, that is, according to germination capacity in vitro and main characters of germs (length and biomass). The cultivars were separated into three groups relative to chloride salinization and according to the complex of characters considered as indicators: into cultivars with high, medium and low resistance.

9.1   INTRODUCTION

The problem of soil salinization and plants resistance to salinization belongs to the problems, which are actual and complicated worldwide. Salinization is considered as abiotic stress factor leading to a considerable decrease of adaptive and productive properties of plants.[1]

The toxic effects of salts are not only experienced by agricultural plants when growing on salt soils but also by the plants applied as greenery in an urban environment. The application of chemical reagents, which decrease the risk of ice-crusted ground, leads to the accumulation of salts along the roads and streets and at closely situated territories. Consequently, the selection of salt-resistant ornamental plants in the roadside plantings is necessary.[2]

When ranking the unfavorable factors according to the measure of effect and danger (risk) for plants in the greenery of Moscow, H. G. Yakubov[3] placed the salinization of soils by anti-ice reagents of older generation at the first place (e.g., salt technical—NaCl). The excess of these compounds in the soil solution is toxic for most of the plants. The most readily soluble salts, which easily penetrate in the cytoplasm, such as $NaCl$, $MgCl_2$, and

$CaCl_2$ are most harmful. Sparingly soluble forms such as $CaSO_4$, $MgSO_4$, and $CaCO_3$ are less toxic. Lower toxicity of sulfate salinization is related, in particular, with that, in contrast with Cl^- ion, the SO_4^{2+} ion is necessary in small amounts for the normal plants nutrition. Only its excess is harmful.

Salts negatively affect the seeds germination, which leads to the appearance of disjointed and weak germs. Frequently, plants do not finish the full development cycle during the vegetation period.[4] It is necessary to take into account the phase of plants development: germs with small roots react stronger than mature plants with the well-developed root system. Plants, which grew on saline soils, are characterized by smaller biomass; sometimes, browning edges and inner areas of the leaves, spotting, crinkle, and beginning of chlorosis are observed, and internal anatomic changes can also occur.[5]

Salt compounds evoke the disturbance of metabolic processes, gas exchange, homeostasis, and use of available energy sources in the plant cells. The disturbance of nitrogen exchange leads to the intensive proteins decomposition; consequently, the accumulation of intermediate metabolic products toxic to plants occurs.[6]

Salt-resistant plants are valued for their ability to tolerate soil salinization and develop active ecological and biological reactions aimed to strengthening of plants vitality in these conditions. The plants, which pass the complete development cycle in the soils with salt ions content above 0.2% of soil mass, are attributed to be salt tolerant.[7]

A number of methods are developed for the diagnostic of the degree of plants' salt resistance: microscopic, counting of seeds germination percentage, evaluation of salt scorches appearance rate of cut plants, and so forth. The experiments in vitro and in situ are usually based on the creation of provocative background.[8–9]

9.2 MATERIALS AND METHODOLOGY

Seventeen cultivars of annual flower plants from the collection of Tyumen State University were applied for the study of salt resistance. The cultivars represented five species: California poppy (*Eschscholzia californica* Cham.)—Papaveraceae family, Common flax (*Linum usitatissimum* L.)—Linaceae family, Tagetes erect (*Tagetes erecta* L.), small flowering tagetes (*Tagetes patula* L.), Chinese Callisaurus (*Callistephus chinensis* (L.) Nees.)—Asteraceae family.

Visually health seeds of the same reproduction year were selected for the experiment. The laboratory germination capacity of the seeds from all samples was preliminarily tested. The seeds with laboratory germination rate above 80% were used for the study of salt resistance.

Seeds germination was performed in Petri dishes on the filter paper moistened by distilled water (control) or by salt (NaCl) solutions in concentrations of 0.15, 1.05, and 1.95%. Water solutions of chemically pure solutions of table salt (NaCl) were prepared with osmotic pressure levels of 0.1, 0.8, and 1.4 mPa, respectively. The sample size was 50 seeds with triple repetition of the experiment.

The Petri dishes were preliminarily sterilized for 3 h in the dry heat oven "ED-15" (BINDER, Tuttlingen, Germany). Before the start of the experiment, the seeds were disinfected with 1% solution of $KMnO_4$ for 10 min. The seeds were germinated in the dry-air thermostat for electrical cooling TSO-1/80 SPU under the temperature of 22°C. On the third and seventh day (in dependence on plant species), the seeds germination energy was calculated. Laboratory germination capacity was determined on the 14th day (State Standard of Russia GOST-12,038–84).[12] After determining the laboratory germination capacity, the length and mass of germs were measured for normally germinated seeds on the same day.

After counting germinated seeds, the salt resistance was calculated according to the equation:

$$P = n1/n* 100,$$

where P is the salt resistance of the sample, n is the number of germs in control, and $n1$ is the number of germs in the salt solution.

Using morphological parameters of the germs, the salt tolerance was also determined according to the degree (percentage) of the decrease of the character under the consideration of the salinization background relative to control:

$$P = a/b * 100\%,$$

where P is the sample salt resistance in % of the control, a is the value of the character under consideration under salinization, and b is the value of the same character under control.

Statistical analysis was performed according to the method of G. F. Lakin.[13] The arithmetic mean of parameters under study (X), the error of arithmetic mean (Sx), and the coefficient of variation (CV) were calculated. All the seedlings were measured—these are individual characteristics.

Laboratory germination was considered as an average of the replicates. The statistics for length and mass were calculated on the basis of individual seedlings. The statistics for germination capacity were calculated for three replicates.

9.3   RESULTS AND DISCUSSION

It is known that salt stress considerably affects the processes of seeds germination. Consequently, the determination of salt resistance was started with the analysis of laboratory germination capacity of the seeds (Fig. 9.1). When germinating under chloride salinization, a considerable decrease in the seeds germination capacity was observed in all cultivars and species under study. The maximum inhibiting effect of stress factor was observed under the salt concentration of 1.95%. The lower sensitivity of germinating seeds to salinization was observed in the variant with the salt solution concentration of 0.15%.

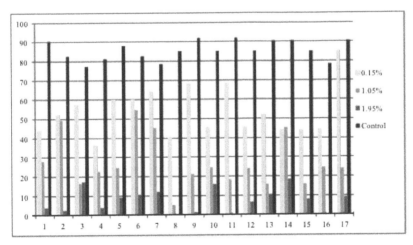

FIGURE 9.1 Laboratory germination capacity of seeds of ornamental plants species and cultivars under study (%): 1—*Tagetes patula* L. 'Yellow', 2—*T. patula* L. 'Double giants', 3—*T. patula* L. 'Oreng prints', 4—*T. patula* L. 'Carmen', 5—*T. patula* L. 'Tiger eyes', 6—*Tagetes erecta* L. 'Solar giants', 7—*T. erecta* L. 'Smile', 8—*Callistephus chinensis* (L.) 'Nees'. 'Winter cherry', 9—*Eschscholzia californica* Cham. 'Violet beam', 10—*E. californica* Cham. 'White castle', 11—*E. californica* Cham. 'Yellow queen', 12—*E. californica* Cham. 'Sparkling carpet', 13—*E. californica* Cham. 'Ballet dancer', 14—*Linum usitatissimum* L. 'Szegedi olaglen', 15—*L. usitatissimum* L. 'Fibriferum', 16—*L. usitatissimum* L. 'Oleiferum', 17—*L. usitatissimum* L. 'Natasja'.

The best results for experimental and control variants were observed for the following species and cultivars: *Eschscholzia californica* Cham. 'Ballet dancer', *L. usitatissimum* L. 'Szegedi olaglen', *T. erecta* L. 'Solar giants'. The slowing and cessation of seeds germination in reference to saline substrate were observed in the following cultivars: *T. patula* L. 'Carmen', *C. chinensis* (L.) 'Nees'. 'Winter cherry', *L. usitatissimum* L. 'Fibriferum' and 'Oleiferum'.

The results of the laboratory germination capacity of the seeds were verified by the data on salt resistance determined by morphometric parameters of seedlings. In the variants with the highest NaCl concentration in the solution (1.95%), very low resistance to salinization was observed. The germs of flower plants in the variant with 0.15% concentration were characterized by the lower sensitivity to the stress factor (Fig. 9.2).

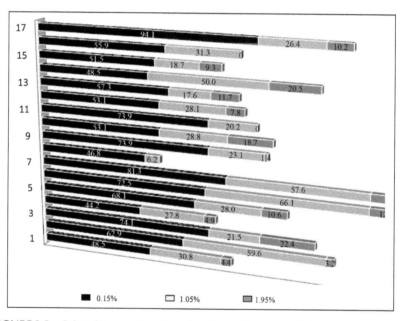

FIGURE 9.2   Salt resistance of species and cultivars of ornamental plants under study (%): 1—*T. patula* L. 'Yellow', 2—*T. patula* L. 'Double giants', 3—*T. patula* L. 'Oreng prints', 4—*T. patula* L. 'Carmen', 5—*T. patula* L. 'Tiger eyes', 6—*T. erecta* L. 'Solar giants', 7—*T. erecta* L. 'Smile', 8—*Callistephus chinensis* (L.) 'Nees'. 'Winter cherry', 9—*E. californica* Cham. 'Violet beam', 10—*E. californica* Cham. 'White castle', 11—*E. californica* Cham. 'Yellow queen', 12—*E. californica* Cham. 'Sparkling carpet', 13—*E. californica* Cham. 'Ballet dancer', 14—*L. usitatissimum* L. 'Szegedi olaglen', 15—*L. usitatissimum* L. 'Fibriferum', 16—*L. usitatissimum* L. 'Oleiferum', 17—*L. usitatissimum* L. 'Natasja'.

The comparative evaluation of the species and cultivars according to the indicators of salt resistance divide them into three groups: with high, medium and low resistance. Such approach is applied in a number of guidelines. Our analysis showed that the representatives of three species were characterized by low resistance to salinization, namely *C. chinensis* (L.) 'Nees'. 'Winter cherry', *T. patula* L. 'Carmen', *L. usitatissimum* L. 'Fibriferum' (third group). High adaptive capacity to salinization was shown by two cultivars of Tagetes erect (*T. erecta* L.): 'Smile' and 'Solar giants' and two cultivars of common flax (*L. usitatissimum* L.): 'Natasja' and 'Szegedi olaglen' (first group). Other species and cultivars were in the intermediate position (second group).

From our point of view, such a distribution based only on the germination capacity could not be completely objective because it does not reflect the state of germs and consequently the possibility of getting the valuable germs in field conditions, which is important for the expression of adaptive and productive properties of plants in ontogenesis. Consequently, in our experiment, the length and mass of the germs were analyzed. The morphological characters of germs are often applied as the biometric indicators of environmental factors that affect the plant organism under formation, in particular, pine,[14] corn,[15] safflower[16], and spring and winter wheat.[17,18]

The reaction of cultivars on stress factor displayed the decrease of germ length with well-expressed dependence on the concentration of the salt solution (Table 9.1). The average value of this indicator among cultivars relative to control amounted 66.6, 37.2, and 14.9% under concentrations of 0.15, 1.05, and 1.95%, respectively.

Under high concentration, the loss of germs was observed in five cultivars. The degree of the character variability in the averaged data in the control variant was weak; and in the variants with salinization, it was medium, with expressed dependence of the coefficient of variation on the NaCl concentration.

The analysis of species and cultivars according to the decrease of germ length in the variant with concentration of 0.15% showed that the cultivars of *T. patula* L. and *T. erecta* L. were characterized by the minimum sensitivity to the stress factor as their salt resistance varied from 72.8 to 109.7%. In the cultivar Oreng prints, a stimulating effect of this character was observed. The 'Yellow queen' *(E. californica* Cham.) and two cultivars of *L. usitatissimum* L. ('Oleiferum' and 'Natasja') showed low resistance. Thus, the group of high resistance amounted to seven cultivars, the

TABLE 9.1   The Effect of Salinization on the Length of Germs of Ornamental Plants Cultivars and Species Under Study (cm).

Species cultivar	NaCl concentration				
	0.15%	1.05%	1.95%	Control	
T. patula L. Yellow	4.91±0.16	3.23±0.43	1.33±0.43	6.73±0.13	
T. patula L. Double giants	4.84±0.52	2.76±0.32	–	5.97±0.63	
T. patula L. Oreng prints	6.31±0.33	3.01±0.19	0.58±0.07	5.75±0.19	
T. patula L. Carmen	6.20±0.93	5.11±0.60	0.79±0.05	7.73±0.47	
T. patula L. Tiger eyes	5.10±0.63	3.08±0.76	0.84±0.06	7.01±0.61	
T. erecta L. Solar giants	6.75±0.56	5.29±0.14	1.31±0.10	7.80±0.40	
T. erecta L. Smile	6.63±0.80	3.37±0.29	1.03±0.60	8.03±0.99	
C. chinensis (L.) Nees. Winter cherry	2.96±0.53	1.91±0.60	–	5.11±0.58	
E. californica Cham. Violet beam	3.77±0.56	1.21±0.47	–	6.38±0.69	
E. californica Cham. White castle	3.98±0.11	1.43±0.02	1.10±0.03	6.55±0.08	
E. californica Cham. Yellow queen	2.39±0.07	1.01±0.03	–	6.13±0.11	
E. californica Cham. Sparkling carpet	3.25±0.05	1.48±0.03	1.11±0.06	6.34±0.06	
E. californica Cham. Ballet dancer	3.13±0.04	1.51±0.04	0.73±0.08	5.86±0.08	
L. usitatissimum L. Szegedi olaglen	3.51±0.03	2.00±0.05	1.08±0.06	5.61±0.03	
L. usitatissimum L. Fibriferum	2.58±0.06	0.87±0.03	0.44±0.06	4.62±0.04	
L. usitatissimum L. Oleiferum	2.16±0.03	1.14±0.05	–	5.55±0.03	
L. usitatissimum L. Natasja	3.08±0.04	1.63±0.05	0.90±0.03	6.32±0.71	
Mean (n=17)	4.21±0.37	2.35±0.32	0.94±0.14	6.32±0.22	
CV, %	10.57	15.13	18.95	5.55	

Note: C.—Callistephus, E.—Eschscholzia, L.—Linum, T.—Tagetes.

group of low resistance amounted three cultivars, and seven cultivars were attributed to the group of medium resistance.

When increasing the concentration of salt solution up to 1.05 and 1.95%, the inhibition of both, aboveground part and root system was observed. The indicator of the ratio of germ lengths in the experiment and control varied from 16.5 to 67.8% under the concentration of 1.05% and from 0 to 19.8% under the concentration of 1.95%. The number of cultivars in the group with low resistance increased. It is necessary to take into consideration that the cultivars *T. erecta* and *T. patula* showed higher tolerance to the stress effect.

In a comparative evaluation of species and cultivars of plants with reference to the biomass of germs, we could find the differences in the reaction of chloride salinization (Table 9.2).

A considerable decrease of germs biomass under salinization was observed under maximum NaCl concentration (1.95%) with minimum indicator of salt resistance (average of 21.1% among all cultivars). Germs of flower plants in the salt substrate with 0.15% concentration developed with lower lag from control. The average salt resistance of all cultivars amounted 62.2% with the variation of this indicator from 52.0 to 110.2%. According to germ biomass, as well as its length, a stimulating effect under low salinization was observed in the cultivar Oreng prints (*T. patula* L.). It is necessary to take into consideration that this cultivar had relatively high resistance to stress factor, where the increase of salt concentration in the solution was up to 1.05%. The salt concentration of 1.95% evoked considerable inhibition of growth processes, but they were less expressed than in a number of other cultivars, the salt resistance amounted 19.5%, which is close to the average population value (21.1%).

When comparing the data on the laboratory seeds germination, germ length, and biomass, it was detected that only two cultivars of *T. erecta* showed high resistance to salinization according to the complex of characters, namely Solar giants and Smile. Two cultivars of *L. usitatissimum*: Szegedi olaglen and Natasja were attributed to the group with high tolerance according to the results of laboratory seeds germination. At the same time, according to the characters of germ length and biomass, these cultivars fall into the group with medium resistance to chloride salinization. The less expressed inhibition of growth processes of the 'Szegedi olaglen' cultivar relatively to other species and cultivars under different salinization levels can be attributed to its advantages.

TABLE 9.2   The Effect of Salinization on the Germ Biomass of Species and Cultivars of Plants Under Study (mg).

Species cultivar	NaCl concentration			
	0.15%	1.05%	1.95%	Control
T. patula L. Yellow	29.00±0.90	22.00±2.10	8.20±1.01	50.03±1.88
T. patula L. Double giants	32.20±5.43	18.05±2.02	–	37.03±4.02
T. patula L. Oreng prints	41.36±2.08	30.25±2.08	7.12±1.12	37.54±2.03
T. patula L. Carmen	35.02±1.81	38.69±1.20	9.08±1.02	64.36±7.57
T. patula L. Tiger eyes	37.26±2.96	21.27±1.09	10.21±1.06	52.36±6.08
T. erecta L. Solar giants	45.18±4.14	43.33±1.33	16.36±0.70	67.02±6.26
T. erecta L. Smile	36.38±3.17	28.33±3.33	9.27±0.33	64.58±1.08
C. chinensis (L.) Nees. Winter cherry	18.02±2.15	10.86±0.44	–	35.23±2.12
E. californica Cham. Violet beam	28.33±1.33	43.53±4.32	–	43.32±6.08
E. californica Cham. White castle	27.00±1.49	15.30±0.70	8.09±0.36	46.20±1.47
E. californica Cham. Yellow queen	22.81±1.02	10.30±0.39	–	49.33±4.44
E. californica Cham. Sparkling carpet	24.40±0.66	14.80±0.35	11.90±0.78	44.40±1.24
E. californica Cham. Ballet dancer	22.90±0.88	12.55±0.37	8.00±0.57	44.06±4.01
L. usitatissimum L. Szegedi olaglen	25.30±0.36	20.10±0.37	12.42±0.36	37.60±0.60
L. usitatissimum L. Fibriferum	21.40±0.49	14.30±0.36	6.20±0.86	36.10±0.48
L. usitatissimum L. Oleiferum	21.80±0.53	12.22±0.49	–	41.00±0.33
L. usitatissimum L. Natasja	26.60±0.54	18.00±0.84	11.33±0.49	45.30±0.71
Mean (n=17)	29.11±1.76	21.99±1.28	9.85±0.72	46.79±2.96
CV (%)	15.85	17.51	19.63	13.85

Note: For the names of genus of plants, see Note of Table 9.1.

9.4 CONCLUSION

It was found that chloride salinization exerts an inhibiting effect on the seeds germination of decorative plants. The dependence of laboratory seeds germination capacity and morphometric parameters of germs on the concentration of sodium salt solutions was observed: the higher the concentration, the stronger was the inhibition of growth processes. The concentrations applied in the study exerted a multidirectional effect on seeds germination and development of germs. The species and cultivars differed considerably according to the decrease of characters under study.

Under general regularity of the growth inhibition under provocative background since the moment of seeds germination, the norm of reaction on the stress factor was individual for each sample. The resistance to chloride salinization at the early stages of ontogenesis is determined by species and genotypic peculiarities of cultivars. More complete and objective evaluation of species and cultivars of ornamental plants can be obtained on the basis of synthesizing data on population (seeds germination capacity) and individual (morphometric parameters of germs) characters. It is necessary to take into consideration that it is possible to apply these results after additional evaluation in natural conditions.

KEYWORDS

- chloride salinization
- germination of seeds
- germ

REFERENCES

1. Baranova, E. N.; Gulevich, A. A. Problems and Prospects of Genetic and Engineering Approach for Addressing Issues of Plant Resistance to Salinity. *Agric. Biol.* **2006,** *1,* 39–56.
2. Goryshina, T. K. *Plant Ecology: Textbook;* Higher School: Moscow, 1979; p 368 (In Russian).

3. Yakubov, H. G.; Yakubov, H. G.; Nikolayevsky, V. S. *New Approaches to Environmental Assessment of Environmental Pollution and the State of Evergreen Plantations in Moscow;* Ecology of Big City: Almanach. Prima-Press: Moscow, 2000; 4, pp 53–58. (In Russian).

4. Bome, H. A. *Soil Science;* Tyumen State University: Tyumen, 2000; p 80 (In Russian).

5. Black, K. A. *The Plant and the Soil;* Kolos: Moscow, 1973; p 503 (In Russian).

6. Loseva, A. S.; Petrov-Spiridonov, A. E. *Plant Resistance to Adverse Environmental Factors;* A.K. Agricultural Academy: Timiryazev Moscow, 1993; p 447 (In Russian).

7. Strogonov, B. P. *The Physiology of Agricultural Plants;* Kolos: Moscow, 1967; p 380 (In Russian).

8. Udovenko, G. V.; Semushkina, L. A.; Sinelnikova, V. N. *Features a Variety of Tolerance Salinity Evaluation Methods. Methods of Evaluating the Resistance of Plants to Adverse Environmental Conditions;* Kolos: Leninrad, 1976; pp 228–238 (In Russian).

9. Polevoj, V. V.; Chirkova, T. V.; Letova, L. A. et al. *Practical Works on Growth and Resistance of Plants. Handbook;* Publisher of the St. Petersburg University: St. Petersburg, 2001; p 212 (In Russian).

10. Belozerova, A. A.; Bome, N. A. Study of Spring Wheat Reaction to Solinity on the Variability of Sprouts Morphometric Parameters. *Fundam. Res.* **2014,** *12*(2)*,* 300–306 (In Russian).

11. Bome, N. A. Intraspecific Diversity of Barley (*Hordeum vulgare* L.) About Resistance to Salinity Chloride. *Agrobiology* **2014,** *2,* 16–22 (In Russian).

12. Agricultural Seeds. Methods for Determining the Germination (GOST 12038–84). Official Publication. Standartinform: Moscow, 2011; pp 36–64 (In Russian)

13. Lakin, G. F. *Biometrics;* Higher School: Moscow, 1980; p 295 (In Russian).

14. Taeger, S.; Sparks, T. H.; Menzel, A. Effects of Temperature and Drought Manipulations on Seedlings of Scots Pine Provenances. *Plant Biol. (Stuttg).* **2015,** *17,* 361–372. DOI: 10.1111/plb.12245.

15. Ma, X. F.; Wang, L. H.; Shi, X.; Zheng, L. X.; Wang, M. X.; Yao, Y. Q.; Gai, H. J. Effects of Water Deficit At Seedlings Stage on Maize Root Development and Anatomical Structure. *Ying Yong Sheng Tai Xue Bao.* **2010,** *21*(7), 1731–1736. (Article in Chinese, Abstract in English).

16. Khodadad, M. An Evaluation of Safflower Genotypes (*Carthamus tinctorius* L.). Seed Germination and Seedlings Characters in Salt Stress Conditions. *Afr. J. Agric. Res.* **2011,** *2*(3), 1667–1672.

17. Bome, A. Ya.; Bome, N. A. Reaction of Varieties of Soft Spring Wheat Domestic and Foreign Selection at Low Temperatures. *Mod. High Technol.* **2006,** *6,* 6–62 (In Russian).

18. Sidorovich, M. M.; Kundelchuk, O. P. Monitoring the Impact of Environmental Factors on the Growth and Developmental Coordination Organs Seedlings of Winter Wheat Method Hytotestirovanie. *Proceedings of the Belarusian State University.* **2016,** *11*(10), 170–178 (In Russian).

CHAPTER 10

BIOLOGICALLY ACTIVE COMPOUNDS SEARCH AMONG *ALOE* SPP.

NATALIA N. SAZHINA[1,*], PETER V. LAPSHIN[2], and NATALIYA V. ZAGOSKINA[2]

[1]*Emanuel Institute of Biochemical Physics, Russian Academy of Sciences, 4 Kosygin St., Moscow, Russia 119334*

[2]*Timiryazev Institute of Plant Physiology, Russian Academy of Sciences, 35 Botanicheskaya St., Moscow, Russia 127276*

*Corresponding author. E-mail: Natnik48s@yandex.ru

CONTENTS

ABSTRACT

One of the pharmaceutical science problems is studying biological activity, including antioxidant activity (AOA) of various herbs. The pharmacological value of plants is substantially connected with various compounds of secondary metabolism including phenolic compounds. The preparations from the components of some succulent plants, particularly some species of the genus *Aloe* are widely applied in medicine.

In present work, the total AOA for ethanol extracts of 15 various species of *Aloe* (*Aloe* L.) were measured by ammetric and chemiluminescence methods which are the most active sources of biologically active components. The comparative analysis of their antioxidant properties was done. For some *Aloe* spp., the considerable difference between AOA values received by these methods is noted. The most active *Aloe* spp., such as *Aloe pillansii*, *Aloe broomii*, and *Aloe spinosissima* which act as potential producers of biologically active substances are exposed. They can be of no less perspective sources of the biologically active compounds than *Aloe arborescen*s and *Aloe vera* that are being used now.

10.1 INTRODUCTION

One of pharmaceutical science problems is studying biological, including antioxidant activity (AOA) of various herbs for the purpose of search among them the most active sources of biologically active substances. The pharmacological value of plants is substantially connected with various substances of secondary metabolism including phenolic connections.[1] These substances play various role in the life of plants such as protection against biotic and abiotic stresses, intensive solar radiation, attacks of pathogens, mechanical damage, and so on. Many succulent plants protect themselves against excessive sunlight by forming a wax raid on leaves, omission development, development of pigments, and compounds of the phenolic nature such as flavonoids and anthocyanins. They absorb short wavelength of the spectrum, and final extent of the damaging action of ultraviolet light often depends on the level of their accumulation in vegetable fiber. It is important to note that polyphenols can also cause AOA of extracts.[2]

In medicine preparations, the components of some succulent plants are widely applied; in particular, some preparations from the species of the genus *Kalanchoe* and *Aloe* are widely applied in domestic medicine.[1] When we carried out screening of 34 *Kalanchoe* sp. according to AOA of their leaves juice, two most active species was revealed,[3] which can appear more perspective producers of biologically active substances in comparison with recently used *Kalanchoe* sp. It is expedient to carry out similar screening among representatives of *Aloe* spp. Modern operational methods of research of antioxidant properties allow them to study at higher level.

Now, the genus *Aloe (Aloe* L.) has more than 270 different species. These plants are originated from South Africa and the island of Madagascar. Representatives of the genus *Aloe* are succulent plants with juicy water reserving leaves in a freakish form.[4] Generally, they are used as decorative, but some species are applied for the medical purposes as their leaves contain useful mineral salts, organic acids, and numerous phenolic compounds. They cause, mainly, biological action, including AOA of this or that species of a plant, that is ability its component to inhibit oxidizing free radical processes. The most known species of *Aloe* are *Aloe arborescens* and *Aloe vera* L. Earlier, these species were used in pharmaceutics for the production of preparations with a broad spectrum of activity.[1] Authors working[5–7] on modern methods investigated a chemical composition and the quantitative content of biologically active substances in juice and extracts of *A. arborescens*. Moreover, antiradical activity of these extracts[6] is defined. In recent years, great interest is taken in the study of *A. vera* because its components show antiradical and antimicrobial activity.[8] For *Aloe ferox*, antioxidant, antimicrobial, anti-inflammatory, and antimalarial action are shown.[9] Moreover, the South African *Aloe barberae* shows antibacterial, fungicide, and anti-inflammatory action.[10] *Aloe mitriformis* and *Aloe saponaria* also showed high antifungal activity that can be alternative *A. vera*.[11] All these data testify to various activities of various representatives of the genus *Aloe*. However, results of purposeful scientific researchs of antioxidant properties of these and other *Aloe* sp. practically are not present.

The purpose of the present work—the AOA definition in 15 species of *Aloe* extracts by ammetric and chemiluminescence (CL) methods and identification among them are the most active producers of biologically active compounds.

10.2   MATERIALS AND METHODOLOGY

Objectives of research were to extract samples of 15 species of the genus *Aloe*, grown in a succulent collection in Timiryazev Institute of plant physiology of the Russian Academy of Sciences in Moscow. Biologically active compounds from leaves were extracted by 70% ethanol, and for *A. arborescence* it was extracted by water, 40% ethanol, and 96% ethanol (for comparison of extraction extent).

Measurements of the total AOA were carried out by electrochemical (ammetric) and CL methods.

The essence of *ammetric method* consists of measurement of the electric current arising from oxidation of investigated substance on a surface of a working electrode at certain electric potential (+1.3V).[12] An oxidation only OH-groups of natural phenolic antioxidants (R–OH) there is at such values of potential. The electrochemical oxidation proceeding under scheme R–OH OH → R–O˙ + e⁻ + H⁺, can be used as model for the measurement of free radical absorption activity. The capturing of free radicals is carried out according to reaction R–OH → R–O˙ + H˙. Both reactions include the rupture of the same bond O–H. In this case, the ability of same phenol type antioxidants to capture free radicals can be measured by value of the oxidizability of these compounds on a working electrode of the ammetric detector.[13] The integral signal (the area under a current curve) is compared with the signal received in same conditions for the comparison of the sample with known concentration. Trolox was used in work as the comparison sample. The total AOA is determined by calibration dependence of the trolox oxidizability (integral on oxidation curve time) on its concentration in mM. Metrological features of AOA are stated by this method in ref 14. The error in determination of the AOA including the error by the reproducibility of results was within 10%.

In *CL method* of AOA the scheme of oxidation system "hemoglobin—hydrogen peroxide—luminol" was used.[15] The detailed measurement technique is given in ref 3. The interaction of hydrogen peroxide (H_2O_2) with "Lum-5373" (DISoft, Russia) participate in reduction of OH˙ -radicals. Besides, as a result of this interaction active ferril–radicals (Hb (˙⁺)–Fe^{4+}=O) are formed.

The formed radicals initiate the luminol oxidation in the process of which a luminol–endoperoxide LO_2^{2-} is formed. Further an aminophthalate anion in excited state $(AP^{2-})^*$ upon which transition to the main state light quantum with a wavelength 425 nm is highlighted. The introduction of antioxidants in "metHb–H_2O_2–luminol" system leads to change of kinetics of its CL and increase in the latent period (*t*), which is directly proportional to the concentration of added antioxidant, a point of a maximum of the first derivative of a CL curve with a time axis.

For the realization of this method in the present work, the device "Lum-5373" (DISoft, Russia) was used. About 50 μl of Hb (15 μM, Sigma, Russia), 100 μl luminol (1 mM, Diaem, Russia), 10 μl H_2O_2 (13 mM, Himmed, Russia), 2.35 ml buffer solution, and different doses of the studied tests (from 0.1 to 30 μl) were added to a cell of the device. The calibration was carried out on dependence of the latent period on concentration of a trolox. A calibration straight line, in μm of a trolox, determined AOA of the studied samples. The error of the AOA determination by this method including the error by the reproducibility of results is not than more 15%. Statistical processing of results was carried out while using standard algorithms of the MS Excel programs. AOA bias by this method did not make more than 15%.

10.3   RESULTS AND DISCUSSIONS

First, we used an ammetric method for definition of ethanol leave extract AOA for different age of several *Aloe* spp. (Fig. 10.1). From the obtained data, the activities of young and old leaves differ, that is, most brightly shown at *A. arborescens*. Proceeding from these data, for the subsequent AOA screening, whenever possible, approximately 1-year-old *Aloe* leaves were used.

Results of AOA measurements for various extracts by two methods are presented in Table 10.1. For AOA, the root mean square deviations from mean values received for four measurements within 2 days for ammetry and three repeated dimensions for CL are indicated.

For *Aloe striata* (No. 10), two probes are specified such as ethanol extract and juice. For *A. arborescens* (No. 1), four probes (No 1a–1d) for different extracts are presented. Results demonstrate that the maximal extent of extraction of the active compounds determined by AOA in both methods was reached 70 and 96% by ethanol; therefore, the comparative analysis of AOA for 15 Aloe types, presented in Table 10.1, was carried out for 70% ethanol extracts.

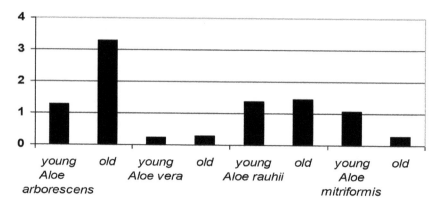

FIGURE 10.1 Antioxidant activity (AOA) (mM trolox) of ethanol extracts for young and old leaves of several *Aloe* spp., measured by ammetry: 0.25 g homogenate of fresh plant was dissolved in 1.5 ml 70% ethanol.

TABLE 10.1 Antioxidant Activity (AOA) of the Extract Probes of *Aloe* spp., Measured by Two Methods (author's modification in ref 16).

No. probe	Plant name	Extract types	AOA, mm trolox, ammetry	AOA, mm trolox, chemiluminescence
1	*Aloe arborescens*	70% ethanol 2 ml, 1 g *Aloe*	3.68±0.26	0.44±0.12
2	*Aloe vera*	–	0.24±0.02	0.04±0.01
3	*Aloe spinosissima*	–	2.16±0.11	1.28±0.15
4	*Aloe delaeti*	–	0.88±0.07	0.19±0.08
5	*Aloe pillansii*	–	5.16±0.23	4.55±0.32
6	*Aloe rauhii hybr.*	–	2.12±0.11	0.29±0.08
7	*Aloe jucunda*	–	0.78±0.05	0.12±0.02
8	*Aloe squarrosa*	–	1.73±0.11	0.65±0.05
9	*Aloe variegata*	–	0.85±0.03	0.43±0.09
10b	*Aloe striata*	Juice	1.05±0.04	1.03±0.08
11	*Aloe dorotheae*	70% ethanol 2 ml, 1 g *Aloe*	0.51±0.02	0.19±0.02
12	*Aloe hemmingii*	–	0.47±0.04	0.07±0.02
13	*Aloe broomii*	–	1.68±0.15	2.65±0.12
14	*Aloe plicatilis*	–	0.92±0.04	0.75±0.15
15	*Aloe brevifolia*	–	0.49±0.03	0.40±0.12

TABLE 10.1 *(Continued)*

No. probe	Plant name	Extract types	AOA, mm trolox, ammetry	AOA, mm trolox, chemiluminescence
1a	*Aloe arborescens*	Water 1 ml, 0.5 g *Aloe*	2.26±0.06	0.13±0.01
1b	*Aloe arborescens*	40% ethanol 1 ml, 0.5 g *Aloe*	2.78±0.12	0.25±0.03
1c	*Aloe arborescens*	70% ethanol 1 ml, 0.5 g *Aloe*	2.99±0.12	0.30±0.03
1d	*Aloe arborescens*	96% ethanol 1 ml, 0.5 g *Aloe*	3.28±0.14	0.29±0.03

Note: In this table and in text numbers of probes are given.

The AOA lowest values appeared for *A. vera* (No. 2), the highest—for *A. pillansii* (No. 5). For different *Aloe* spp., the strong distinction in AOA values measured by two methods (at 5–10 times) is observed; therefore, correlation of results of all 15 exemplars turned out low ($r=0.704$ with reliability of approximation $R^2=0.496$ and a significance level for unilateral criterion ≤ 0.001). In case of AOA for juice, of a Kalanchoe of similar kind, it was not observed.[3] It can demonstrate the significant difference in composition of the studied *Aloe* spp.

Kinetic CL curves, typical for separate *Aloe* sp. are given in Figure 10.2.

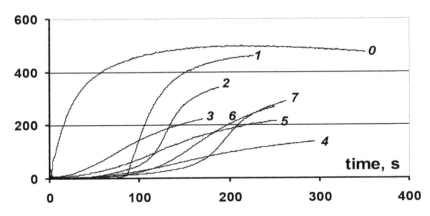

FIGURE 10.2 Dependence of chemiluminescence (CL) intensity (arbitrary units) on time (sec) at addition in oxidizing mix various samples[16]: *0*—blank; *1*—1 μl of trolox; *2*—0,5 μl of probe (No. 13); *3*—1 μl (No. 1), *4*—2 μl (No. 1); *5*—20 μl (No. 2); *6*—2 μl (No. 10a), *7*—2 μl (No. 10b). Designation number in brackets—see Table 10.1.

For comparison, the CL-gram for 1 μl of trolox (curve 2) is given on which the latent period (t) is legibly expressed and CL-amplitude practically does not decrease. Such CL-grams are characteristic for many low-molecular phenol compounds such as ascorbic, gallic acids, catechols, quercetin, and so forth. The decrease of CL-amplitude is, as a rule, bound to the presence of samples of protein and ferment structures. For kinetic curves of some *Aloe* extracts (curves 2, 6, 7) rather expressed latent period and the decrease of CL-amplitude depending on an extract dose is observed. Such behavior of CL curves is observed in samples No. 3, 5, 6, 9, 10, 13–15 and testifies about high phenol antioxidant content in them. That is confirmed by their high oxidizability recorded by ammetry (Table 10.1). A. *striata* juice and extract behave differently (No. 10— curves 6 and 7). Juice gives the larger latent period than ethanol extract that is bound, probably, with various inhibition of luminol oxidation by juice and spirit extract components. The checked ethanol influence on CL-parameters showed that in the used concentrations this influence is practically not shown.

The brightest representative of the considered groups of *Aloe* extracts is the sample No. 5 (Fig. 10.3).

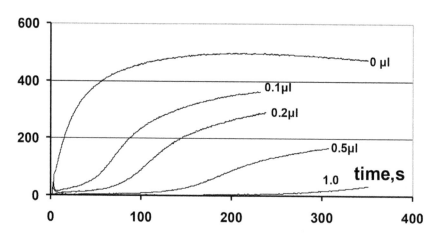

FIGURE 10.3   Kinetics of CL intensity (arbitrary units) at addition in oxidizing mix of various doses (μl) of sample No. 5 (see Table 10.1).[16]

Both methods for it show the high AOA values. CL is extinguished almost completely already at a small extract dose (1 μl) and due to increase in the latent period that is confirmed by the high AOA value on ammetry, and due to decrease of CL-amplitude that demonstrates existence in leaves of any proteins and enzymes.

A. pillansii is not studied and is originated from South Africa such as deserts of Angola, desert Namib. *A. pillansii* prefers flat stony tops of low hills. *A. pillansii* is strongly branched trees (height of 6–9 m, at a trunk thickness to meter). Branches are whitish-gray, naked, smooth, brilliant, and on the ends of the socket the leaves are very juicy. Leaves are 25–35 cm long, 5–6 cm wide, and bluish-green. Inflorescences are branchy, upright, flowers tubular, and yellow. Components of the second group of extract samples (No. 1, 2, 7, 11, 12) suppress CL in a different way (Fig. 10.2, curves 3, 4, 5): badly expressed small latent period even at high doses (a curve 5), slow luminescence development, and considerable decrease of amplitude. These extracts have the AOA small values measured by ammetry. It demonstrates low content of active low-molecular antioxidants in leaves of these species, but there are components, most likely, proteins or ferments which strongly suppress CL-amplitude, but are not oxidized on the anode of ammetric detector. For a sample No. 2 (*A. vera*), there are polysaccharides, as shown in ref 8.

For the most popular *A. arborescens* (No. 1), the high AOA values were received by ammetry that, apparently, is caused by the presence of enough phenol compounds (eloenin, aloin, aloe-emodin and so forth), organic acids and some amino acids.[5,6] However, the AOA values measured by CL have appeared 10 times less. As shown in ref 6, *A. arborescens* juice causes an inactivation of the H_2O_2 molecules due to the presence of enzymes in it (a catalase, peroxidase) and ions of heavy metals which potentiate the process of H_2O_2 decomposition. Besides, polysaccharide complexes as a part of *A. arborescens* juice helatirut Fe^{2+} and connect ions of heavy metals.[6] Therefore, these components of juice or extract do not allow normal development of CL, connecting hydroperoxides and Fe^{2+} from hemoglobin, "(Hb)—(H_2O_2)—luminol" which are present at CL oxidizing mix "(Hb)—(H_2O_2)—luminol," decreasing latent period values defining AOA. It is also possible for other species having the low AOA values (No. 6, 7, 11, 12).

In Figure 10.4a,b, comparative diagrams for 15 extract samples are given.

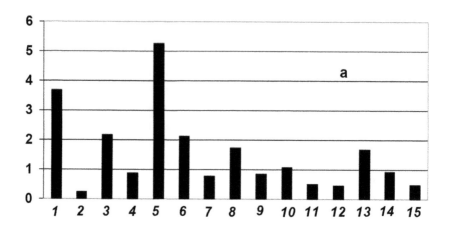

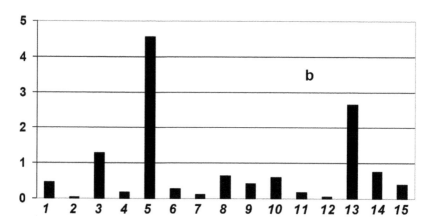

FIGURE 10.4   antioxidant activity (AOA) (mM trolox) diagrams for 15 *Aloe* extracts (probe numbers from Table 10.1), obtained by methods: *a*—ammetric, *b*—chemiluminescence.[16]

A. pillansii extract (No. 5 from Table 10.1) was indisputable "outsider." Both methods have shown for him almost identical high AOA values, and components interfering CL development, as in *A. arborescens*, are present in insignificant quantity. For a sample No. 3 (*A. spinosissima*), ammetric AOA values were almost twice higher, than for CL, and for a sample No. 13 (*A. broomii*) on the contrary.

Certainly, it is desirable to study a chemical composition of juices or extracts of these *Aloe* spp. to understand at the expense of what it occurs.

For possible expansion using studied abovementioned *Aloe* sp. as sources of biologically active compounds, carrying out additional researches of a chemical composition of these plant juice and extracts, and also the antibacterial, antimicrobial, phytoregulating, and other properties of their components are necessary. Perhaps, they will be not less perspective than the known *Aloe* sp., and for their use in medicine.

10.4 CONCLUSION

In the present study, the total AOA for ethanol extracts of 15 various species of *Aloe* (*Aloe* L.) by ammetric and CL methods were measured. The comparative analysis of their antioxidant properties for the purpose of search among them, the most active sources of biologically active components was made. The most active representative of *Aloe* spp. from the point of AOA view is *A. pillansii* having the high AOA values received by both methods is revealed. In addition *A. pillansii, A. arborescens* also *A. broomii* and *A. spinosissima* were rather active.

Thus, as a result of our researches, it is possible to note that the variety of forms of morphological adaptations for succulent plants is interconnected with a variety of their metabolism which is shown at the level of essential distinctions in accumulation of different classes of polyphenols and their AOA. For species of the genus *Aloe* by results of preliminary analyses new perspective taxons for further studying of their biochemical properties are allocated. These species can be not less perspective in pharmacy and medicine sources of the biologically active compounds than used now *A. arborescens* and *A. vera*.

KEYWORDS

- **ammetry**
- **chemiluminescense**
- **trolox**
- **polyphenols**

REFERENCES

1. Zaprometov, M. N. *Fundamentals of Biochemistry of Phenolic Compounds;* Higher School: Moscow, 1974; p 213 (In Russian).
2. Lapshin, P. V.; Sazhina, N. N. *Succulent Plants: Content of Phenolic Compounds and Antioxidant Activity;* Materials of the IX International Symposium Phenolic Compounds: Fundamental and Applied Aspects; Institute of Plant Physiology RAS: Moscow, 2015; pp 337–342 (In Russian).
3. Sazhina, N. N.; Lapshin, P. V.; Zagoskina, N. V.; Korotkova, E. I.; Misin, V. M. Comporative Study of Antioxidante Properties for Various Kalanchoe Kind Juices. *Chem. Veg. Raw mater.* **2013,** *3,* 13–119 (In Russian).
4. Blinova, R. F.; Yakovlev, G. P., Eds. *Botanical Pharmacognosy Quality Dictionary;* Vysshaja Shkola: Moscow, 1990; p 272 (In Russian).
5. Olennikov, D. N.; Zilfikarov, I. N.; Ibragimov, T. A. Research of Chemical Composition Aloe Treelike (*Aloe arborescens* Mill.). *Chem. Veg. Raw Mater.* **2010,** *3,* 77–82 (In Russian).
6. Olennikov, D. N.; Zilfikarov, I. N.; Ibragimov, T. A.; Toropova, A. A.; Tanhatva, L. M. Chemical Composition of Aloe Treelike Juice (*Aloe arborescens* Mill.) and its Antioxidatic Activity (in vitro). *Chem. Veg. Raw Mater.* **2010,** *3,* 83–90 (In Russian).
7. Luccia, B. D.; Manzo, N.; Vivo, M.; et al. A Biochemical and Cellular Approach to Explore the Antiproliferative and Prodifferentiative Activity of *Aloe arborescens* Leaf Extract. *Phytother. Res.* **2013,** *27*(12), 1819–1828.
8. Kaithwas, G.; Singh, P.; Bhatia, D. Evaluation of *in Vitro* and *in Vivo* Antioxidant Potential of Polysaccharides from Aloe Vera (*Aloe Barbadensis* Miller) Gel. *Drug Chem. Toxicol.* **2014,** *37*(2), 135–143.
9. Chen, W.; Van Wyk, B.-E.; Vermaak, I.; Viljoen, A. M. Cape Aloes—A Review of the Phytochemistry, Pharmacology and Commercialisation of Aloe Ferox. *Phytochem. Lett.* **2012,** *5,* 1–12.
10. Ndhlala, A. R.; Amoo, S. O.; Stafford, G. I.; Finnie, J. F.; Van Staden, J. Antimicrobial, Anti Inflammatory and Mutagenic Investigation of the South African Tree Aloe (*Aloe Barberae*). *J. Ethnopharmacol.* **2009,** *124,* 404–408.
11. Zapata, P. J.; Navarro, D.; Guillen, F.; Castillo, S.; Martinez-Romero, D.; Valero, D.; Serrano, M. Characterisation of Gels from Different Aloe sp. as Antifungal Treatment: Potential Crops for Industrial Applications. *Ind. Crops Prod.* **2013,** *42,* 223–230.
12. Yashin, A. Y. Inject-Flowing System with Ammetric Detector for Selective Definition of Antioxidants in Foodstuff and Drinks. *Russ. Chem. Mag. LII* **2008,** *2,* 130–135 (In Russian).
13. Peyrat-Maillard, M. N.; Bonnely, S.; Berset, C. Determination of the Antioxidant Activity of Phenolic Compounds by Coulometric Detection. *Talanta* **2000,** *51,* 709–715.
14. Biryukov, V. V. Features of Antioxidant Concentration Determination by an Ammetric Method. *Chem. Veg. Raw Mater.* **2013,** *3,* 169–172 (In Russian).

15. Teselkin, Y. O.; Babenkova, I. V.; Lyubitsky, O. B.; Klebanov, G. I.; Vladimirov, Y. A. Inhibition of Luminol Oxidation in the Presence of Hemoglobin and Hydroperoxide by Serum Antioxidants. Questions of Medical *Chemistry* **1997,** *43*(2), 87–93 (In Russian).
16. Sazhina, N. N.; Lapshin, P. V.; Zagoskina, N. V.; Misin, V. M. Comporative Study of Antioxidate Properties for Various Aloe Kinds Extracts. *Chem. Veg. Raw Mater.* **2015,** *2,* 179–186 (In Russian).

PART IV
Fruit Growing and Breeding

CHAPTER 11

CULTIVATION OF APPLE (*MALUS DOMESTICA* BORKH.): MAJOR GROWING REGIONS, CULTIVARS, ROOTSTOCKS, AND TECHNOLOGIES

MYKOLA O. BUBLYK*, LYUDMYLA O. BARABASH, LYUDMYLA A. FRYZIUK, and LYUBOV D. BOLDYZHEVA

Institute of Horticulture of the National Academy of Agrarian Sciences of Ukraine, 23 Sadova St., Novosilky, Kyiv-27, 03027, Ukraine

Corresponding author. E-mail: mbublyk@ukr.net

CONTENTS

ABSTRACT

This chapter considers the historical information about the apple production in Ukraine since the middle of the 20th century to the present day and determines the major regions for the industrial apple growing.

At present, the total area of apple orchards in Ukraine is 145,600 ha, that is, about 41% of which lies on the most suitable lands in Western Forest-Steppe and Prydnistrovia. On such lands, approximately 25% of the apple orchards are cultivated, including 5.5% in Transcarpathia, 9.4% in Western and Central Steppe, and 5% in Donbass. During the last 5 years, the manufacturing of fruit planting trees increased from 4.25 to 9.7 million t, among them of apple, it was from 2.47 to 7.42 million t. The main production of planting stock is concentrated in four regions: 38% in Podillia and Prydnistrovia, 15% in Transcarpathia, 17% in Mykolaiv region, and 20% in Dnipropetrovsk, Zaporizhia, and Donetsk regions.

The average annual apple output in the mentioned period was about 1.1 million t; among which almost three-quarters were in individual holdings—mostly to meet their own requirements for fresh fruits.

In this chapter, the characteristics of the major and promising apple cultivars and rootstocks are presented. The technologies for the fruit growing and the scientific institutions in Ukraine, which deal with researches on horticulture and the advisory services structure, are also presented.

11.1   INTRODUCTION

By the beginning of 2015, the Ukraine's population was 42.9 million. In this country, the intensive urbanization process lasted for many years. Therefore, the urban population (69%) was constantly increasing, whereas the rural one (31%) was decreasing. At present, the highest urbanization level is observed in the industrial regions. For instance, the urbanization level in Donetsk region was over 90%, and in Luhansk, Dnipropetrovsk, and Kharkiv regions, more than 80% of the population live in towns and town-type settlements. The rural population exceeds the town population in Transcarpathia, Ivano-Frankivsk, Rivne, Ternopil, and Chernivtsi regions only.

The natural and resource potential of the Ukrainian agriculture includes land, water, climatic and biological resources. Among which, land is of the greatest importance. It is the essential means of production in this field.

Over the last decades, its development was based only on the maximum drawing of arable lands into production.

Among the European countries, Ukraine is characterized by the high area of agricultural lands.[1] At the beginning of 2015, it was 41.5 million ha, among which, cultivated land was 32.5 million ha. During the last decade, it decreased by 210,000 ha, mainly because of being transferred to the category of nonagricultural lands on the basis of the ecological considerations and as a result of conducting land protection measures. Partly, the lands were allotted to different users for the nonagricultural work.

In Ukraine, the part of the agricultural lands in the general land structure is about 70%, and arable land is 54%. Ukraine possesses reserves of productive lands, which exceeds approximately four times than its modern inner requirements. The area per capita in Europe averages 0.43 ha, among which, for cultivated lands, it is 0.24 ha, and in Ukraine it is 0.97 and 0.76 ha, respectively.

Horticulture is of great importance for the country. In Ukraine, two categories of horticultural farms were historically formed: industrial orchards of agricultural enterprises and those of individual owners. The production of the latter is assigned mainly to be consumed by their owners.

At present, the total area of orchards in our country is approximately 239,000 ha, among which 79,000 ha is farm orchards of agricultural enterprises, the rest is of individual owners. The annual volume of the fruit production averages about 2.0 million t during the last 5 years.

11.2 HISTORY OF APPLE GROWING

Under the annals, the first information about the fruit crops cultivation on the territory of modern Ukraine goes back to the 11th century. The industrial fruit production began in the 18th century, and in the second half of the 19th century, it achieved a wide range already.

By the beginning of the 20th century, the total orchards area was approximately 230,000 ha, among which about 40% was farm area and apple orchards area was approximately 135,000 ha.[2] At the middle of that century, the total area increased by more than two times and achieved 609,000 ha, among which industrial orchards area was 57%. Simultaneously, the apple orchards area was 343,600 ha (57%), among which the farm area was 74%.

In 1960, the fruit orchards area in Ukraine exceeded 1 million ha and achieved maximum (1.3 million ha), among which farm area of agricultural enterprises was approximately 67%. Apple orchards then occupied 881,600 ha, including 77% of farm area.

Subsequently, because of the crisis in the economy of the Soviet Union and then in Ukraine, the orchard area began to decrease sharply as early as in 2000 and achieved 425,000 ha, among which 256,000 ha of the apple area, and in 2005, 300,000 ha and 155,000 ha, respectively. The farm orchards were 65% in 2000 and in 2005 their part decreased to 48%, apple ones were 82 and 67%, respectively.

The maximum level of the fruit production was achieved in 1984–1990, which was almost 3 million t, and in the farm orchards about 43% of this volume. The apple production in that period achieved 2.1 million t, among which in industrial farms, it was 67%. At the end of the 1990s, the production level decreased to 1 million t owing to the crisis. Since 2000, it has increased to 1.5–2.0 million t, the apple part fluctuating from 0.8 to 1.1 million t. However, the part of fruits grown in the agricultural enterprises was not more than 18% at that time, and for apples, it was 27%.

11.2.1   MAJOR APPLE GROWING REGIONS

The territory of Ukraine is known for various climatic conditions. It has been subdivided into 11 fruit-growing regions for the scientifically substantiated distribution of fruit crops: Polissia, Eastern Forest-Steppe, Western Forest-Steppe, Prydnistrovia, Western Steppe, Central Steppe, North-Eastern Steppe, Southern Steppe, Prycarpathia, Transcarpathia, and Crimea[3] (Fig. 11.1).

The climatic conditions of most of the regions are favorable for the apple cultivation. However, the most suitable are Forest-Steppe (especially western districts), Central Steppe, and Southern Polissia. Those are zones where temperately continental climate dominates. In the most severe winters, the air temperatures lower considerably (−35 to −38°C). However, the absolute temperature minimum and the indications similar to it are observed very rarely—for example, during the previous century, it was observed eight times. Therefore, more significant characteristic is middle minimum among the absolute ones. The lowest among them (−27 to −28°C) is observed in the Northern-East Ukraine (Kharkiv—Sumy), and in the western and central regions, it is −23 to −25°C.

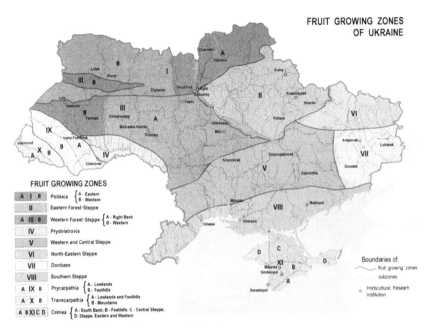

FIGURE 11.1 Fruit growing zones of Ukraine.

Source: Adapted from M. Yu. Gushchin.

In Ukraine, apple orchards more often suffer not from critical tempera-tures but from their fluctuations, especially in the second half of winter. At the beginning and in the middle of January, most of the cultivars end the period of the physiological dormancy—under-thaw trees get active and lose protective functions.

The weather conditions in that period are characterized based on a considerable temperature fluctuation in February and March when apple trees end exogenous dormancy and may be damaged by low temperatures. For example, in March, the middle minimum temperature can decrease in the range of −10 to −12°C, and the absolute maximum one is achieved at +25°C.

The majority of the cultivars can be damaged considerably when the air temperature drops to −20 to −25°C, and non-winter-hardy, West Euro-pean ones freeze slightly below −18 to −20°C.

In winter, the root system of fruit trees is susceptible to low soil temper-atures. Even their short-time effect in top strata influences the plants' vital activity. In some regions, the soil temperatures of −10 to −14°C are critically low for the roots of the majority of apple clonal rootstocks. Taking this into consideration, scientists have established the northern limit of the industrial apple cultivation on the mentioned rootstocks. On M.9, this boundary coincides with the position of the soil temperature isometric line: at a depth of 20 cm (−10°C), on MM 106 (−12°C), on 62–396 and 54–118 (−14°C) (Fig. 11.2).

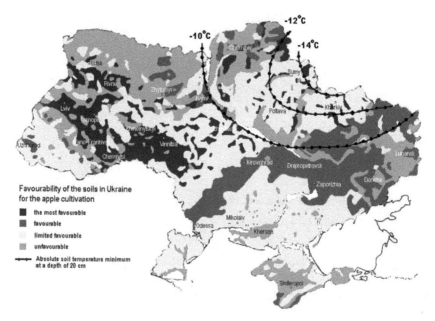

FIGURE 11.2   Map of apple growing regions in Ukraine.

Source: Adapted from M. O. Bublyk, P. V. Kondratenko, L. O. Barabash, L. A. Fryziuk.

Late spring frosts are also dangerous for the apple cultivation. The analysis of weather conditions shows that during the flowering period, the frosts (−2.5 to −5°C) occur 50–80% of the year. They are able to obstruct normal fructification.

The sum of active temperatures above 10°C in the northern regions is 2400–2600°C, in the central ones, it is 3000–3200°C, and in the southern regions, 3200–3800°C. It is enough for the most of apple cultivars.

One of the most important ecological factors that determines the potential productivity of apple orchards is moisture-providing: the amount of precipitations and their distribution during the year as well as the reserves of productive moisture in the root-habitant soil layer before the beginning of the vegetation period and in autumn.

The analysis of the moistening conditions in Forest-Steppe and Steppe has shown that during the vegetation, about 3500–5000 m³ of water from 1 ha of a fruiting orchard depending on its type is expended for the water provision and the precipitation amount in this period in Western Forest-Steppe is 400–450 mm; in Eastern Forest-Steppe and Steppe, it is 300–325 mm, that is, 3000–4500 m³/ha. A certain amount of precipitations is compensated by winter reserves of moisture which in the 1-m soil layer before the beginning of vegetation are in Western Forest-Steppe 175–200 mm, in Eastern Forest-Steppe and Steppe 150–175 mm of total moisture. For the compensation of those reserves, apple orchards need irrigation. Only the orchards on seedling rootstocks in Western Forest-Steppe can grow without it.

In the Institute of Horticulture of National Academy of Agrarian Sciences (IH of NAAS), special methods were elaborated for assessing the suitability of regions for the industrial cultivation of certain fruit crops.[4] The researches resulted in establishing four groups of soils in the territory of Ukraine that were determined concerning the degree of their favorability for this crop cultivation—the most favorable, favorable, limitedly favorable, and unfavorable.

The *most favorable soils* are light-gray podzolized ones, gray podzolized, dark-gray podzolized, and podzolized chernozems; *favorable soils* are soddy medium-podzolized sandy loam, common deep chernozems, common deep low in humus, common deep chernozems medium in humus, soils brown podzolized, korichnezem mountain with rubble chipping; *limitedly favorable* soils are soddy-podzolized sandy on deep sands, meadow carbonate, which demand hydromeliorative practice, southern chernozems, and chestnut soils that can be used with watering only; *unfavorable* soils are alkaline and salined gleyed carbonate soils.[5,6]

The most favorable and favorable soils are distributed compactly in Western Forest-Steppe, Central Steppe, and Steppe and rarely in Polissia. In Ukraine, the area of the most favorable and favorable soil for apple is 15.8 million ha (see Figure 11.2); whereas, at present, the total area of these crop orchards occupies is 145,600 ha.[7] About 41% of orchards are on the most favorable soils (in Western Forest-Steppe and Prydnistrovia), nearly 25% on favorable ones, among them in Transcarpathia (5.5%), Western and Central Steppe (9.4 and 5.0%, respectively), and North-Eastern Forest-Steppe (4.0%).

Approximately 34% of the apple orchards are on the limitedly favorable soils, among them in the Right-Bank part of Western Forest-Steppe (9.0%), Southern Steppe (7.7%), Crimea (6.2%), and Eastern Forest-Steppe (7.7%).

It should be noted that the productivity of this crop depends on the degree of the soil favorability. The most productive soils are the most favorable ones (41% of the total orchards area), where for the last 5 years, 48% of the total apple output is obtained, a quarter of it on the favorable soils (25% of the area); however, on the limitedly favorable ones (34% of the area), only 27% of the total output. Such dependence is explained by the fact that the soils of the worse quality demand additional expenses on their cultivation, but farms not always have costs for it.

11.2.2   PRODUCTION OF PLANTING STOCK

For the last 5 years, the production of fruit planting trees increased from 4.25 to 9.7 million t, among which for apple, from 2.47 to 7.42 million t.[8] The planting stock manufacturing is concentrated mainly in four regions: 38% in Ilya and Prydnistrovia; 15%in Transcarpathia; 17% in Mykolaiv region; and 20% in Dnipropetrovsk, Zaporizhia, and Donetsk regions (Fig. 11.3). The dimensions of nursery farms are very different. Most of the nursery farms grow 20,000–50,000 planting trees yearly, whereas about 10 farms grow 300,000–500,000 trees.

11.3   APPLE PRODUCTION

In Ukraine, the average annual output of apples for the last 5 years was about 1.1 million t, including almost three quarters produced by private holdings mainly to meet their own requirements for fresh fruits. The role of these orchards in the formation of the supply of high quality and great parties of production on the inner market is not large and in prospect is very likely to decrease.

Under the statistic data, the yield of the apple orchards is very low. However, it was not taken into consideration that in the 1990s, after the attempts of the transition to the market relations in agriculture, the management of the great part of the apple orchards was practically not carried out. At present, they are unfavorable for the fruit production, but for the economic reasons, they are not stubbed out.

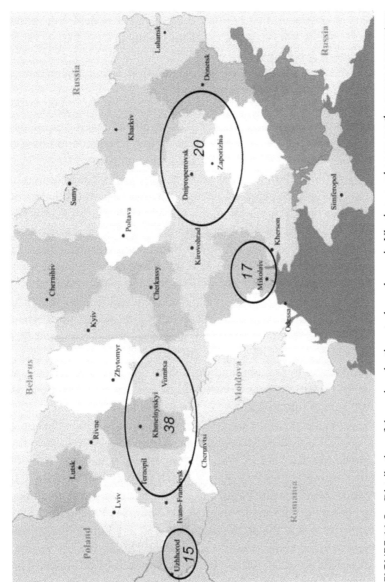

FIGURE 11.3 Distribution of the major planting stock producers in Ukraine: number means the average percentage of planting trees produced in the region.

Source: Ref. [5].

In the modern farm orchards, the apple yield is within 20–25 t/ha, whereas 45–80 t/ha is achieved in the advanced farms.

At present, the increase of the intense orchards part in the structure of fruit orchards is observed, that is, their establishment is held at a rather rapid rate (rank one is Vinnytsia region). This results in a gradual increase of the part of marketable apples.

For the last 10 years, their yearly consumption in our country rose averagely from 13 to 18 kg per capita.

One of the characteristics of the present-day state of the industrial fruit production in Ukraine is great versatility of its organizational forms—from farmsteads with areas of 15 ha, where three to five constant and not less than 10 seasonal workers are employed for great companies. Nowadays, approximately 80% of the industrial apple production is concentrated in almost 400 agricultural enterprises. They are the only major producers of marketable apples that come to the market. The average area of orchards in such farms is 50–100 ha and more, the number of constant workers being 12–20 and seasonal ones employed in harvesting not less than 25 depending on the yield.

Over the last years, the process of joining of separate horticultural enterprises to big agroholdings began and the considerable investments of which made it possible to restore the existing orchards as well as to establish new ones. For example, in "Svarog West Group"—a company that has brought together companies which are located in the Khmelnytskyi, Chernivtsi, and Zhytomyr regions and work in all the sectors of agriculture (plant growing, livestock rearing, horticulture, and processing) orchards take over 500 ha and further increase of their areas is planned.

In 2010–2014, the export of fresh pome fruits, among which the apple percentage is 97.5%, decreased by 4.9 times (from 99,000 to 20,200 t).[9] The main buyer was Russia (97.1%). The same trend was observed with those of fruits import (apples 90.9%), which fell from 207,100 to 51,800 t, which is by four times. In spite of the fact that the trade surpluses of apples are in many countries, their major supplier in Ukraine is Poland (93.1%).

At the same time, it is the apple juice produced in Ukraine that enjoys a considerable demand at the international market. During the last 5 years, its export increased from 62,500 to 103,900 t, which is by 1.7 times. The products were sold mainly in Poland (40.6%) and Russia (36.1%), less in Austria (10.7%), Germany (5.2%), and Belarus (2.8%). Simultaneously, its import to Ukraine reduced from 8900 to 2500 t that is by 3.6 times. The major suppliers to the market were Moldova (36.2%), China (30.5%), and Russia (23.7%).

11.3.1 CULTIVARS

According to the data of the listing of the perennial orchards, Melba and Papirovka refer to summer varieties of apples; 'Antonovka Zvychaina' and 'Slava Peremozhtsyam' are autumn ones; and 'Reinette Symyrenka', 'Caleville Neige', 'Jonathan', 'Golden Delicious', 'Banana Winter' refer to winter cultivars that dominated in the inland orchards over the last 25 years. In general, although in apple orchards numerous cultivars are grown, only 13–15 of them dominate. For example, the Melba trees occupy 1.5% of the total amount of this crop trees,' Papirovka'—1.6%, 'Antonovka Zvychaina'—3.0%, 'Slava Peremozhtsyam'—3.6%, 'Idared'—4.5%, 'Jonathan'—9.1%, 'Caleville Neige'—10.2%, 'Golden Delicious'—10.8%, and 'Reinette Symyrenka'—13%. Such correlation between cultivars will evidently change considerably in the nearest future since in accordance with the data of studying the great number of new inland and foreign cultivars conducted for the last 15 years quite a number of those were determined which in keeping with the complex of the characteristics valuable for economy and economic indexes are the best for the establishment of modern orchards. The cultivars 'Williams' Pride', 'Julia', 'Quinty', 'Papirovka', 'Redfree', 'Yamba' (summer ones), 'Amulet', 'Delicia', 'Delbarestival', 'Slava Peremozhtsyam', 'Teremok' (autumn cultivars), 'Idared', 'Ascolda', 'Gala Mast', 'Golden Delicious', 'Jonagold', 'Peredgirne', 'Radogost', 'Reinette Symyrenka', 'Royal Red Delicious', 'Rumyany Al'pinist', 'Spartan', 'Tavriya', and 'Champion' (winter) can ensure the products of high market and taste qualities for the realization as fresh at the inner market and 'Ascolda', 'Peredgirne', 'Radogost', 'Reinette Symyrenka', 'Slava Peremozhtsyam' for the foreign market. 'Alkmene', 'Antonovka Zvychaina', 'James Grieve', 'Imrus', 'Liberty', 'Priam', 'Red Boskoop', and 'Rovyesnik' are the most favorable for the orchards where apples are grown for processing.

The breeders of the Ukrainian research institutions created a number of new apple cultivars in recent years that are just acquiring spread in production.[10] In particular, the scab immune cultivars 'Edera', 'Perlyna Kyyeva', 'Guarant', 'Amulet', and 'Skifske Zoloto' were entered into the register of the plants which are suitable for the dissemination in Ukraine. New scab immune cultivars bred at the IH of NAAS are tested, namely, summer 'Nastya'; winter 'Todes', 'Dmiana', 'Beregynya', and 'Solomiya'. Along with the disease resistance, they distinguish themselves for the suitability for intensive horticulture, a harmonious combination of the fruits taste and flavor, their long shelf life, and attractive appearance. These cultivars provide a possibility for a manufacturer to use a smaller number of crop protection agents, thus providing a low production cost. In addition, they can be grown by using the organic plant protection system.

Here, a brief description of these cultivars is provided.

11.3.1.1 NASTYA

A summer cultivar obtained by crossing cultivars 'Vista Bella' and 'Prima'. It is recommended for the dissemination in the Forest-Steppe, Steppe, and Polissia.

The tree is middle sized, forms not a large oval crown, and on a middle rootstock begins fruiting in the third to fourth year.

The cultivar is scab immune, mildew resistant, winter-hardy, and productive.

Fruits are of medium size, mass 130–170 g, flattened out orbicular, weak ribbed, greenish-yellow, with red blurred blush of strokes on the largest part of the fruit surface. The flesh is cream colored, fine grained, very dense and juicy, sweet-sour (8.2–8.5 points). Picking maturity and table ripeness come in the first decade of August. Ripening on a tree is not simultaneous and lasts for a month.

11.3.1.2 TODES

A winter cultivar obtained by crossing the cultivar Idared with the hybrid form X-2034 (F_2 *Malus floribunda* 821 × 'Golden Delicious'). It is recommended for the dissemination in Forest-Steppe, Steppe, and Polissia.

The tree is middle sized, forms a thin obconical crown, early ripening, and on a middle rootstock begins fruiting in the second to third year after planting.

The cultivar is scab immune, resistance to powdery mildew is average, and winter-hardiness, and yield is at the Idared level.

The fruits are of medium and large size, mass 170–200 g, elongated–conical, slightly ribbed, greenish-yellow, with maroon-red blush on almost the entire surface. The flesh is cream colored, fine grained, very dense, juicy, sweet-sour (8.0–8.2 points). Picking maturity comes at the end of September, the table ripeness in December. The fruits are stored up to May not losing taste qualities.

11.3.1.3   BEREGYNYA

A winter cultivars were obtained by crossing cultivars 'Askolda' and 'Florina'. The trees are small; on a middle rootstock begin fruiting in the third to fourth year after planting.

The cultivar is scab immune; in the Forest-Steppe, resistance to powdery mildew is average, winter-hardy, and very productive.

Fruits are medium and large, mass 165–220 g, orbicular–conical, with vague dark-red striped blush over the almost entire surface. The flesh is dense, juicy, aromatic, sweet sour (8.4 points). The fruits are transportable, and are stored in a refrigerator until May.

11.3.1.4   DMIANA

A winter cultivar, obtained from the open pollination of cultivar, 'Todes'. The trees are small, compact, and on a middle rootstock begin fruiting in the second to third year after planting.

The cultivar is scab immune, in the Forest-Steppe not damaged by powdery mildew, winter-hardy, very productive.

Fruits are medium, mass 150–170 g, orbicular-conical, greenish-yellow, with red blurred blush on the sunny side of the surface. The flesh is dense, juicy, with expressed brilliant aroma, sweet-sour (8.5–8.7 points). The fruits are transportable, and are stored in a refrigerator until May.

11.3.1.5   SOLOMIYA

It is a winter cultivar, obtained by crossing of cultivars 'Florina' and 'Mayak'. The trees form a small broad-orbicular crown; on a middle rootstock begin fruiting in the second to fourth year after planting.

The cultivar is scab immune; in Forest-Steppe, powdery mildew resistant, winter-hardy, and very productive.

The fruits are medium, mass 160–170 g, oblong–conical, slightly ribbed, green and yellow with a blush on the sunny side of the surface and bluish bloom. The flesh is medium dense and medium juicy, sour-sweet (7.8 points). The fruits are transportable, and are stored in a refrigerator until the end of April.

Picking maturity of all the described winter cultivars begins at the end of September and the delay with harvesting causes overmaturity, and as a result poor fruits storage. The Todes fruits are located mainly in the periphery, have the same coloring and size, without overloading with them. Those of 'Dmiana' impress with their taste, but they are very fruitful therefore require setting of norms of ovary rating.

11.3.2   ROOTSTOCKS

Unfortunately, in Ukraine, there are no data about the types of rootstocks in the existing apple orchards. One can only observe which types are used for the planting stock production during the last decade. Most of planting trees are grown on the middle rootstocks: on MM 106 about 40%, on 54–118 Idared—8%, the part of the formers decreasing year after year and that of the latter increasing. Almost one-third of the total planting stock volume is produced on the dwarf rootstock M.9. The number of planting trees on the clonal ones M.26 (7.1%) and 62–396 (1.7%) goes up. Annually, in Ukraine, about 5% of planting trees are grown on seedling rootstocks, namely, on the seedlings of 'Antonovka Zvychaina' and other apple cultivars mostly in nursery farms of Polissia, Central and Eastern Forest-Steppe. On seedling rootstocks, summer cultivars are propagated as well as 'Slava Peremozhtsyam', 'Teremok', 'Antonovka Zvychaina', 'Spartan', 'Caleville Neige', 'Zymove Lymonne', and 'Rosavka'. Their planting stock is used, as a rule, for the establishment of orchards in Polissia, Eastern Forest-Steppe, and Donbass.

The Ukrainian breeders have created a number of new apple rootstocks, which are valuable, first of all, due to their high frost resistance. Those are super-dwarf rootstocks 'Malyuk' and KD5, dwarf D1071, D3017, D3038, 'Konotopska', 'Sambirska'; half-dwarf 'Baturynska', 'Nadiya', 'Nizhynska'; and the middle ones D1904, D471, and 'Slobozhanska'.[11]

At present, about 5% of planting trees are grown on these rootstocks, and the mass establishment of production apple orchards on them is expected in 3–5 years.

Here is a brief description of these rootstocks.

11.3.2.1   MALYUK

A super-dwarf rootstock bred at the Sumy Research Station of Horticulture of the IH of NAAS.

The mother shrub is upright and low. Shoots are straight, not hairy, red-brown, with a small amount of branches.

The root system of layers has no main root and the separation from the mother shrub is easy. The yield of standard layers is seven to eight per shrub. The root system is high frost resistant, it endures the lowering of the temperature up to − 12 to − 14°C, and takes roots well. The rootstock is well compatible with grafted cultivars, and the damage by pests and diseases is slight.

The trees on 'Malyuk' begin fruiting in the second- to third year after planting, and they are highly productive and drought resistant. Anchoring is insufficient; therefore, the obligatory condition of growing is support.

11.3.2.2   KD5

A super-dwarf rootstock bred at the Krasnokutsk Research Station of Horticulture of the IH of NAAS. The mother shrub is middle sized and slightly branchy. The shoots are slightly curved at the base, thickened, without branching, slightly hairy, and light brown. The root system of layers has no main root; the yield of standard layers is 70,000–75,000 per hectare. The root system is high frost resistant, it endures the temperature lowering up to − 6°C, and takes roots well when irrigating.

The rootstock is well compatible with grafted cultivars, full scab and powdery mildew field resistant and aphids high resistant.

The trees on KD 5 begin fruiting in the first to second year after planting, they are highly productive, but anchoring is insufficient. The obligatory conditions of growing are support and irrigation. The height of mature trees is easily maintained within 1.5–1.8 m, the crown diameter is about 1 m.

11.3.2.3   D1071

A dwarf rootstock bred at the Artemivsk Research Station of Nursery Practice of the IH of NAAS.

The mother shrub is small, 60–72 cm high, and slightly branchy. The shoots are straight, thick, rather hairy, and slightly brown.

The root system of layers has no main root. The yield of standard layers is 8–11 per shrub. The rootstock is winter-hardy; it endures the temperature decrease up to − 13°C.

The rootstock is well compatible with apple cultivars. The trees on it are planted by the planting plan 4 × 1–1.5 m and need support. They are drought resistant, early ripening, and the yield is up to 40.0 t/ha depending on a cultivar.

11.3.2.4   D3017

A dwarf rootstock bred at the Artemivsk Research Station of Nursery Practice of the IH of NAAS.

The mother shrub is compact, low, up to 75 cm high, and not branchy. The yield of standard layers is 300,000 per hectare, rooting 4.4 points. Winter hardiness is high; the root system endures the temperature lowering up to − 14°C. In a nursery, the rootstock is compatible with the cultivars 'Reinette Symyrenka', 'Jonathan', 'Askolda', 'Jonagold', 'Champion', and 'Ligol'. The trees on D3017 are planted by the planting plans 5.0 × 1.5 and 4 × 1.5–2.0 m, lower than those on M.9 by 15% and need support. The yield in the fourth year is 26.4–30.5 t/ha, depending on the cultivar.

11.3.2.5   D3038

A dwarf rootstock bred at the Artemivsk Research Station of Nursery Practice of the IH of NAAS.

The mother shrub is compact, low, up to 60–75 cm high, and slightly branchy. In the parental garden, D3038 ends vegetation in time. The yield of standard layers is seven per shrub, that is, 270,000 per hectare. Winter hardiness is high, the root system endures the temperature lowering up to

−14°C. In a nursery, the rootstock is compatible with the spread cultivars 'Reinette Symyrenka', 'Jonathan', 'Askolda', 'Jonagold', 'Champion', and 'Ligol'. The trees on D3038 are planted by the planting plans 5.0×1.5 and 4×1.5–2.0 m, lower than those on M.9 by 20% and need support. The yield in the fourth year is averagely 14.4–19.3 kg per tree, depending on a cultivar.

11.3.2.6 SAMBIRSKA

A dwarf rootstock bred at the Sumy Research Station of Horticulture of the IH of NAAS.

The mother shrub is upright and of medium height. The shoots are straight, moderately hairy, and brown, with a small amount of branches.

The rootstock roots easily. The yield of standard layers is seven to eight per shrub. Frost resistance is high, and it endures the temperature lowering up to −12 to −14°C. Cultivar Sambirska is well compatible with grafted cultivars, and the damage by pests and diseases is slight.

The trees on this rootstock begin fruiting in the second to -third year after planting, they are highly productive, drought resistant, but anchoring is insufficient; therefore, the obligatory condition of growing is support.

11.3.2.7 KONOTOPSKA

A dwarf rootstock bred at the Sumy Research Station of Horticulture of the IH of NAAS.

The mother shrub is upright, and of medium height. The shoots are straight, not hairy, and brown, with a small amount of branches.

The rootstock roots easily. The yield of standard layers is seven to eight per shrub. The separation from the shrub is easy. The root system is high frost resistant. It endures the temperature lowering up to −14 to −16°C.

Konotopska is well compatible with grafted cultivars, and the damage by pests and diseases is slight.

The trees on this rootstock begin fruiting in the second to third year after planting; they are highly productive, drought resistant, but anchoring is insufficient; therefore, the obligatory condition of growing is support.

11.3.2.8   BATURYNSKA

A half-dwarf rootstock bred at the Sumy Research Station of Horticulture of the IH of NAAS.

The mother shrub is upright, and of medium height. The shoots are straight, moderately hairy, and brown, with a small amount of branches.

The rootstock roots easily. The yield of standard layers is seven to eight per shrub. The separation from the shrub is easy. The root system's frost resistance is high. It endures the temperature lowering up to -12 to $-14°C$.

The rootstock is well compatible with grafted cultivars; the damage by pests and diseases is slight.

The trees on this rootstock begin fruiting in the third to fourth year after planting, they are highly productive, drought resistant, and anchoring is good.

11.3.2.9   NADIYA

A half-dwarf rootstock bred at the Sumy Research Station of Horticulture of the IH of NAAS.

The mother shrub is upright, and of medium height. The shoots are straight, not hairy, and red-brown, with a small amount of branches.

The rootstock roots easily. The yield of standard layers is eight to nine per shrub. The separation from the shrub is easy. The root system's frost resistance is high. It endures the temperature lowering up to -14 to $-16°C$.

The rootstock is well compatible with grafted cultivars.

The trees on this rootstock begin fruiting in the third to fourth year after planting; they are highly productive, drought resistant, and anchoring is good.

11.3.2.10   NIZHYNSKA

A half-dwarf rootstock bred at the Sumy Research Station of Horticulture of the IH of NAAS.

The mother shrub is upright, and of medium height. The shoots are straight, not hairy, and maroon-brown, with a small amount of branches.

The rootstock roots easily. The yield of standard layers is seven to eight per shrub. The separation from the shrub is easy. The root system's frost resistance is high. It endures the temperature lowering up to -12 to $-14°C$. The rootstock is well compatible with grafted cultivars.

The trees on this rootstock begin fruiting in the third to fourth year after planting; they are highly productive, drought resistant, and anchoring is good.

11.3.2.11   D1904

A medium rootstock bred at the Artemivsk Research Station of Nursery Practice of the IH of NAAS.

The shrub has a pyramidal shape, the average height is 70–75 cm. The root system has no main root. D1904 is winter-hardy, the roots endure the temperature lowering up to $-16°C$. In a nursery, this rootstock is well compatible with all the grafted apple cultivars, and the yield of standard planting trees is up to 90%.

The trees on this rootstock begin fruiting in the third to fourth year after planting; they are highly productive, drought resistant, and anchoring is good.

11.3.2.12   D471

A medium rootstock bred at the Artemivsk Research Station of Nursery Practice of the IH of NAAS.

The shrub has a pyramidal shape and medium height. The rootstock is winter-hardy, it endures the temperature lowering up to $-16°C$. The root system has no main root (4.0 points). The yield of standard layers is within six to eight per shrub. The separation from the shrub is hard.

The compatibility of D471 with grafted cultivars is good. The trees on this rootstock begin fruiting in the fourth to fifth year after planting; they are highly productive, drought resistant, and anchoring is good.

11.3.2.13   SLOBOZHANSKA

A middle rootstock bred at the Sumy Research Station of Horticulture of the IH of NAAS.

The mother shrub is upright, and of medium height. The shoots are straight, moderately hairy, red-brown, and averagely branching.

The rootstock roots easily. The yield of standard layers is seven to eight per shrub; the separation from the shrub is easy. The root system's frost resistance is high. It endures the temperature lowering up to -14 to $-16°C$. The rootstock is well compatible with grafted cultivar. The trees on it begin fruiting in the third to fourth year after planting; they are highly productive, drought resistant, and anchoring is good.

11.4   TECHNOLOGIES FOR CULTIVATION

In Ukraine, apple is grown using the following important technologies.[12]

11.4.1   APPLE ORCHARDS ON DWARFING ROOTSTOCKS

On average, 2000–2500 planting trees are planted on 1 ha. The obligatory elements of this technology are installation of supports for every tree separately or espaliers consisting of concrete pillars and wire, drop irrigation, and fertilizer. The width of interrow spaces and density of the distribution of the trees in a row depend on the soil fertility, vegetation period duration, and crown habit. Middle cultivars are planted under a plan of 4–5 × 1–1.5 m, low ones 3.5–4 × 0.75 m. The most widespread crown form is well-proportioned spindle. The single-stage combined palmette is also used. Its central axis is formed like spindle and lateral branches like palmette. The crown height is 2.2–2.5 m.

11.4.2   APPLE ORCHARDS ON MEDIUM ROOTSTOCKS

On average, 500–666 planting trees are planted on 1 ha. The planting plan is 5 × 3–4 m and the crown is spindle-shaped form. The total height of the trees is up to 3–3.5 m, the width of the crown across the row is from 2.2–2.5 at the bottom to 0.3–0.5 m on the top.

In the orchards on seedling rootstocks and for vigorous cultivars on MM.106, the half-flat crown is mostly used for the planting plans being 5 × 4 and 6 × 4 m. Such crown is formed with five to six main branches on the trunk, above the bole, in two stories, orientated in the line of a row in pairs. The other one to two branches are formed storeyless. Besides, for the half-flat crown, the thinned-storied one is formed in the orchards on seed rootstocks.

In Forest-Steppe, the orchards on medium and seedling rootstocks are grown, as a rule, without irrigation; and in Steppe, only a part of areas is irrigated.

11.5   STORAGE AND MARKETING OF APPLES

At present, 60–70% of the apple fruits in Ukraine are consumed as fresh, and the rest is processed. The main type of processing is making concentrates. They are then used for making restored juices.

For the apple storage, fruit stores are used with a total capacity of approximately 300,000 t, where up to 10% of fruits being stored in the regulated atmosphere, which ensures 100% conservation of their qualitative characteristics irrespective of a cultivar, 18–20% in reequipped fruit stores under the conditions of the normal atmosphere, about 15% in old non-hermetical stores by using obsolete energetically capacious cooling systems; the losses can exceed 40%.

In Ukraine, wholesale markets of the fruit and vegetable productions function in Lviv, Kherson, and Khmelnytskyi regions. The greatest and the most modern wholesale market of the agricultural products is the regional agromarketing center Shuvar in Lviv. The agricultural commodity producers from 18 regions realize their products at this market. Over the last years, 1000–1200 t of the fresh fruit and vegetable and other agricultural products are sold here, transparent conditions of pricing are formed, and the price stabilization measures are carried out if necessary.

11.6   POMOLOGICAL INSTITUTES: EXPERIMENTAL STATIONS

The main institution which coordinates and carries out researches on the problems of horticulture is IH of NAAS of Ukraine which lies in the suburbs of Kyiv (http://sad-institut.com.ua/, e-mail: sad-institut@ukr.net), founded in 1930. The researches are conducted under four main directions: breeding and investigation of the cultivars of fruit, small fruit, and nuciferous crops; development of the technologies for the production of the sound planting stock and genetic control over the characteristics of fruit, small fruit, and ornamental crops valuable for economy; elaboration of technologies for growing fruits and berries for different Ukraine's zones of horticulture as well as technologies for the fruits and berries storage and receipts and technologies for making new products of their processing. The institute also develops and produces machines for the nurseries, orchards, and small fruit plantations management.

The scientific network of the IH of NAAS includes six research stations that are distributed in different zones of fruit growing. Their specialization has been determined depending on the crops that are important in the zones of these institutions activity. The Extension Service functions in the regions have been charged to them.

Prydnisrovska Research Station of Horticulture of the IH of NAAS (E-mail: *bospisnaan@gmail.com*) is situated near Chernivtsi. The major research directions are breeding, strain investigation, and elaboration of regional technologies for the pear and walnut production.

Podillia Research Station of Horticulture of the IH of NAAS (E-mail: *prydnistrovska@ukr.net*) lies near Vinnytsia. The major research directions are strain investigation and development of regional technologies for the apple and pear production.

Crimean Research Station of Horticulture of the IH of NAAS (E-mail: *sadovodstvo@ukr.net*) is situated in the neighborhood of Simferopol. Its specialization is breeding and investigation of apple and pear cultivars and development of technologies for the fruit and small fruit crops cultivation in the conditions of Crimea.

M. F. Sydorenko Melitopol Research Station of Horticulture of the IH of NAAS (E-mail: *iosuaan@zp.ukrtel.net*) lies in Melitopol of the Zaporizhia region. The main activity directions are studying the problems of the orchards irrigation in the South Ukraine as well as breeding and investigation of apple, cherry, sweet cherry, apricot, and peach cultivars.

Artemivs'k Research Station of Nursery Practice of Nursery (E-mail: *bospisnaan@gmail.com*) is situated in Bakhmut of Donetsk region. It elaborates regional technologies for the production of the fruit crops planting stock and breeds rootstocks for apple and cultivars of cherry, sweet cherry, and myrobalan plum. The experimental farm, which produces 500,000 planting trees yearly is subordinated to the station.

Sumy Research Station of Horticulture of the IH of NAAS (E-mail: *sumy_dss@ukr.net*) lies in the neighborhood of Konotop of Sumy region. The main activity directions are breeding of apple rootstocks; testing of apple, pear, plum, and cherry cultivars in the severe weather conditions of Northern-East Ukraine; and development of the regional technologies for the fruit and small fruit crops cultivation.

L. P. Symyrenko Research Station of Pomology of the IH of NAAS (E-mail: *mliivis@ukr.net*) is situated near Cherkasy. It is involved in the accomplishment of the State Horticultural Program as well. The major activity direction is conservation of the fruit and small fruit plants genetic bank. In parallel, the institution carries out breeding and investigation of apple, pear, plum, cherry, and sweet cherry cultivars and elaborates regional technologies for the black currant growing.

11.7 CONCLUSION

11.7.1 EXTENSION SERVICE

There are 150 subdivisions of the advisory service in Ukraine. In the majority of regions, they are in each district, and in the others, only a few per region. As a rule, a subdivision includes up to three persons who organize work and involve experts and scientists to provide concrete services. The main of them are:

- Consultations on the participants in state and branch programs of the agroindustrial production development
- Informing agricultural enterprises and rural population about novelties in science and engineering, introduction of modern technologies and achievements of the advanced experience in production and adherence to standards of quality and security of agricultural products
- Development and realization of standard module projects with a full cycle of the agroindustrial production by big, medium, and small enterprises in different natural climatic zones
- Ensuring the development of the agrarian market infrastructure as well as of the rural localities and solving of the rural population social problems

The advisory activity is financed by using costs of the state and local budgets, technical aid, and so forth.

11.7.2 OUTLOOK OF DEVELOPMENT OF APPLE ORCHARDS

The Program of the Horticulture Development in Ukraine is envisaged to 2025 to bring the total apple orchards area to 140,500 ha, the fertile ones taking 105,400 ha, the average yield increasing to 14.5 t/ha and the total output achieving 1.5 million t.

The realization of the program will result in increasing the efficiency of using labor resources and the number of jobs in 2025 by seven times as compared to 2015.

KEYWORDS

- **horticultural farms**
- **planting stock**
- **chernozem**
- **podzol**
- **marketing**

REFERENCES

1. FAOSTAT (FAO Statistics Division) 2015. http://faostat.fao.org/site/567/Desktop-Default.aspx
2. Shestopal, O. M.; Rulyev, V. A.; Kondratenko, P. V.; et al. *Economics and Organization of the Industrial Horticulture of Ukraine;* National Research Centre "Institute of Agrarian Economy": Kyiv, 2010; p 334 (In Ukrainian).
3. Kondratenko, T. Y. Apple in Ukraine. Cultivars. World Scientific Publishing: Kyiv, 2001; p 297 (In Ukrainian).
4. Bublyk, M. O.; Fryziuk, L. A.; Levchuk, L. M. Methodical Fundamentals of the Fruit Crop Distribution in Ukraine. *J. Nat. Sci. Sustainable Technol.* 2014, *8*(4), 635–642.
5. Bublyk, M. O.; Barabash, L. O.; Fryziuk, L. A.; Chorna, G. A. Efficient Distribution of the Main Fruit Crops Plantations in Ukraine: Apple and Pear. Vegetables and Fruits. December 2009; 34–37 (In Ukrainian).
6. Bublyk, M. O.; Fryziuk, L. A.; Levchuk, L. M. Scientifically Substantiated Soil and Climatic Regions for the Industrial Cultivation of Fruit Crops in Ukraine. In *Biological Systems, Biodiversity, and Stability of Plant Communities;* Weisfeld, L. I., Opalko, A. I., Bome, N. A., Bekuzarova, S. A., Eds.; Apple Academic Press: Toronto, New Jersey 2015; 161–173.

7. Plant Growing in Ukraine: Statistical Collection. (2015). State Statistics Service of Ukraine: Kyiv, 2014; p 180 http://www.ukrstat.gov.ua/ (In Ukrainian).

8. List of the Planting Stock of the Fruit, Small Fruit, Nuciferous, Minor Crops, Grapevine and Hop Grown in Ukraine. State Inspectorate of Agriculture of Ukraine. Department of the Control over Seed Production and Nursery Practice: Kyiv, 2013; p 39 (In Ukrainian).

9. Export and Import of Certain Goods by Countries. State Statistics Service of Ukraine. 2015, http://www.ukrstat.gov.ua/ (In Ukrainian).

10. Lytovchenko, O. M.; Pavlyuk, V. V.; Omelchenko, I. K. *The Best Cultivars of Fruit, Small Fruit and Nuciferous Crops of the Ukrainian Breeding;* Press of Ukraine: Kyiv, 2011; p 143 (In Ukrainian).

11. Pomology. In *Apple;* Kondratenko, P. V., Kondratenko, T. Y., Eds.; Nilan–Ltd: Vinnitsa, 2013; pp 594–615 (In Ukrainian).

12. Grynyk, I. V.; Omelchenko, I. K.; Lytovchenko, O. M. *Inland Technologies for the Production, Storage and Processing of Fruits and Small Fruits in Ukraine;* Press of Ukraine: Kyiv, 2012; p 117 (In Ukrainian).

FIGURE 8.1   The flowering of dandelion (*Taraxacum officinale* Webb.)

FIGURE 8.2   *Dracaena fragrans* 'Massangeana'.

FIGURE 8.3   *Monstera deliciosa* Liebm.

FIGURE 8.4   *Monstera deliciosa* 'Borsigiana'.

FIGURE 8.5 *Spathiphyllum wallisii* 'Quatro'.

FIGURE 8.6   *Tetrastigma voinierianum* (Baltet) Gagnep.

FIGURE 13.3   Fruits of hazelnut 'Sofiyivsky 1'.

FIGURE 13.4   Fruits of hazelnut 'Sofiyivsky 2'.

FIGURE 13.5   Fruits of hazelnut 'Sofiyivsky 15'.

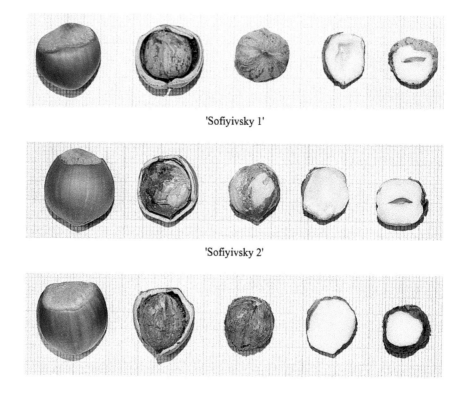

'Sofiyivsky 1'

'Sofiyivsky 2'

'Sofiyivsky 15'

FIGURE 13.6 Fruits of new hazelnut cultivars: 'Sofiyivsky 1' 'Sofiyivsky 2', 'Sofiyivsky 15'.

(b)

(a)

FIGURE 15.1   Raceme formation in black currant; (a) 'Shalun'ja' and (b) 'Divo Zvjaginoj'.

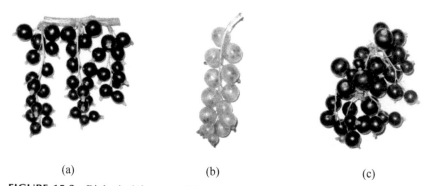

(a)   (b)   (c)

FIGURE 15.2   Biological features of berry formation in a raceme of the black currant cultivars; (a) Pandora, (b) 'Izumrudnoe ozherel'je', and (c) 'Talisman'.

(a)   (b)

FIGURE 15.3   Differences in cluster length of black currant cultivars; (a) Chjornyi zhemchug and (b) Karmelita.

CHAPTER 12

THE PRODUCTIVITY OF APPLE ORCHARDS ON TERRACED AND GENTLE SLOPES

ZHAMAL H. BAKUEV[1,*] and LIANA CH. GAGLOEVA[1]

[1]North Caucasian Research Institution of Mountain and Foothill Gardening, 23, Shardanova St., Nalchik, Kabardino-Balkarsk Republic, Russia, 360004, E-mail: liana.kabisova@mail.ru

*Corresponding author. E-mail: kbrapple@mail.ru

CONTENTS

ABSTRACT

In this chapter, results of experimental data on the study of productivity of intensive gardens of an apple tree are given in conditions on the terraced and gentle slopes of North Caucasus foothills in the Kabardino-Balkar Republic.

12.1 INTRODUCTION

The central part of the North Caucasus is one of the leading regions of agricultural production of the south of the Russian Federation where the big development program of intensive gardening has been planned. In the solution of these tasks, the special place is allocated to Kabardino-Balkar Republic. It is characterized by favorable agroecological conditions. In spite of the fact that in the past the republic was one of the main zones of production of various fruit products, now industrial gardening is not always and not everywhere rather effective. One of the reasons is that many extensive gardens in foothill and mountainous areas require replacement by plantings of more intensive type on clonal stocks. The productivity and cost efficiency of such obsolete plantings do not provide expanded reproduction of an industry. Because of complex nature of creation of intensive fruit plantings in the conditions of vertical zonality of the central part of the North Caucasus requires accurately developed recommendations to scientific reasons for a possibility of cultivation of an apple tree on slopes, it should be noted that research of these questions is very actual and has great national economic value that is confirmed by outstanding cost efficiency of implementation of the received results in the production.

12.2 PURPOSE OF RESEARCH PROBLEM

The main objective of researches consisted in theoretical reasons and development of effective elements of the technology of cultivation of intensive gardens on slope lands in the central part of the North Caucasus, and also to develop recommendations about increase of productivity of an apple tree in relation in Kabardino-Balkar Republic conditions.

The main objective was solved—study influence of specific conditions on terraced slopes of various steepness and also on flat, poorly sloping slopes on growth, productivity, and quality of fruits of an apple tree.

12.3 MATERIALS AND METHODOLOGY

The production of plantings of research is performed according to a technique of expeditionary (route) inspection in mountain conditions on Dragavtsev (1956).[1]

Laying of experiences, supervision, and accounting of the studied factors were carried out according to techniques of the All-Russian research institution of gardening[2] and the All-Russian research institution of selection of fruit crops,[3] and also with requirements of agro-technical researches in gardening.[4]

Handling of experimental data is carried out with use of methods of mathematical statistics.[5]

12.4 RESULTS AND DISCUSSION

12.4.1 STUDYING OF PRODUCTIVITY OF INTENSIVE GARDENS OF AN APPLE TREE ON THE TERRACED SLOPES

In Kabardino-Balkar Republic in connection with limitation of arable lands, development under gardens of low-convenient slope lands is actual. It is known that on slopes, condition of moisture providing, mineral food, the thermal and light modes for fruit plants on different sites of slopes develops inadequately.[6–9] With respect thereto, we were in a forest and mountain fruit zone of Kabardino-Balkar Republic, at the height of 550–650 m above sea level in village Kenzhe during the period from 2011 to 2015 and have conducted researches of productivity on an apple tree in the conditions of a difficult mountainous terrain on the terraced slopes. Soils in experience gray forest spread by heavy loams. The amount of precipitations in a year is the 650–720th. The total area of pilot sites of 60 ha was occupied by an orchard on slope lands.

Researches are conducted on two sites, which are slopes of hills and spurs of northern slopes of Black (woody) mountains. The area occupied by plantings makes more than 60 ha.

Slopes in southern and adjacent expositions in mountainous areas receive more solar radiation, are better lit, are warmer, and at the same time less, humidified than the opposite slopes. Depending on continentality of climate, the steepness, the dominating winds, and other factors of distinction between opposite slopes amplify. It, in turn, defines conditions of growth of fruit trees.

On the slopes, the fluctuation of temperature factor has significant effect on fruit plants. On the slopes turned to the south and the east, damageability of boles and skeletal branches of solar burns considerably increases. On the slopes of the southern exposition, trees are stronger and damaged by frosts. An important condition of growth of an apple tree on slopes is moisture security and mineral nutrition. In the southern zone, fruits growing on the slopes of northern, northeast, and northwest expositions are optimum for an apple tree.

The configuration of slopes also plays an important role in providing the necessary mode for growth and development of fruit crops. On concave slopes, condition for growth of fruit is better than on convex. The favorable combination of natural and climatic factors for growth of an apple tree, plum, and a cherry plum on slopes substantially is defined by vertical zonality. In this regard, certain, more optimum fruit crops slopes are also recommended for various zones.

In the conditions of the forest and the mountain zones, slopes of the southern and western expositions are warmer. The temperature of ground air in summer months on the stated slopes is above 10°C than that on northern and east slopes. The difference of temperatures on a soil surface between slopes of northern and southern expositions reaches to 6–80°C in the afternoon and at night to 1.5–20°C in favor of the southern slope. Soil temperature at a depth of 5 cm in middle part of the southern slope is slightly higher, in comparison with a northern slope, on 20°C. Similar regularity, but with a smaller difference, by comparison of temperature of the soil of the western and east slopes. In the first case, the temperature was 1 and 0.80°C higher than that on a slope of east exposition.

Therefore, on slopes of the western and especially southern expositions, the effect of a temperature factor is much higher than on opposite slopes.

Important condition for the growth of fruit crops on slopes is to provide moisture and elements of mineral food in the soil. The moisture content in the soil on the southern slope is less than on the northern slope (Table 12.1).

In 5 years on a slope of a northern exposition in a meter layer of earth on a cloth of a terrace of productive moisture was more, than on the southern slope, on 33 mm. In drought years, when rainfall for the period May to September dropped out 170–181 mm less the required amount, the difference in content of productive moisture in a meter layer of earth reached 51 mm. Slopes of eastern and western exposition on moisture security of the soil are intermediate between northern and southern slopes.

TABLE 12.1 Humidity and Elements of Mineral Nutrition on the Terraced Slopes of a Different Exposition in a Meter Layer of Earth.

Slope exposition	Productive moisture (mm)	Elements of mineral food of soil (mg/kg)		
		NO_3	P_2O_5	K_2O
Northern	165	26.3	36.3	284.0
Southern	132	17.2	31.0	256.0
Eastern	153	20.6	39.1	258.0
Western	148	24.6	36.1	262.0

During the vegetative period, moistening of the soil on the terraced slopes is uneven. As a rule, in the spring and at the beginning of summer (May to June), when plentiful rainfall drops out, moisture in the soil is more than in the second half of summer and an early autumn (July to September). During this period, the temperature and moisture in the soil decreases, especially on slopes of the southern exposition. Similar data are obtained for the fructifying garden, which contains a non-terraced slope under grassing with the steepness of 12–14°.

Therefore, the humidity mode for growth of an apple tree more evenly develops on the slopes turned to the north. On the southern slope, especially in the second half of vegetation, moisture content in the soil considerably goes down.

In the valleys between slopes, the humidity mode in the soil is slightly better than on slopes. It is explained by water drainage of both superficial and subsoil during loss of rainfall.

The soil slopes between themselves generally differ on the content of nitrate nitrogen. The soil on slopes of northern and western expositions contains more nitrates (see Table 12.1). The soil on the slope of the southern exposition contains less nitrates than on the other slopes.

In the soil content, mobile forms of phosphorus and potassium of an essential difference between slopes is not established. In the upper layers of the earth, the mobile forms of phosphorus is less (5–13 mg on 1 kg of the soil), and from depth of 60 cm, its content increases.

The content of a mobile form of potassium in the soil on all slopes is rather high. On an equal site, moisture and mineral food for an apple tree are distributed on the square rather regularly; on terraces, the best place for growth is the cloth.

Terracing in all zones of bogharic gardening promotes accumulating of moisture in the soil. Moisture inventories can increase by 1 ha of a cloth of a terrace on 600–700 m³ in a year. On terraces, almost all dropping-out rainfall is absorbed by the soil of a cloth, unlike deep draining plowing where the part flows down on a slope, causing an erosion.[6]

The mode of humidity of the soil on a cloth of terraces develops unequally. The extraction part of a cloth is a smaller berm and bulk part is most humidified. It is also promoted by bigger accumulation on a cloth of terraces of snow and its slow thawing.

Improvement of the water content of the soil on terraces promotes more active mobilization of available forms of mineral food to plants; especially nitrate nitrogen (Table 12.2).

TABLE 12.2   Content of Productive Moisture and Mobile Forms of Elements of Mineral Food in a Meter Layer of Earth on Terraces.[6]

Slop exposition	Terrace elements		Productive moisture (mm)	NO_3 (mg/kg) soil	P_2O_5 (mg/kg) soil
Northern	Road bed	Holy part	176	Trails	53.7
		Middle	156	14.7	53.0
		Bulk part	145	9.0	32.9
	Berm		150	Trails	61.7
Southern	Road bed	Holy part	142	Trails	47.2
		Middle	139	18.8	27.3
		Bulk part	94	19.1	30.9
	Berm		88	No	32.3
Eastern	Road bed	Holy part	159	No	41.2
		Middle	164	13.1	37.3
		Bulk part	112	Trails	29.8
	Berm		127	No	32.0
Western	Road bed	Holy part	164	Trails	36.8
		Middle	166	17.4	31.1
		Bulk part	129	17.4	28.4
	Berm		No	No	32.8

The nature of distribution of elements of mineral in the food of fruit plants depending on an exposition is not similar for the terraced slopes. On the slopes of the southern exposition under a grassing of nitrates, they are not enough, and in the top layers of the earth, they are absent more often. In comparison with the slopes of other expositions, the nitrates are two times less here. In the valley, the amount of nutrients in the soil is more in comparison with adjacent slopes.[6–12]

The obtained data on the growth and productivity of an apple tree on slopes of different expositions are presented in Tables 12.3 and 12.4.

TABLE 12.3 Growth of an Apple Tree on MM106 Stock on Different Expositions of Slopes by the Steepness 10–120° (Middle Part), a Garden is Put in 2003 According to the Scheme 5 × 2.4 m (2011–2015).[8]

Slop exposition	Cultivar	Length circles of shtamb (cm)	Total gain of a circle of a shtamb (cm)	Escape length (cm)
Southern (control)	Aydared	19.2	10.2	32.3
	Golden Delishes	17.4	8.7	27.5
	Rennet Simirenko	23.3	11.4	38.6
	Florina	26.4	11.6	42.3
	Prima	24.2	10.3	37.7
	Red Free	24.5	10.4	34.3
	Melba	18.5	9.6	34.5
	Average on cultivars	21.9	10.3	35.3
Eastern	Aydared	24.7	14.3	43.3
	Golden Delishes	22.3	12.5	35.5
	Rennet Simirenko	26.6	16.4	48.2
	Florina	28.2	16.6	51.4
	Prima	26.8	15.2	47.3
	Red Free	27.0	14.4	44.5
	Melba	23.2	13.5	43.2
	Average on cultivars	25.5	14.7	37.7
Western	Aydared	22.3	12.5	38.5
	Golden Delishes	19.7	10.3	31.3
	Rennet Simirenko	23.8	13.2	37.5

TABLE 12.3   *(Continued)*

Slop exposition	Cultivar	Length circles of shtamb (cm)	Total gain of a circle of a shtamb (cm)	Escape length (cm)
	Florina	27.3	15.3	48.4
	Prima	25.5	14.3	42.7
	Red Free	25.7	12.4	39.8
	Melba	20.5	10.8	38.0
	Average on cultivars	23.5	12.7	39.5
Northern	Aydared	26.7	15.6	47.5
	Golden Delishes	24.5	13.7	43.6
	Rennet Simirenko	29.7	17.8	52.7
	Florina	30.2	18.5	57.8
	Prima	28.8	16.8	52.3
	Red Free	28.6	15.7	49.7
	Melba	25.5	15.2	47.5
	Average on cultivars	27.7	16.2	50.2
LSD_{50}		2.4	1.6	4.0

Note: LSD is Least Significant Difference

Data given in Tables 12.3 demonstrate that apple tree on a slope of a northern exposition grow more actively; for example, the average length of a circle of a shtamb at a 10-year apple tree of a cultivar 'Golden Delishes' equaled 24.5 cm or 7.1 cm more than that of control. It was confirmed by productivity of an apple tree and mass of a fruit (Table 12.4).

As one would expect on productivity of an apple tree, the slope of a northern exposition was allocated. For example, productivity of cultivar 'Golden Delishes' for 3 years has averaged 26.0 t/ha and was higher than on a slope of the southern exposition on 5.6 t/ha. The productivity of an apple tree on the western and eastern slopes authentically did not differ on account of years, and on average for 3 years. However, on productivity, the apple tree on a slope of the western exposition is closer to northern, than to the east.

The mass of a fruit of an apple tree of a cultivar of 'Golden Delishes' on a slope of a northern exposition was larger and equaled to 155.0 g, against 128.5 g on the southern slope.

TABLE 12.4 Productivity and Mass of a Fruit of an Apple Tree on MM106 Stock on Different Expositions of the Terraced Slopes. Scheme of Planting of Trees 5 × 2.4 m, 833 trees per hectare. (2011–2015).[8]

Slop exposition	Cultivar	Total gain of a circle of a shtamb (cm)	Escape length (cm)
Southern (control)	Aydared	21.3	130.5
	Golden Delishes	20.4	128.5
	Rennet Simirenko	22.2	120.5
	Florina	21.0	120.0
	Prima	19.0	110.5
	Red Free	20.8	122.0
Eastern	Melba	24.7	155.3
	Average of cultivars	25.3	145.5
	Aydared	25.2	140.4
	Golden Delishes	22.5	125.5
	Rennet Simirenko	21.2	112.5
	Florina	23.8	135.8
Western	Prima	22.3	145.5
	Red Free	23.2	137.0
	Melba	24.3	128.4
	Average on cultivars	20.7	120.0
	Aydared	20.5	112.0
	Golden Delishes	22.2	128.6
Northern	Rennet Simirenko	27.0	175.5
	Florina	26.0	155.0
	Prima	26.2	145.0
	Red Free	24.5	135.0
	Melba	21.8	115.5
	Average of cultivars	25.1	145.2
LSD_{50}	2	2.6	15.4

The mass of a fruit on slopes of the western and eastern expositions is intermediate between northern and southern slopes.

The productivity of an apple tree on slopes of the western and eastern exposition is also more than on slopes of the southern exposition. At the same time, a reliable difference in productivity of an apple tree between these slopes is not established. Similar regularity has been noted on other cultivars of an apple tree.

The example of an apple tree of winter cultivars 'Aydared', 'Golden Delishes', and 'Florina' and summer cultivars 'Melba' and 'Red Free' landed on bulk part of a cloth of terraces has made observation over their productivity in different belts of slopes of various expositions (Fig. 12.1).

FIGURE 12.1   Gardens on the terraced slopes of the foothills of Kabardino-Balkar Republic, the settlement of Kenzhe.[8]

This was confirmed data about the productivity of the studied varieties. On all cultivars, productivity in the lower parts of slopes is reliable. At the same time, the more noticeable difference in productivity between the specified parts of a slope has appeared on winter cultivars. Therefore, the productivity of a winter cultivar 'Aydared', average for 3 years, in the lower part of a slope has made 293.8 t/ha and was higher than on the top tier (k) by 1.35 times.

The productivity of a cultivar 'Aydared' in a middle part of a slope was less than in the lower belt of a slope, but is significantly higher than in the top part of a slope. Here, the productivity on average for 3 years equaled to 269.8 t/ha and surpassed control almost on 76.7 t/ha.

On a summer cultivar, 'Melba' distinctions in productivity were less noticeable. The productivity in the lower part of a slope surpassed control on 30 t/ha.

All cultivars observed distinction and the average mass of a fruit. The mass of a fruit from top to down on a slope slightly increased on cultivars 'Melba', 'Red Free', 'Florina', and 'Golden Delishes' and also considerably increased in a cultivar 'Aydared'. Therefore, in a cultivar Aydared, the difference on the mass of a fruit between the top and lower parts of a slope was 68 g.

Various conditions of moistening and mineral food on the slopes exert considerable impact on growth of root system of an apple tree. The root system of 10-year trees of an apple tree in the lower part of a slope by more than 2–3 times surpasses root system of the trees placed in the top part of a slope (Table 12.5) in the extent and weight.

TABLE 12.5 The Sizes of Root System at a 10-Year Apple Tree of a Cultivar Aydared in Various Parts of a Terraced Slope of a Northwest Exposition with the Steepness 10–12°.

Part of slope	Length of roots		Mass of roots	
	(m)	In (%) to the top part of a slope	(g)	In (%) to the top part of a slope
Top	97	100	346	100
Average	353	230	656	226
Lower	351	229	913	330
LSD_{50}	15.0	–	20.8	–

Note: LSD is Least Significant Difference

On terraces, the considerable part of roots (40.3–45.0%) of length sum in volume of excavation is placed on the processed part of a cloth and only 20.1–23.9% in a zone of a bulk slope. In the direction of a bulk slope, growth of roots (on length and weight) restrains.

The maximum removal of roots at a 10-year apple tree from a shtamb toward a slope is of 93 cm, while at trees in a middle part of a slope at distance of 145 cm and in the lower part of 185 cm. The bulk of roots on terraces is placed with a 10-year apple tree in a radius of 1 m in the lower and average parts of a slope and in a radius 0.5 m in the top part, in a layer by the 0–60th. Here, trees in the top part of a slope 79% of length and 86.7% of mass of roots of their sum in volume of excavation, respectively, at trees in a middle part of a slope of 86.2 and 90.9% and in the lower part of a slope of 86.2 and 89.3% have been concentrated. The maximum depth of penetration of vertical roots into the soil has made the 65–70th.

Therefore, roots of an apple tree on slopes develop more powerfully in conditions with the best moisture security. Here, they branch more weakly than roots on slope sites with smaller security with moisture. At the top part of a slope at 10-year trees of an apple tree length of roots in unit of their weight (1 g) was 80.1 cm, and in the lower part of a slope 55.4 cm. Features of growth of roots of an apple tree remain on the not terraced sites of slopes. In the fructifying gardens, the root system of an apple tree in the lower part of slopes is more developed than at the top.

In conclusion, we shall note that apple trees on the terraced slopes, irrespective of a cultivar, grow at the steepness in the fructifying age more actively and are more fruitful in the lower tier of northern slopes. Distinction degree in the growth and productivity of an apple tree between belts of slopes depends on the steepness, extent, and a cultivar. In summer cultivars 'Melba' and 'Red Free', these distinctions are less noticeable than in winter cultivars 'Aydared', 'Golden Delishes', and 'Florina'. It is also revealed that cultivars, immune to a scab, 'Red Free' and 'Florina' differed in bigger adaptability to cultivation microzones, and within a cultivar, they had no essential difference in indicators of productivity and mass of a fruit on different parts and expositions of slopes. Thus, superiority in growth and productivity slopes can be listed in the following decreasing sequence on expositions: northern, western, eastern, and southern; and in parts slopes: lower, average, and top.

12.4.2   PRODUCTIVITY-INTENSIVE BUTTRESSNESS AND SUPER-INTENSIVE LANE-DWARFISH GARDENS OF AN APPLE TREE IN THE CONDITIONS OF VERTICAL ZONALITY OF KABARDINO-BALKAR REPUBLIC

The placement of the super-intensive and intensive gardens of an apple tree in the conditions of vertical zonality of the republic does not meet the modern requirements and interferes with rational use of land resources.[6,10] With respect thereto, we have conducted researches in intensive plantings of various designs. At the beginning, agroecological features of four fruit zones of the republic (Table 12.6) are considered.

There are certain risks in case of cultivation of gardens in the republic. Along with risks of social and economic nature, we will consider the risks connected with an environment of cultivation of plantings.

A number of environmental factors affect the growth apple trees in a certain period that leading to reduction in the productivity of gardens are: snowless winters with air temperatures below −20°C; intensive hail damages; the long air droughts accompanied with hot dry winds with relative humidity of air that is lower than 45–50%; plentiful freezing of trees; and return of subzero temperatures to the early-spring period (Table 12.7).

The accounts which are carried out by us have shown that higher rates of growth of trees of an apple tree took place in a forest and mountain zone, in a borderland with a foothill zone. Much more weakly trees grow in mountain-steppe and steppe zones.

One of the important indicators of growth of an apple tree is the size of a shtamb. These Tables 12.7 demonstrate that on a thickening of shtamb there are essential distinctions between fruit zones. Most actively growth of trees happens in a forest and mountain zone, and more weakly trees develop in a mountain-steppe zone. Thus, the length of the circle of the stem in all grades in the first case averaged 15.2 cm, whereas in the second—only 9.2 cm. This is evidenced by the data of the account during the whole year of testing (Table 12.8).

The greatest length of surpluses was in a forest and mountain zone 44.3 cm (on average on all cultivars), and the smallest in a mountain-steppe zone—25.8 cm between the studied options (fruit zones) of distinction on the specified indicators have a reliable difference.

TABLE 12.6   Agroecological Characteristic of Fruit Zones of Kabardino-Balkar Republic.

Fruit zone	Administrative areas	Height over sea level (m)	Soil varieties	Climate type	Rainfall (mm)	Air moistening coefficient	Sum of active temperatures (°C)	Average annual isotherm (°C)
Steppe zone	Tersky, Prokhladnensky, Maisky, Baksansky (northeast part), Leskensky, Urvansky (northern and northeast parts)	220–450	Meadow, meadow chernozem, dark chestnut, secondary carbonate, meadow and marsh	Continental	330–450	0.6–0.8	3200–3422	+9.5
Foothill zone	Baksansky (the central part), Chegemsky (northeast part), Zolsky (northeast part without mountain), Nalchik	350–550	Gray forest, meadow steppe lixivious, so-so powerful pre-Caucasian chernozems, pebble soils	Continental, mild-continental	400–615	1.0–1.1	2417–3200	+9.0
Forest and mountain zone	Baksansky (southwest part), Chegemsky (the central part), Urvansky (southern and southeast parts), Zolsky (southeast part), Chereksky (southern and southwest parts), Nalchik	550–850	Gray forest, mountain, and forest opodzolenny, chernozems the pre-Caucasian poor alkali	Continental, mild-continental	750–850	1.2–1.4	2010–2800	+8.5

TABLE 12.6 *(Continued)*

Fruit zone	Administrative areas	Height over sea level (m)	Soil varieties	Climate type	Rainfall (mm)	Air moistening coefficient	Sum of active temperatures (°C)	Average annual isotherm (°C)
Mountain-steppe zone	Chereksky (southern and southwest mountain parts), Chegemsky (southeast mountain part), Zolsky (central part), Baksansky (upper courses)	900–1800	Mountain meadow, mountain chernozems: carbonate and residual and carbonate, so-so powerful, ordinary powerful, clay on the alyuvio-delyuvialnykh limy sandstones	Continental	350–436	1.1–1.3	1800–2600	+7.5

TABLE 12.7   Approach Frequency in Kabardino-Balkar Republic of the Adverse Natural Phenomena and a Share of Yield Losses from Their Impact (2006–2015).

Types of the adverse phenomena	Average values in 10 years			
	Steppe	Foothill	Forest and mountain	Mountain-steppe
Hail damages	0.5/15	1.6/40	2.1/60	0.3/10
Snowless winters with $t° < -20°C$	0.5/30	0.0/0	0.0/0	0.0/0
Hot dry winds	1.1/15	0.2/5	0.0/0	0.6/7
Freezing	0.3/5	0.4/8	0.4/10	0.3/5

Note: In the numerator—the frequency of adverse events in the 10 years of time; and in the denominator—commercial crop losses, %. In relation to culture of the apple tree cultivated on the irrigated lands possible risks as a result of approach of all harmful phenomena on zones make, respectively: in steppe—13.8%; in foothill—13%, in forest and mountain—24.5%, and in mountain steppe—18.5%. Minimal risks of growing apple orchards are noted in the steppe and in the foothill zones.

TABLE 12.8   Influence of Height Above Sea Level on a Circle of Shtamb and a Year Gain of Escapes of an Apple Tree, (Landing of 2003, the scheme 5×2 m, an average for 2011–2015).[8]

Fruit zone/height above sea level (m)	Cultivar	Average length of a circle of a shtamb (cm)	Average length of a year gain of escapes (cm)
Steppe/188–250 m	Melba	12.4	27
	Red Free	10.3	35
	Gala Must	12.6	28
	Prima	12.8	37
	Golden Delishes (control)	11.5	24
	Florina	12.8	35
	Rennet Simirenko	11.6	32
	Average on cultivars	12.0	31.1
	LSD_{50}	1.0	2.5
Foothill/430–485 m	Melba	14.5	38
	Red Free	12.2	42
	Gala Must	14.8	35
	Prima	14.9	41
	Golden Delishes (control)	13.4	31
	Florina	15.2	44
	Rennet Simirenko	13.3	42
	Average on cultivars	14.0	38.6
	LSD_{50}	1.2	2.8

TABLE 12.8 *(Continued)*

Fruit zone/height above sea level (m)	Cultivar	Average length of a circle of a shtamb (cm)	Average length of a year gain of escapes (cm)
Forest and mountain/533–830 m	Melba	15.5	43
	Red Free	13.6	44
	Gala Must	15.8	42
	Prima	16.3	47
	Golden Delishes (control)	14.7	35
	Florina	16.4	51
	Rennet Simirenko	14.5	48
	Average of cultivars	15.2	44.3
	LSD$_{50}$	1.4	3.2
Mountain-steppe/1160–1200 m	Melba	9.2	21
	Red Free	8.7	27
	Gala Must	8.4	20
	Prima	9.8	32
	Golden Delishes (control)	8.5	22
	Florina	10.4	33
	Rennet Simirenko	9.2	26
	Average of cultivars	9.2	25.8
	LSD$_{50}$	1.0	2.5

Note: LSD is Least Significant Difference

Weaker growth rates and escapes are characteristic of fruit trees in a mountain-steppe zone. Here, it is much less, than in forest and mountain and foothill zones and with a smaller difference than in a steppe zone.

The important indicator determining productivity of an apple tree is the area of leaves. Favorable conditions for moistening in a forest and mountain zone promote forming of well-developed sheet device. At a cultivar apple tree of 'Simirenko' of 10–12 summer age, the area of a sheet plate of escapes on zones has constituted the Rennet: in steppe—20.5 cm², in foothill—23.3 cm², in forest and mountain—24.2 cm², and in a mountain-steppe zone—17.2 cm² (Table 12.9).

In forest and mountain and foothill zones in connection with steady rains, quite often there are great difficulties in preservation of the sheet device of an apple tree from defeat by mushroom diseases. In steppe and especially in a mountain-steppe zone, mushroom diseases in the apple tree are surprised to a lesser extent.

TABLE 12.9 The Size of a Sheet Plate of an Apple Tree of a 'Cultivar Rennet', 'Simirenko' Depending on Height Above Sea Level (Landing of 2003, the Scheme 5 × 2 m, an Average for 2011–2015).[8]

Fruit zone/height above sea level (m)	Average area of a leaf (cm^2)
Steppe/188–250	205
Foothill/430–485	23.3
Forest and mountain/533–830	24.2
Mountain-steppe/1160–1200	17.2
LSD$_{50}$	2.2

Note: LSD is Least Significant Difference

Thus, in the central part of the North Caucasus, the apple tree grows in a forest and mountain zone, in a borderland with the foothills better.

The most big crops of fruits of an apple tree are also noted in foothill and forest and mountain zones, regions of an ecological optimum of conditions (Table 12.10).

TABLE 12.10 Influence of Height Above Sea Level on Productivity of an Apple Tree (2003 of Landing, the Scheme 5 × 2 m, an Average for 2011–2015).[8]

Group on terms of maturing of fruits	Cultivar	Average productivity at different heights above sea level on zones (t/ha)				The smallest essential difference
		188–250 m/ steppe	430–485 m/ foothill	533–830 m/ forest and mountain	1160–1200 m/ mountain-steppe	
Summer	Melba	19.2	22.6	23.8	17.8	2.0
	Red Free	22.5	25.8	27.6	20.5	2.3
Autumn	Gala Must	24.4	28.7	33.6	22.3	2.2
	Prima	18.5	22.5	26.5	17.8	2.0
Winter	Golden Delishes	23.6	31.4	37.3	17.2	2.6
	Rennet Simirenko	21.4	28.7	35.6	15.4	3.0
	Florina	19.3	26.8	32.5	16.3	2.8
LSD$_{50}$		2.2	2.5	3.1	2.0	

Note: LSD is Least Significant Difference

Compliance of natural factors for requirements of an apple tree in a zone of an ecological optimum also promotes that here fruits of the cultivated cultivars of an apple tree differ in the large sizes, the correct form, and extremely bright beautiful coloring that puts them out of competition in comparison with fruits from other zones.

Long-term supervision of a number of varieties has shown that varieties in conditions of forest and mountain zones, have the period of growth and maturation of fetuses is shorter and their fruits do not differ significantly in weight (Table 12.11) there.

TABLE 12.11 The Mass of a Fruit of an Apple Tree in Plantings at Different Heights Above Sea Level (2011–2015).

Terms of maturing of fruits	Cultivar	Average productivity at different heights above sea level on zones (t/ha)				The smallest essential difference
		188–250 m/ steppe	430–485 m/ foothill	533–830 m/ forest and mountain	1160–1200 m/ mountain-steppe	
Summer	Melba	103	110	115	105	10.0
	Red Free	127	132	140	125	10.8
Autumn	Gala Must	115	125	145	110	9.5
	Prima	120	135	145	115	8.8
Winter	Rennet Simirenko	135	155	180	105	12.2
	Florina	125	150	165	110	12.0
LSD_{50}		8.5	10.0	12.5	9.6	

Note: LSD is Least Significant Difference

For example, a summer cultivar 'Melba' for growth and maturing of fruits in a forest and mountain zone required on average 86 days. At the height of 1200 m, the period from setting to a deadline of cleaning of fruits equals on average to 151 days. At the height of 1200 m, a summer cultivar 'Melba' yields the fruits which do not significantly differ on weight from the fruits which are grown up at the height of 550 m above sea level (forest and mountain zone).

Ripening of fruits of an apple tree of a winter cultivar of the 'Rennet' of 'Simirenko' at the height of 550 m requires on average 153 days. The period from setting before cleaning at the height of 1200 m makes 143 days. As a result, fruits of the specified cultivar in the first case in years with occurrence of early cold weather do not grow ripe, and at the height of 1200 m reach ripeness only in years with warm and long fall, and the sizes much more small than in optimum conditions of cultivation.

We have considered questions of cultivation of the super-intensive gardens of an apple tree in various fruit zones of Kabardino-Balkaria.

Landings of the super-intensive gardens (Fig. 12.2) according to the scheme 3.5×0.9 m (with placement of 3170 trees on 1 ha) have been carried out by the saplings of different cultivars of an apple tree brought from Italy, Belgium, Holland, and Austria (saplings "knip-baum").

FIGURE 12.2 Super-intensive lane-dwarfish gardens of an apple tree in Kabardino-Balkar Republic.[6]

They differ from usual saplings coevals in more powerful development, existence of 5–6 and more side branches (5+, 7+) located at the height from 70 to 100 cm and also existence of a kolchats (floral kidneys). Length of side branches is from 30 to 50–60 cm.

The sizes of young apple trees in the spring, ages of 3 and 4 years, are shown in Table 12.12. Apparently from the provided data on the thickness of a shtamb (circle length) and height of a tree, there are noticeable distinctions. On the basis of growth force, the studied cultivars can be distributed to three groups: strongly tall is Red Delishes Hapke and Grenni Smith, so-so tall—'Breburn', 'FudzhiKiku', and 'Golden Delishes', 'Golden Delishes' a clone of B, 'Golden Raingers' and a cultivar poorly tall—'Red Chif Kamspur', 'Super Chif Sandidzh', 'Zheromin', and 'Early Red Wang'.

TABLE 12.12 Biometric Indicators of Young Apple Trees in the Super-Intensive Gardens Depending on a Cultivation Zone (Landing of 2009, the Scheme 3.5×0.9 m, a Stock of M9, the 3175 trees per hectare, an Average for 2012–2013.

Fruit zone/ height above sea level m)	Cultivar	Length of a circle of a shtamb (cm)	Length of a year gain of escapes (cm)	Height of a tree (m)
Steppe/188–250	Golden Delishes (control)	11.4	31.5	2.5
	Golden Delishes, clone B	12.2	33.4	2.6
	Golden Raingers	11.7	32.6	2.5
	Red Delishes Hapke	13.8	41.4	3.0
	Red Chif Kamspur	9.3	25.5	2.1
	Super Chif Sandidzh	9.0	24.0	2.1
	Early Red Wang	9.5	27.4	2.3
	Zheromin	10.6	29.0	2.3
	Breburn	11.4	31.0	2.3
	FudzhiKiku	12.7	34.8	2.4
	Grenni Smith	13.6	40.2	3.0
	Average on cultivars	11.4	31.9	2.5
	LSD_{50}	1.0	3.0	0.3
Foothill/ 430–485	Golden Delishes (control)	13.5	38.5	3.0
	Golden Delishes, clone B	14.4	41.5	3.0
	Golden Raingers	13.5	40.3	3.0
	Red Delishes Hapke	15.4	48.8	3.5
	Red Chif Kamspur	11.2	30.5	2.0

TABLE 12.12　*(Continued)*

Fruit zone/ height above sea level m)	Cultivar	Length of a circle of a shtamb (cm)	Length of a year gain of escapes (cm)	Height of a tree (m)
	Super Chif Sandidzh	11.1	28.6	2.1
	Early Red Wang	11.7	33.1	2.1
	Zheromin	12.3	36.9	2.2
	Breburn	13.6	37.5	2.5
	FudzhiKiku	14.5	41.0	2.6
	Grenni Smith	14.8	45.4	3.5
	Average on cultivars	13.3	38.3	2.7
	LSD$_{50}$	1.2	3.4	0.3
Forest and mountain/533–830	Golden Delishes (control)	14.5	45.1	3.2
	Golden Delishes, clone B	15.6	43.5	3.2
	Golden Raingers	14.4	42.4	3.2
	Red Delishes Hapke	17.1	55.7	3.5
	Red Chif Kamspur	12.3	35.5	2.2
	Super Chif Sandidzh	12.2	33.2	2.3
	Early Red Wang	12.5	38.3	2.3
	Zheromin	12.6	38.9	2.5
	Breburn	15.3	42.5	2.9
	FudzhiKiku	15.6	45.2	2.8
	Grenni Smith	16.2	48.8	3.5
	Average on cultivars	14.5	42.6	2.9
	LSD$_{50}$	1.3	4.0	0.3
Mountain-steppe/1160–1200	Golden Delishes (control)	8.2	28.7	2.0
	Golden Delishes, clone B	9.3	29.3	2.1
	Golden Raingers	9.7	27.0	2.0
	Red Delishes Hapke	11.4	34.6	2.4
	Red Chif Kamspur	8.4	20.2	1.9
	Super Chif Sandidzh	8.0	18.8	1.9
	Early Red Wang	8.3	24.9	2.0
	Zheromin	9.5	26.0	2.0
	Breburn	10.5	30.5	2.1

TABLE 12.12 *(Continued)*

Fruit zone/height above sea level m)	Cultivar	Length of a circle of a shtamb (cm)	Length of a year gain of escapes (cm)	Height of a tree (m)
	FudzhiKiku	9.4	22.8	1.9
	Grenni Smith	9.3	25.4	2.0
	Average on cultivars	9.3	26.2	2.0
	LSD$_{50}$	1.0	2.5	

Note: LSD is Least Significant Difference

The height of 4-year trees of cultivars Red Delishes Hapke and Grenni Smith in foothill and forest and mountain zones reaches 3.5 m, which is the most admissible for this garden, taking into account an arrangement at this height of a antihaily grid. The height of tree of cultivars 'Golden Delishes' and its clones reaches 3.2 m and also approached admissible border where restriction was required (see Table 12.12).

Poorly tall cultivars 'Red Chif Kamspur', 'Early Red Wang', 'Super Chif Sandidzh', and 'Zheromin' had tree height of 2.2–2.5 m which is significantly less than the abovenamed cultivars. In steppe and mountain-steppe zones, the height of trees and other biometric indicators were lower than in foothill and forest and mountain zones (see Table 12.12).

Data on fructification of apple trees in fourth and fifth are presented to vegetation (2012 and 2013) in Table 12.13. The settlement number of fruits after thinning of an ovary for the fourth year has 70–80 pieces, and in the fifth year, 90–100 pieces.

After a June normalization of an ovary on each tree at cultivars with plentiful blossoming, 'Golden Delishes' and its clones and 'Grenni Smith' in foothill and forest and mountain zones remained within 70–100 fruits for the fourth year and 90–110 fruits for the fifth year. At the poorly tall cultivars 'Red Chif Kamspur', 'Super Chif Sandidzh', 'Zheromin', 'Early Red Wang' in these zones, there was a moderate quantity of fruits (40–60sht). Steppe and especially mountain-steppe zones conceded in fructification and quality of fruits to foothill and especially forest and mountain zones (see Table 12.12).

During vegetation in a garden, it was supported high soil care: a drop irrigation, 2–3 times treatment by fertilizers (nitroammofosk), spraying against wreckers and diseases (to only 20 times), application of herbicides in a near-stem strip.

TABLE 12.13 Productivity of Cultivars of an Apple Tree in Super-Intensive Planting Depending on a Zone of Cultivation (2012–2013).[6–8]

Cultivar	Average productivity at different heights above sea level on zones (t/ha)				The smallest essential difference
	188–250 m/ steppe	430–485 m/ foothill	533–830 m/ forest and mountain	1160–1200 m/ mountain-steppe	
Golden Delishes (control)	40.4	47.9	48.6	33.2	3.8
Golden Delishes, clone B	42.3	45.6	45.9	35.4	3.5
Golden Raingers	45.8	51.2	53.8	37.3	4.0
Red Delishes Hapke	36.9	38.4	36.5	30.4	3.2
Red Chif Kamspur	26.1	28.6	29.8	21.1	2.2
Super Chif Sandidzh	28.5	33.0	35.3	22.5	2.4
Early Red Wang	27.5	31.4	33.8	21.8	3.0
Zheromin	25.6	30.1	31.6	20.4	3.2
Breburn	36.4	39.7	41.2	27.3	3.8
FudzhiKiku	45.4	41.3	40.3	29.5	4.0
Grenni Smith	42.6	45.2	44.6	25.7	4.4
LSD_{50}	4.2	3.8	4.4	3.0	–

Note: LSD is Least Significant Difference

All this provided fruits of the large size with a diameter of 70 mm and more. The harvest of fruits at the majority of cultivars made within 13–18 kg from a tree in a year. For example, average productivity of cultivar 'Golden Raingers' in a forest and mountain zone made 53.8 t/ha. In a foothill zone, the big crop of this cultivar which made 51.2 t/ha (Table 12.13) was also received.

Thus, on average, productivity from hectare depending on height above sea level (cultivation zones) in 2 years of cultivars 'Golden Delishes', 'Golden Reaingers', 'Golden Delishes' clone of B, 'Grenni Smith', 'Red Delishes Hapke', 'FudzhiKiku', and 'Breburn' was more. Cultivars 'Red Chif Kamspur', 'Super Chif Sandidzh', 'Early Red Wang', and 'Zheromin' among themselves do not considerably differ on a harvest, but significantly concede to cultivars of the first group. The last has the smaller sizes of kroner; therefore, they should be placed more bodying, or to apply taller semi-dwarfish stocks to them (for example, M26, CK2, MM106). Moreover, in intensive buttressness plantings of the super-intensive lane-dwarfish gardens, more big crops are received in a forest and mountain zone, then in the decreasing sequence in foothill, steppe, and mountain-steppe zones (see Table 12.13).

Having studied the commercial qualities of fruits, it should be noted that, as on traditional and intensive plantations, in gardens of superintense content, a regularity revealed, under which the best results of the average weight of fruits and their commercial quality were obtained in foothill, forest and mountain areas. Here the average mass of fruits has reached 170–205 g depending on a cultivar (Table 12.14).

TABLE 12.14 Influence of Height above Sea Level on Commodity Qualities of Fruits of Young Apple Trees in the Super-Intensive Plantings (2012–2013).[6-8]

Cultivar	Average mass of a fruit (g)	Marketability of fruits (%)		
		The highest and first cultivar	Second cultivar	Third cultivar
Steppe/188–250				
Golden Delishes (control)	145	87.0	8	5
Golden Delishes, clone B	157	88.0	9	3
Golden Raingers	155	85.0	7	8
Red Delishes Hapke	185	90.0	5	5
Red Chif Kamspur	165	87.0	7	6
Super Chif Sandidzh	170	90.0	6	4
Early Red Wang	160	85.0	9	6
Zheromin	165	87.0	9	4
Breburn	180	96.0	3	1
FudzhiKiku	173	95.0	4	1
Grenni Smith	195	94.0	4	2
LSD$_{50}$	15.6			
Foothill/430–485				
Golden Delishes (control)	175	90.0	7	3
Golden Delishes, clone B	182	92.0	6	2
Golden Raingers	180	91.0	5	3
Red Delishes Hapke	190	95.0	3	2
Red Chif Kamspur	180	92.0	5	3
Super Chif Sandidzh	175	90.0	7	3
Early Red Wang	178	87.0	8	5
Zheromin	171	89.0	9	2
Breburn	182	93.0	5	2
FudzhiKiku	178	94.0	5	1
Grenni Smith	205	95.0	3	2
LSD$_{50}$	17.0			
Forest on the mountain/533–830				
Golden Delishes (control)	185	92.0	6	2

TABLE 12.14   *(Continued)*

Cultivar	Average mass of a fruit (g)	Marketability of fruits (%)		
		The highest and first cultivar	Second cultivar	Third cultivar
Golden Delishes, clone B	187	94.0	5	1
Golden Raingers	180	92.0	5	2
Red Delishes Hapke	195	95.0	3	2
Red Chif Kamspur	185	94.0	3	3
Super Chif Sandidzh	185	92.0	7	1
Early Red Wang	180	88.0	8	4
Zheromin	177	90.0	8	2
Breburn	180	92.0	5	3
FudzhiKiku	170	90.0	5	5
Grenni Smith	195	92.0	5	3
LSD_{50}	17.8	–	–	–
Mountain-steppe/1160–1200				
Golden Delishes (control)	140	85.0	8	7
Golden Delishes, clone B	150	87.0	8	5
Golden Raingers	145	85.0	7	8
Red Delishes Hapke	165	90.0	4	6
Red Chif Kamspur	155	85.0	6	9
Super Chif Sandidzh	160	87.0	7	6
Early Red Wang	153	84.0	9	7
Zheromin	155	82.0	9	9
Breburn	157	80.0	11	9
FudzhiKiku	160	85.0	9	6
Grenni Smith	155	80.0	9	11
LSD_{50}	15.0			

Note: LSG is Least Significant Difference

The largest, one-dimensional fruits are received at a cultivar 'Grenni Smith', green fruits reminding 'Rennet Simirenko'.

Other studied apple tree cultivars also have high-quality fruits. Therefore, fruits in the super-intensive gardens are characterized by high commodity rates.

Apparently, fruits, estimated by weight and commercial qualities, exceed the fruits of varieties grown in foothill, forest and mountain areas, where usually about 90% or higher fruits have the highest and first levels of assessment and less than 10% have a third quality. Here, it should be noted that late winter cultivars such as 'Breburn', 'Grenni Smith', and 'FudzhiKiku' in a mountain-steppe zone had much more fruits with low quality, whereas in a steppe zone, they have not badly proved.

We have conducted researches on identification of influence of vertical zonality in terms of cleaning of fruits of an apple tree in the super-intensive lane-dwarfish gardens by method of determination of the density of pulp. Table 12.15 demonstrates that terms of removing of fruits of the same cultivars at different heights above sea level were different and depended on a growth zone that was also confirmed by the carried-out tests for pulp density. We fixed date—numbers of months when the measured density of pulp coincided with density index recommended for a removing, which served as a signal in farms to start harvesting.

Distinctions in start dates of a removable maturity of the studied winter cultivars are essential, both on cultivars within one zone and within a cultivar depending on a cultivation zone. The gradient of delay of the beginning of maturing of fruits on all cultivars has made 3.1 days on each 100 m of rise above sea level. Therefore, for example, the cultivar 'Golden Delishes' ripened in a foothill zone in 7–8 days after a steppe zone, in forest and mountain in 5–6 days after foothill, and in a mountain-steppe zone in 16–17 days after forest and mountain.

It is established that the difference in terms of maturing of this cultivar between the lowest in the conditions of vertical zonation concerning height above sea level in a steppe zone and the highest mountain-steppe makes about 1 month (see Table 12.15). In separate years, no ripening in a mountain-steppe zone at the height of 1200 m above sea level of fruits of an apple tree of cultivars 'FudzhiKiku' and 'Grenni Smith' has been noticed.

For comparative study of cost-effectiveness analysis of cultivation of three types of gardens of an apple tree of various intensity, we will provide aggregated data on the example of a cultivar 'Golden Delishes' who met all types of plantings.

Table 12.16 confirms earlier noted regularity and specify preferential performance indicators of cultivation of all types of gardens in forest and mountain and foothill zones. Here it should be noted that in Kabardino-Balkarsk Republic, intensive gardens on the terraced slopes are found only in a forest and mountain fruit zone and data on productivity are taken as averages on expositions and parts of slopes.

TABLE 12.15 Approach of Terms of a Removing of Fruits of an Apple Tree in the Super-Intensive Plantings Depending on Height Above Sea Level (Average Data for 2011–2013).[6-8]

Cultivar	Cultivar indexes of density of pulp recommended for a removing of fruits (kg/cm2)	Dates of approach on zones of optimum terms of a collecting of fruits according to a pulp density index; number, month			
		188–250 m/ steppe	430–485 m/ foothill	533–830 m/forest and mountain	1160–1200 m/ mountain-steppe
Golden Delishes (control)	7.5–8.0	12.09	19.09	25.09	11.10
Golden Delishes, clone B	7.5–8.0	15.09	22.09	28.09	14.10
Golden Raingers	7.5–8.0	15.09	22.09	28.09	14.10
Red Delishes Hapke	7.0–7.5	4.09	11.09	17.09	3.10
Red Chif Kamspur	7.5–8.0	3.09	10.09	16.09	2.10
Super Chif Sandidzh	7.5–8.0	4.09	11.09	17.09	3.10
Early Red Wang	7.5–8.0	4.09	11.09	17.09	3.10
Zheromin	7.5–8.0	4.09	11.09	17.09	3.10
Breburn	8.0–8.5	22.09	29.09	5.10	21.10
FudzhiKiku	8.5–9.0	5.10	12.10	18.10	4.11
Grenni Smith	10.0–10.5	1.10	7.10	13.10	29.10

TABLE 12.16 Comparative Economic Figures of Production of Apples of a Cultivar 'Golden Delishes' in Gardens of Various Type of Intensity in the Conditions of Vertical Zonality of Kabardino-Balkar Republic.[8]

Planting type, number of trees on 1 ha	Average productivity with 1 ha (t)	Average selling price of 1 t (thousand rubles)	The cost of products from 1 ha (thousand rubles)	Production expenses, everything (thousand rubles)	Net income (thousand rubles)	Prime cost of 1 cwt of production (rubles)	Level of profitability (%)
			Steppe zone				
Intensive 1000	23.6	17.0	401.2	122.0	279.2	516.9	228.9
Super-intensive 3175	40.4	23.0	929.4	250.0	679.4	618.8	271.8
			Foothill zone				
Intensive. 1000	31.4	19.0	596.6	126.0	470.6	401.3	373.5
Super-intensive 3175	47.9	25.0	1197.5	255.0	942.5	532.4	369.6
			Forest and mountain zone				
Intensive 1000	37.3	20.0	746.0	131.0	615.0	351.2	469.5
Super-intensive 3175	48.6	25.0	1215.0	255.0	960.0	524.7	376.5
Intensive on terraced slopes 833	29.7	22.0	653.4	123.0	530.0	414.1	431.0
			Mountain-steppe zone				
Intensive 1000	17.2	15.0	258.0	108.0	150.0	628.0	139.0
Super-intensive 3175	33.2	20.0	664.0	245.0	419.0	737.9	171.0

Note: cwt—hundredweight center.

On gained net income (profit) from 1 ha of a garden in all fruit zones to Kabardino-Balkar Republic, the leading position is taken by super-intensive lane-dwarfish gardens, followed by intensive buttressness gardens and in a forest and mountain zone on the third place intensive gardens on the terraced slopes. On profitability level, super-intensive gardens exceed intensive only in steppe (for 43.0%) and mountain-steppe (for 32.0%) zones and slightly yield to intensive gardens in foothill (for 4.0%) and intensive (for 93.0%) and terrace gardens (for 54.5%) in forest and mountain zones (see Table 12.15).

12.5 CONCLUSION

1. For rational use of favorable agroecological conditions and optimization of placement of fruit crops taking into account vertical zonality in the territory of Kabardino-Balkarsk Republic, it is expedient to use for increase of productivity of plantings winter cultivars of an apple tree which should be put on northern and western expositions of slopes in their lower tiers and autumn and summer cultivars of an apple tree in the top parts of eastern and southern slopes. In the conditions of vertical zonality, it is more preferable to make development under gardens in the following sequence: to put generally winter cultivars of apples in foothill and forest and mountain zones, in a steppe zone— autumn cultivars of apples, and in a mountain-steppe zone, summer and autumn cultivars of apples.

2. For receiving the greatest profit from unit area in all types of gardens, it is necessary to give to crowns the spindle-shaped form at the studied optimum density of landing. 'Red Chif Kamspur', 'Super Chif Sandidzh', 'Early Red Wang', and 'Zheromin' with compact kroner are offered to put spurovs, poorly tall cultivars of an apple tree on a dwarfish stock of M9 with consolidation to 0.5–0.7 m, or at the scheme 3.5 × 0.9 m on semi-dwarfish stocks of M26, CK2, and MM106.

3. When laying gardens on the slopes, including on the terraces of the foothills, forest and mountain areas of the Kabardino-Balkaria Republic, you should first of all use Florina, Golden Resistance varieties, which have high resistance to diseases, including scab.

KEYWORDS

- vertical zonality
- potassium

REFERENCES

1. Dragavtsev, A. P. Apple-Tree of Mountain Dwellings. Ecology and Features of Cultivation on the Example of ZailiyskyAla Tau. Moscow, 1956; p 252 (in Russian).
2. *Program and Technique of a Cultivar Investigation of Fruit, Berry and Nut Bearing Crops.* Michurinsk, 1973; p 492 (in Russian).
3. Dyadchenko, D. G.; Sheykina, T. V. Economic Evaluation of Cultivars. Program and Technique of a Cultivar Investigation of Fruit, Berry and Nut Bearing Crops. Orel, 1999; pp 235–246 (in Russian).
4. Potapov, V. A.; Faustov, V. V.; Pilschikov, F. N. *Fruit Growing Kolos*; Moscow, 2000; p 340 (in Russian).
5. Potapov, V. A. A Technique of Researches and Variation Statistics in Scientific Fruit Growing, Problems and Solutions: Collection of Reports. The International Scientific and Methodical Conference on March 25–26, 1998. Michurinsk. 1998, *1*, 7–15 (in Russian).
6. Bakuyev, ZhH. *A Gardening Intensification in the Foothills of Kabardino-Balkar Republic.* Print Centre Publishing House: Nal'chik, 2012; p 360 (in Russian).
7. Berbekov, V. N.; Bakuyev, ZhH. Ways of an Intensification of Gardening in the Conditions of the Foothills of the Central Part of the North Caucasus. Collection of Scientific Works, Volume XX Fruit growing and Berry Growing of Russia. Moscow, 2008; pp 274–277 (in Russian).
8. Berbekov, V. N.; Bakuyev, ZhH; Gagloyeva, LCh. Intensive Gardens in the Conditions of Vertical Zonality of the Central Part of the North Caucasus. *Monograph.* Print Center Publishing House: Nal'chik, 2016; p 153 (in Russian).
9. Luchkov, P. G.; Kudayev, R. H.; Bakuyev, ZhH.; Kalmykov, M. M.; Beslaneev, B. B. Clonal Stocks in an Apple-Tree Intensification on Slopes of the Central Part of the North Caucasus. *Gard. Wine Gro*w. 2003, 3, 4–5 (in Russian).
10. Luchkov, P. G.; Kudayev, R. H.; Bakuyev, ZhH.; Beslaneev, B. B. Increase of Efficiency of Mountain and Foothill Gardening in the North Caucasus. *Bull. Russ. Acad. Agric. Sci.* **2005,** *4,* 36–39 (in Russian).
11. Gon, K. M.; Pearson, D. R.; Daly, M. J. Effects of Apple Orchard Production Systems on Some Important Soil Physical, Chemical and Biological Quality Parameters. *Biol. Agric. Hortic.18,* 2001, *3,* 269–292.

CHAPTER 13

HAZELNUT (*CORYLUS DOMESTICA* KOS. ET OPAL.) RESEARCH AND BREEDING AT NATIONAL DENDROLOGICAL PARK (NDP) "SOFIYIVKA" OF THE NATIONAL ACADEMY OF SCIENCES (NAS) OF UKRAINE

IVAN S. KOSENKO[1], ANATOLY I. OPALKO[1,2,*], OLEKSANDR A. BALABAK[1], OLGA A. OPALKO[1], and ALLA V. BALABAK[2]

[1]*National Dendrological Park "Sofiyivka" of NAS of Ukraine, 12-a, Kyivska St., Uman, Cherkassy Region, 20300, Ukraine, E-mail: ndp.sofievka@gmail.com*

[2]*Uman National University of Horticulture, 1, Instytutska St., Uman, Cherkassy Region, 20305, Ukraine*

[*]Corresponding author. E-mail: opalko_a@ukr.net

CONTENTS

ABSTRACT

The necessity of assortment enrichment of the hazelnut, also known as filbert (sometimes called cobnuts), as one of the most valuable nut crops, is associated with the need of search and selection of sources and donors of scarce signs for breeding and stable crop capacity of industrial plantations as well as the improvement of the quality of domestic hazelnut cultivars, especially concerning the form of nut kernel. As the result of the genetic collection *Corylus* spp. of the National Dendrological Park (NDP) "Sofiyivka" of National Academy of Sciences (NAS) of Ukraine screening and according to economic-valuable features, a set of samples was taken, and the best of the samples were included in the hybridization program, in particular with representatives *Corylus chinensis* Franch. A number of hybrid seedlings were received, and the best of them were prepared to be submitted to the state registration of the State Veterinary and Phytosanitary Service of Ukraine, in particular, new hazelnut cultivars: 'Sofiyivsky 1'; 'Sofiyivsky 2'; and 'Sofiyivsky 15', which develop fruits with a nearly globular shape, are characterized by an increased winter hardiness and drought tolerance, and lack of nut-bearing periodicity, compared with Turkish and Azerbaijan cultivars. Having summed up the results of the research, a new breeding scheme was made, which helped increase the passing of breeding material through the stages of a breeding scheme for 5–8 years, namely, at the growing stage of F_1 seedlings for 3–4 years, in a hybrid orchard for 1–2, and at the propagation stage of the best seedlings with layering for 1–2 years.

13.1 INTRODUCTION

With the development of intensive fruit production, cultured forms of *Corylus* spp., which are grown under the name of hazelnut, also known as filbert (sometimes called cobnuts), become more important as nut crops. Although, hazelnut takes an important place among nut-bearing crops in the world, in Ukraine it is not grown widely, which is why the need in the nuts is satisfied only by 12–15%. The rest is covered by the import, which causes high prices of the nuts themselves and their processed products, as a whole and/or crushed hazelnuts are in various confectionaries: cakes, cookies, deserts, baked things, chocolate, candies, halvah, and other tasty output. Besides, hazelnut kernels contain phospholipids in the amount enough for its economically efficient use in a technological process of getting lecithin, liposomal nano-emulsions and nano-dispersions for functional foodstuff and feed, as well as for the manufacturing of liposomal medicinal preparations. Although, so far, walnut is the most popular crop in Ukrainian orchards and homesteads, and in the regions with warm winters, almonds are grown in small areas. Due to the breeding efforts, the tendency of the considerable increase in hazelnut cultivation becomes quite real.

Simultaneously, and sometimes even faster than in fruit production, representatives of *Corylus* spp. are significant in ornamental horticulture. At present, in the green plantation of the cities, *Corylus colurna* L. and *Corylus maxima* Mill. are widely used; other species are kept mostly in collections of botanical gardens, dendrological parks, and arboretums. However, the experience of the cultivation of many ornamentally valuable forms of North American and East Asian species confirms promising opportunities for the introduction of *Corylus* representatives in ornamental horticulture. Due to the resistance of *Corylus* spp. to unfavorable environmental factors, some of its species and cultivars can be successfully used to reinforce slopes and in various protective plantations.[1]

In natural flora of Ukraine genus *Corylus* L. is represented with only one species *Corylus avellana* L.[2] The researches of classification/taxonomy, phylogeny, and biogeography of *Corylus* spp. carried out at the end of the previous century with the use of deoxyribonucleic acid (DNA) successions of some *Corylus* representatives, enabled to specify the connections between the localities of Old World and North American areas.[3–6]

The genetic affinity of the species of subsection *Siphonochlamys,* namely North American species *Corylus cornuta* Marsh. and *Corylus californica* (A. DC.) Rose, the latter is classified as a subspecies *C. cornuta* (*C. cornuta* ssp. *californica* (A. DC.) A. E. Murray), with East Asian species *Clematis mandshurica* Maxim., which is considered to be a synonym to *Corylus sieboldiana* var. *mandshurica* (Maxim.) C. K. Schneid., allows to assume ancient migration between East Asia and North America in the Paleogene and Neogene Periods probably through Bering Strait (Fig. 13.1).

Not less significant is the data concerning the migration of *avellana*-like species, which resulted in the formation of current areal of *Corylus* (Fig. 13.2). The generalization of the published results gives every ground to confirm the availability of three localities of *Corylus* spp.: North

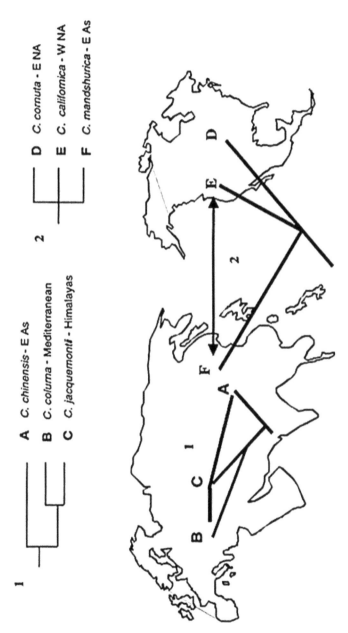

FIGURE 13.1 Key scheme of biogeographic interspecific connections in section *Corylus*: **A**—East Asia, **B**—Mediterranean area, **C**—Himalayas, **D**—East of North America, **E**—West of North America, **F**—East Asia, **1**→biogeographic connections between East Asian, Mediterranean, and Himalayan species of section *Corylus* and subsection *Colurnae*, **2**→intercontinental bridge of the connections of species of section *Corylus* and subsection *Siphonochlamys*

Source: Adapted from ref 6.

American, Minor-Asiatic–European, and East Asian.[2] The last two are connected by a narrow isthmus along the mountains of Iran, Afghanistan, and the Himalayas. Thus, the areal of the genus is not disconnected in Old World. Similar regularities are typical for some species of *Corylus*. The results of studying Neogenic flora of East Asia confirm that subtropical vegetation existed in Neogene in the Manchurian floristic region, as well as everywhere in Europe and Siberia. One can assume that *Corylus* areal in Neogen was much wider and covered large areas of the whole Europe, Siberia, North America, and East Asia, including Japan. At present, 20 excavated *Corylus* spp. are known. Excavated remains of *Corylus* spp. were found in Pliocene deposits in Middle Europe and in preglacial deposits of anthropogenesis in North Germany and England.[1]

C. avellana is the youngest and the most widely spread species among all the contemporary species of *Corylus*. Its habitat covers the whole Europe, except for far north and partially a steppe part, and also the Asia Minor and Caucasus. It is possible that Transcaucasia and Black Sea coast of Asia Minor were the center of *C. avellana* formation, as Transcaucasia is characterized with the diversity of its forms, and Asia Minor is home for this species cultivation. Besides, numerous varieties of *C. avellana* are originated from the abovementioned regions.[1,2]

Latin name of *Corylus* genus has been known since ancient times and has been used in Vergil's works. That was the way how ancient Romans called wild hazelnuts. Pliny was the first to call hazelnuts as *avellana*. He used a specific name *avellana* as a derivative from the name of an Italian city Avellino in Campania, where *Corylus* spp. was grown widely from ancient times. In 1751, Carl Linnaeus used *C. avellana* as a scientific name for hazelnuts.[2]

The extinction of Neogenic *Corylus* spp. during a glacier period explained the existence of *vicarious species* of hazelnut in Manchurian floristic region. For instance, *C. avellana* is widely spread in Europe and Asia Minor, but not in Siberia. In Northeast China, Amur region, Korea, and Japan, it is replaced with *Collinsia heterophylla* Fisch. ex Trautv, which has a very similar leaf and fruit shape to *C. avellana*. Manchurian hazelnut *C. sieboldiana* var. *mandshurica* (former name *C. mandshurica*) is cultivated in Northeast China, coastland, Korea, and Japan, and it is *Neogenic relict*, which is poorly adjusted to its current areal. Based on this fact, L. A. Smolianinova[7] considers Manchurian hazelnut to be an endangered species. Contrary to this, *C. heterophylla* is a frost-resistant

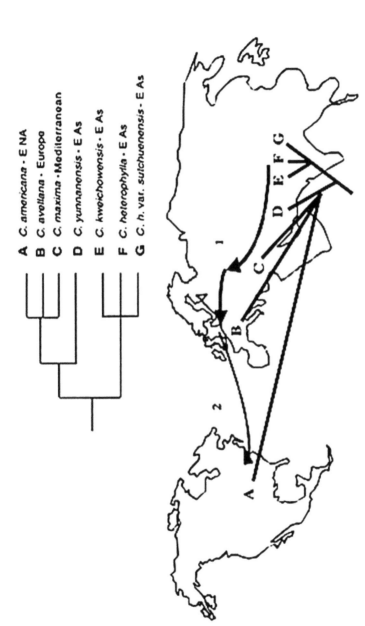

FIGURE 13.2   Key migration scheme of the representatives of section *Corylus* and subsection *Corylus*: A—North America, B—Europe, C—Mediterranean area, D, E, F, G—East Asia, 1→migration route from Asia across the Mediterranean to Europe, 2→migration route from Europe across North Atlantic to North America.

Source: Adapted from ref 6.

oligotroph. One can assume that *C. heterophylla* belongs to those woody plants which, after the first and the second icing, migrated from Arctic regions to northern parts of Asia.[1]

In Northeast China, there are several *Corylus* spp. close to Japanese and North American ones; this fact confirms a direct connection of America with Asia in the geological past. For example, the abovementioned Manchurian hazelnut is close to North American *C. cornuta, C. brevituba* Kom.—to *C. californica* (now *C. cornuta* ssp. *californica*), and *C. heterophylla*—to *Corylus americana* Walter.

Hence, it has been proved that current *Corylus* spp. habitat is presented by the remains of a large area of this species in Neogene.[1,2]

Taking into consideration the data concerning genesis and habitation of *Corylus* spp., it is possible to define phylogenic connections of *Corylus* spp. within the genus. Until recently, the scheme of Ye. H. Bobrov[8] has been considered to be the most complete as 19 *Corylus* spp. were divided into three subgenera: *Acanthochlamys, Phyllochlamys,* and *Siphono-chlamys,* typifying the genesis stages of *Corylus.* According to the scheme, the oldest species—*Corylus ferox* Wall. and *C. tibetica* Batalin—belong to subgenus *Acanthochlamys.* Their localities are separated from other Asian species, which makes an impression of phylogenetic not-belonging of *C. ferox* and *C. tibetica* to the rest of the representatives of the Asian group of *Corylus.*

The cited data,[6] received at the end of the previous century as the result of the analysis of DNA sequence, were used to specify the issue of classification and molecular phylogeny of *Corylus* spp. In general, these novelties do not deny the above-cited conclusions of Ye. H. Bobrov—they amplify interspecific connections of *Corylus* spp. However, to single out primitive (from the point of view of domestication) *C. ferox* species into a separate section *Acanthochlamys,* as well as to include it in subsection of *Colurnae* spp. with a life form "tree" help better understand species phylogeny and make more reasonable selection of initial material for hazelnut breeding.

Due to this, the availability of two sections within a genus was found out:

- *Acanthochlamys,* which includes a base species *C. ferox* and probably *C. tibetica* (*C. ferox* var. *tibetica* (Batalin) Franch.)
- *Corylus,* which incorporates the rest of the species

At present, three subsections are classified in *Corylus* section:

- *Corylus* (bush species with leafy, more or less bell-shaped involucres)
- *Colurnae* (species of a life form "tree" with deeply dissected involucres)
- *Siphonochlamys* (bush species with tubular involucres)

Populations of *C. avellana*, *C. maxima*, and *C. avellana* var. *pontica* (former *Corylus pontica* K. Koch) have been grown as nuts since the ancient times. *C. americana* was introduced into a culture much later; however, in cultivar diversity of this species, there are forms which are similar to *C. avellana*, *C. maxima*, and *C. avellana* var. *pontica* representatives (by size and shape), cultured long time ago. Further crossings between them as well as with *C. colurna*, *C. Americana*, and other *Corylus* spp. gave quite a number of modern hazelnut cultivars, which is a contemporary name for cultured cultivars of *Corylus* spp. Yet, it is difficult to conclude the share of heredity of each abovementioned species in the formation of full hazelnut genotype;[1,9] therefore, it will be logically correct to combine all the cultivated cultivars in one general species.[2]

In the national registry of the cultivars suitable for cultivation in Ukraine in 2016 (as in force of February 22, 2016),[10] there are only three Italian hazelnut cultivars—'Halle', 'Cosford', and 'Barcelona'. We find them in chapter "Fruits and small fruits" in a group of cultivated plants under general title "*Corylus maxima* Mill.," in English—"Hazelnut," which cannot be considered to be correct. Hazelnut is a general Americanized name popular in the United States, which was used for local *Corylus* spp. from the times of first settlements; it received its official status after 1981 when "Filbert board" in the state of Oregon decided to standardize it. And in biological dictionaries, the Ukrainian word "Funduk (*C. maxima*)" is translated into English as "hazelnut" and "filbert." In Great Britain, filbert is used to denominate both a tree and a nut. In addition, Turkey, which is the world's major supplier of hazelnuts, generally uses "Findik" as the name of hazelnut and "Hazelnut" is used in published works. Such nomenclative divergence confirms the necessity to unify it, which can be accomplished by overall introduction of general specific name *Corylus domestica* Kos. et Opal, suggested by us in 2007[11] for all cultivated cultivars of *Corylus* spp., and which became more popular in scientific publications.[2,4,12–14]

Several researchers believe that *C. colurna* (Ukrainian name "Bear nut wood," also known as Turkish Hazel) takes an intermediary place between European and American–Asian species. Some species with leafy involucres most likely originate from this woody species, and they grow in Asia, America, and Europe—*C. heterophylla, C. americana,* and some species, spread in Europe—*C. maxima, C. avellana* var. *pontica* (former *C. pontica* K. Koch), and also, probably *Corylus colchica* Albov.[1]

All *Corylus* spp. that were studied by the cytologists usually have diploid number of chromosomes ($2n=22$); however, there is an information about other chromosome numbers in somatic cells of some representatives of *Corylus* sp.[1] First of all, it is about the publication of Robert H. Woodworth,[15] who, having studied meiotic preparations *C. americana, C. colurna, C. cornuta, C. heterophylla* var. *sutchuensis* Franch., *C. pontica* (now it is a synonym of *C. avellana* var. *pontica* (K. Koch) H. J. P. Winkl.), *C. sieboldiana* та *C. vilmorinii* Rehd., counted 14 chromosomes in them, which proves $2n=28$. The article, published in prestigious botanical magazine "Botanical Gazette" in 1929,[16–18] caused long-term uncertainty as to diploid chromosome number in the mentioned *Corylus* sp. Some authors, referring to R. Woodworth, stated possible $2n=28$ together with normative $2n=22$,[16,17] others emphasized the necessity to experimentally confirm the calculation as to $2n=28$,[19] whereas Veli Erdogan,[20] from the very beginning, called R. Woodworth's calculations false because of wrong interpretation of meiotic preparations. The error could occur when imperfect methodology of cytological research was used, and the difficulties were explained by very small *Corylus* spp. chromosomes. At present, $2n=2x=22$ is accepted as a norm for *Corylus* spp.[1,20–24] except for some cases of chromosome aberrations and polyploidy,[21,22] Almost all representatives of *Corylus* are diploids, but Robert Botta with coauthors (1986) informed about spontaneous tetraploid *C. heterophylla.*[21] Among other examples of unusual number of chromosomes, it is worth mentioning aneuploids with $2n=18$ in somatic tissues of *C. colurna* and *C. maxima* f. *atropurpurea* (Dochnahl) H. J. P. Winkl.[18,23] and also tetraploids induced by colchicine treatment, and spontaneous triploid seedlings *C. avellana.*[17,21]

For a long period of time, hazelnuts were harvested mainly in wild forests. Cultivars of *Corylus* spp. were first domesticated in Turkey over 2000 years ago. At present, hazelnut production in this country is well developed. It is important to state that 400,000 ha of hazelnut plantations support 250,000 Turkish families, and production, processing, and exporting of nuts provide 8 million jobs. Export traditions of this country are also well known. Turkey has been a world leader of hazelnut markets for over 600 years.[25] At the beginning of the 21st century, world hazelnut production was over 900,000 metric t. It ranks second after almond production (*Prunus dulcis* [Mill.] D. A. Webb). The share of Turkey in this quantity of hazelnut is 70–75%, annual production of unshelled nuts being 450,000–800,000 metric t. Italy ranks second with the production of 100,000–130,000 metric t, the United States and Azerbaijan rank third and the fourth, respectively, and their annual production is 20,000–35,000 metric t. Then comes Georgia—25,000–30,000 t, China and Iran—18,000–25,000 t, and Spain; the latter decreased its production to 12,000–17,000 t. Contrary to this, France has doubled the production recently, and it reached 8000–10,000 t, compared with the indicator of the previous 5-year period—4000–5000 t; Poland and Kirghizstan produce 3000–4000 metric t and Croatia—about 2000 metric t.

All hazelnuts can be shelled and sold as raw kernel. Kernels of hazelnut are sold better at the world market, and their quantity ranges within 200,000–300,000 t per year depending on the supply and demand. At present, world production of shelled nuts amounts to 500,000 metric t. Turkey is the leader in this industry—360,000 t, which makes 76% of the world production, then comes Italy—40,000–45,000 t, Azerbaijan—about 20,000 t, Georgia—15,000–17,000 t, the United States—12,000–15,000 t, Spain—less than 10,000 metric t; the total share of other producers is slightly more than 12,000 metric t.

Ukraine ranks 30th in the list of producers of hazelnuts and its production is 20 t,[26] whereas, in the 1990s, this indicator was six to eight times higher. Taking into account the fact that our long-term researches prove the feasibility of successful hazelnut production almost all over the country, provided proper organizational measure are taken, including the development of processing industry and the improvement of local cultivar genotypes,[1,25,27] the current condition of hazelnut production in Ukraine has to be classified as unsatisfactory. In Ukraine, the area of hazelnut plantations at the farms of all categories and forms of ownership does not exceed 1000 ha. Hazelnut yield on these plantations is 0.18–0.43 t/ha on an average, including 0.01–0.13 at farm enterprises, 1.10–3.15 t/ha in households. In recent years, the number of farmers who try to develop hazelnut plantations has increased. The progress in these positive tendencies is hindered by the lack of planting material. In the near future, the deficit will be overcome, but insufficient development of processing industry is expected to be a bigger problem.[27]

Annually, Ukraine imports 2800–3100 t of hazelnuts, having average annual production of about 20–40 t. There is no official statistics regarding the harvesting of wild *Corylus* spp. nuts. In other words, at present, only 1% of internal needs of Ukraine in hazelnuts is satisfied (domestic production and import combined together), that of internal demand—25–30%. Hazelnut deficit is partially compensated with peanut, which worsens the quality of local confectionaries and reduces the chance to enter foreign markets. Having analyzed the dynamics of general nutrition culture, and in turn, the consumption of hazelnut products, one can predict the increase of internal demand ranging 2–3% per year, which will result in average annual demand equal to 11,000 metric t in 2017–2021, and in 2022–2026—over 12,000 t of hazelnut nuts.[25,27]

Based on the abovementioned data, the aim of the research is to make screening of the existing hazelnut cultivars of local and foreign breeding as to their adaptation to the cultivation in agroclimatic conditions of Ukraine and its adequacy to producers' and food processors' requests and those of consumers of nuts and processed products, to involve the best of them into a breeding process aimed at developing new cultivars, and in turn, to work out a breeding scheme, which will make it possible to speed up the passing of breeding material through the stages of the breeding scheme by 5–8 years.

13.2　MATERIALS AND METHODOLOGY

Economic-valuable characteristics of 165 cultivar samples and form hybrid composition of genetic collection of *Corylus* L. of National Dendrological Park (NDP) "Sofiyivka" of National Academy of Sciences (NAS) of Ukraine, which are tested in the plots of a collection and hybrid orchard and the orchard of primary cultivar studying, as well as kernels in terms of standard 14% humidity, were studied with help of common methodology.[28,29] The choice and separation of branches with female flowers, interspecific hybridization, and crossing of hazelnut cultivars (*C. domestica* Kos. et Opal.) with representatives of Chinese hazelnut *Corylus chinensis* Franch. were done at the beginning of spring development of male inflorescence (before pollen release). Pollination was performed without removing the separated, untying only an upper part of a sleeve and tying it again after pollination and/or without untying the separated, injecting pollen into the separated with insufflator MO-03, and repeated pollination was performed 2–3 days later. The technology of hazelnut cultivation in a harvest and hybrid orchard and the orchard of primary cultivar testing is the one commonly used in the Forest Steppe zone of Ukraine. The number of general lipids was determined by extracting from kernels with diethyl ether in Sokslet device (GOST 10,857-64),[30] acid and peroxide value of oil— by conventional methodology.[31] Phospholipid content was determined by gravimetric method, which consisted in settling down phospholipids with acetone from lipid extract according to Folch method.[32] A statistical analysis of the received data was made with help of a disperse analysis[33] by using statistical software of Microsoft Excel.

13.3   RESULTS AND DISCUSSION

The evaluation results of hazelnut cultivars of domestic and foreign breeding with reference to the complex of economic-valuable characteristics proved the advantages of cultivars 'Dohidnyi', 'Funduk-85', and 'Bolhradska novynka', which increased their yield by 4.5–6.4 times (from 61.6–179.2 to 3948–8092 kg/ha) during the first 3 fruiting years, and oil content in kernels was rather high (68.7–75.8%). In addition, the yield capacity of the studied hazelnut cultivars depended on cultivar peculiarities as to the start of stable fruiting rather than on meteorological conditions of a vegetative period. In addition to the mentioned cultivars such as 'Dohidnyi', 'Funduk-85', and 'Bolhradska novynka', early fruiting was typical for 'Dar Pavlenka', 'Zorynskyi', 'Zuidivskyi', 'Karamanivskyi', 'Stepovyi', and 'Shedevr'; their yield capacity increased by 1.7–4.5 times during the first 3 fruiting years, while high oil-bearing cultivar 'Hrandioznyi' appeared to be inferior to high-yielding 'Dohidnyi' by 171.4 kg/ha in the first fruiting year, and in the third year—by 414.4 kg/ha.

The indicators of oil content in hazelnut kernels of the studied cultivars changed slightly depending on meteorological conditions of the experimental years, but general tendency of cultivar specificity remained similar. With reference to oil content, cultivars such as 'Urozhainyi-80', 'Funduk-85', and 'Hrandioznyi' were the best with the average indicators of 74.5, 74.3, and 74.2% in 2012–2014, respectively. Thus, all the studied cultivars can be included into the group of high oil-bearing crops, except for 'Lozivskyi urozhainyi' and 'Dar Pavlenka'.

In 2012–2014, on an average, 'Dohidnyi' had the highest oil content in its kernels (197.1 kg/ha). High indicators of average oil output per hectare were recorded in 'Funduk-85'—165.4, and 'Bolhradska novynka'— 148.1 kg/ha. The indictors of the rest of the cultivars were much lower with interspecific variability ranging from 39.5 ('Hrandioznyi') to 110.3 kg/ha ('Stepovyi').

Having considered the evaluation results and the findings of previous researches,[1,25,27,34] a breeding task with respect to hazelnut for Ukraine for the near future was determined by the following parameters:

- Potential yield capacity—3.5 t/ha
- Nut mass/weight—3.0–3.5 g
- Kernel yield—at least 50% of the total fruit mass
- Shell thickness—not more than 1.0 mm
- A number of nuts in a collective fruit—more than four fruits.

Taking into account these and other important parameters of the assortment improvement, the following major trends in hazelnut breeding were identified:[34]

- To increase winter resistance by developing forms with earlier transition to winter dormancy in the fall and longer deep dormancy in winter
- To increase winter resistance of generative buds (in particular, male inflorescences) and to develop cultivars with late flowering terms
- To develop cultivars with annual high yield of large nuts
- To make harvesting easier, to separate pericarp from the shell, and to take kernel out of the shell without damaging it
- To increase a kernel share in a nut mass
- To improve kernel quality (clean surface, resistance to shrinkage, pleasant taste and flavor, suitability for long shelf life)
- To develop cultivars with programmed fruit characteristics, suitable for precooking, roasting, and making some confectionaries
- To improve resistance to pathogens of bacterial and fungal diseases and pests, first of all, powdery mildew and bud mite
- To develop self-fertile cultivars, which can be grown in one-cultivar plantations

- To develop cultivars with various harvesting terms but with synchronous flowering (to ensure cross-pollination)
- To develop cultivars whose bushes do not need thinning

The filling of a nut increases the share of kernel, which is a positive sign; however, fitting density to a shell also increases, and crushing will most likely occur when taking a kernel out. It is expedient to develop cultivars with good filling, but there should be some space between a shell and a kernel for better separating of the latter.[1]

As vegetative/clonal propagation of nuts is preferable than seed one, plant breeders face the task to develop scions, which will be compatible with rootstocks and resistant to disease and pest agents that destroy roots.

Along with general trends, there are some specific tendencies in breeding. It is the development of the cultivars which form fruits of a nearly globular shape, with smooth clean kernel without pellicle shucks. To reduce the number of empty (without kernels) nuts is a breeding and agrotechnical task. The selection of nut filling and self-fertility, on the one side, and the choice of good pollinators, on the other side, will facilitate the decrease of empty nut percentage in newly developed and existing hazelnut cultivars.[1]

The shell thickness of hazelnut is important (not more than 1 mm) as well as the easiness of taking a nut out of involucre. It becomes even more important when breeding is aimed at developing cultivars suitable for mechanical harvesting. To perform breeding and to plan the crossing, the preference was given to the samples with early maturation of involucres, which open and release nuts in time.

Hazelnut cultivars are successfully propagated by old, well-tested ways—layering, root sprouts, and bush division. Therefore, they are own-rooted plants in most cases. However, it becomes a frequent practice to graft hazelnut on woody hazelnut, which makes the issue of rootstock breeding significant.[12]

In hazelnut breeding, there is a necessity to develop special cultivars-pollinators, which would form big catkins with a large amount of pollen with such set of *S*-genes that will not block pollination; this will allow reducing the number of plants-pollinators on the plantation.[1] Besides, there is statistics that nut quality depends on the peculiarities of pollinators, for instance, xenia (metaxenia—the influence of pollen on maternal tissue of the fruit) occurs on pellicle on kernels formed from fertilization with *C. cornuta* pollen.[2,35,36]

Until recently, the choice of natural populations has been the major method in world hazelnut breeding practice, due to which most of the cultivars were developed, and they constitute the industrial assortment of many regions of hazelnut cultivation. The cultivar of folk breeding, 'Cherkeskyi-2', is grown on 90% of the industrial plantation of Russia's coastland. Its usage in the crossing of both mother and parent component led to receiving such high-quality cultivars such as 'Zorynskyi', 'Bythyn-skyi', 'Tuapsynskyi', 'Yuvileinyi', and others. Hazelnut 'Tombul' (round), as well as 'Fosa', 'Mincane', and 'Karafindik' dominate in Black Sea coast of Turkey. 'Tonda Gentile delle Langhe' is the most important cultivar in the Piedmont region of Italy; cultivars 'Gentile Romana' and 'Tonda di Giffoni' originated from it. 'Mortarella' belongs to universal Italian cultivars. In France, the most valuable old cultivars would be 'Fertile de Coutard' and 'Merveille de Bollwiller'. Hazelnut breeders from the United States used French cultivar 'Fertile de Coutard', named 'Barcelona', in wide crossings with local cultivar 'Rush' (*C. americana*) and also 'Italian Red'. The fact that 'Negret' dominates the industrial plantations in Spain is explained by its self-fertility.[2] In the last decades, the basis of industrial hazelnut culture in South Russia includes such cultivars as 'Cherkeskyi-2', 'Panaheskyi', 'Adyheiskyi-1', 'Ata-Baba', 'Romana', 'Hachapura', and 'Nemsa'.[1]

In a view of the fact that valuable cultivars such as 'Borovskyi', 'Dar Pavlenka', 'Koronchastyi', 'Lozivskyi kuliastyi', 'Raketnyi', 'Sere-brystyi' (Ukraine), 'Adyheiskyi 1', 'Tuapsynskyi', 'Yuvileinyi' (Russia), 'Gunslebert', 'Gustav's Zellernuss', 'Kadetten Zellernuss', and 'Louise' (Germany), and others were developed by using the method of progeny analysis, which resulted from sowing free pollinated seeds. In our breeding practice, we also used seedlings, received from seeds which were formed from free pollination of the best alien cultivars.

The seeds to be sown were taken from healthy plants of the most valuable cultivar samples, which were singled out for a set of charac-teristics corresponding to the parameters of a hazelnut breeding task for Ukraine. Mother plants were planted next to the cultivars without serious defects. Those were large fruit-west-European hazelnut cultivars which had both higher winter resistance and good yield capacity: 'Louise', 'Gunslebert', 'Gustav's Zellernuss', 'Cannon Ball', 'Merveille de Boll-willer', and others. The attempts were made to receive the combination of winter resistance with the resistance to disease and pest agents by making the choice in the progeny when seeds of free pollinated cultivars were planted—'Garibaldi', 'Gubener Barcelloner', and 'Mels'. Seedlings adapted to the conditions of Ukraine are received when planting the seeds from a large collection, created by a well-known Ukrainian plant breeder F. A. Pavlenko and also the cultivars of R. F. Kudasheva—'Pervenets', 'Moskovskyi rubin', 'Akademik Yablokov', and other cultivars of the Caucasian group—'Adyheiskyi-1', 'Cherkeskyi-2', 'Ata-Baba', 'Nemsa', and 'Chhykvistava'. According to F. A. Pavlenko's findings/experience, to plant seeds of the best West European cultivars and samples of Sochi's hazelnut population proves to be promising.

Although the majority of the existing hazelnut cultivars were selected and developed from free wind pollinated populations, intraspecific hybridization, which involves different cultivars, varieties, and forms into crossing, has always been and still is the main and the most effective induction method of genetic diversity, that is, the creation of initial material for selection.

As the requirements to the quality and productivity of new hazelnut cultivars increase, the importance of controlled crossings will increase as well. The hazelnut breeding scheme, worked out and suggested by us, includes such stages:

- 1st year—to choose pairs for hybridization and crossing of parental forms selected
- 2nd year—to grow seedlings of first generation F_1 in controlled conditions
- 3rd–5th years—to grow and evaluate seedlings of first generation F_1 in a hybrid orchard
- 6th year—organoleptic evaluation of nuts
- 7th year—integrated evaluation of nuts
- 8th year—planting layers, integrated evaluation of nuts
- 9th year—to regrow layers in a nursery, integrated evaluation of nuts
- 10th year—to lay the orchard of preliminary cultivar studying, integrated evaluation of nuts
- 11th–13th years—integrated evaluation of the chosen seedlings in the orchard of preliminary cultivar studying, preparation for the expertise with the aim to be included into the National registry of the cultivars suitable for cultivation in Ukraine

- 14th year—to give a name to a cultivar and to enter it into the national registry of the cultivars suitable for cultivation in Ukraine, widespread propagation

An accelerated passing of the selected material through the stages of the scheme for 3–4 years occurs at the stages of growing seedlings F_1 in a hybrid orchard (1–2 years) and speedup propagation of the layering of the best seedlings (1–2 years). Cultivar 'Sofiyivsky 15' started fruiting in the third year after crossing. Although in the first year, a seedling formed only female flowers, it did not prevent from processing organoleptic evaluation of the fruits, which developed from crossed free pollination.

Due to the fact that a wide spectrum of splitting as to economic-valuable characteristics was recorded in seed progeny from both controlled crossing and free pollination, promising forms from 44 seedling populations were chosen for next hybridization cycles and direct cultivar testing aimed at further use in polyclonal plantations where mutual pollination would be possible.

During the crossing, recombination of genes takes place in hybrid organisms, that is, characteristics and features of mother and parent specimens. However, the formative mission of hybridization is not limited by this. New features can result from crossing due to the interaction between allele and nonallele genes within a hybrid core, as well as the interaction between the complexes of core-inherited structures with plasmagenes. Among new formations, which occur in hybrid seedlings and progeny from splitting in the second and the following generations, transgressive genotypes are the most desirable ones as they surpass both initial forms and existing cultivars by a set of anthropo-adaptive features.[1,36] Transgressive segregation or highly favorable multiallelic segregation take place as a result of total effect of favorable polymer genes. However, there are no clear criteria that will help to predict the combination in which transgressive splitting is expected; therefore, it is a proper choice of crossing components that leads to hybridization success.

Biological features of *Corylus* spp. predetermine specific aspects of hazelnut breeding. Hazelnut, as well as other representatives of *Corylus,* belongs to typical monoecious intersex plants, which facilitates cross-pollination, and, in turn, ensures regular heterozygosity of specimens and heterogeneity of populations.[1] In creating initial material for hazelnut breeding, intravarietal hybridization is used. However, different species and varieties of *Corylus* spp. took part in the development of numerous cultivars; therefore, when crossing hazelnut, it is not always easy to differentiate between intraspecific and interspecific hybrids. To make a well-thought choice of pairs for hybridization, it is necessary to consider specific aspects of flowering and sexual reproduction in addition to the information about the hierarchy of allele interactions of incompatibility genes.

Among different mechanisms in the structure and functioning of generative organs which prevent self-fertilization, various forms of dichogamy are the most effective,[1,34] namely:

- Non-simultaneous maturation of anthers and pistils of androgynous/ hermaphrodite flowers and/or their non-simultaneous separation of generative organs within one plant (monoecism) or formation of intersex flowers on different plants (dioecy), and so forth
- Genetic systems of incompatibility, which block self-fertilization on gene level even in case of self-pollination

As a result, for good pollination, the flowering of female and male flowers should take place at the same time. The availability of natural clones in many *Corylus* spp. reduces the efficiency of monoecism as a mechanism of ensuring allogamy, but their inherent genetic system of self-incompatibility adds to monoecism and impedes geitonogamy within both one plant and clone.[1]

Thus, all wild *Corylus* spp., as well as most of the hazelnut cultivars, cannot set fruits from autogamy or geitonogamy.[18] There are partially self-fertilized genotypes, like Turkish cultivar 'Tombul', but 'Tombul' and other partially self-fertilized cultivars require cross-pollination to fully implement their productive potential. Geitonogamy can be observed in some introduced species which are represented mostly by single specimens in many parks. In 2002, we found a 140-year-old *C. colurna* fruiting tree (not regularly, though) in the village of Drabiv, Cherkassy region.[1]

Cross-pollination itself does not guarantee successful fertilization and good fruits. Cross-pollination with pollen of proper genotype is necessary to develop reproductive nuts. Sporophyte incompatibility is typical for the plants of all species of *Corylus*, which means that pollen germination is determined by the interaction of genotype of a plant source of pollen (sporophyte) and genotype of style tissue with stigma (also sporophyte).

Hazelnut incompatibility is controlled by S-gene, which can be in many allele conditions. At present, over 30 alleles of this gene are known. Two forms of allele interactions were identified—domination and codomination, and the hierarchy of the interaction between individual alleles of S-gene was also built.[1,37]

The availability of self-incompatibility genes, which not only blocks self-fertilization but can also cause cross incompatibility, predetermines some difficulties in working out a hazelnut crossing program and makes the procedure of choosing cultivar pollinators in hazelnut orchards very important, emphasizing mutual responsibility levels of alleles of self-incompatibility genes of the cultivars. To identify S-genes of any hazelnut cultivar, it is necessary to use cultivars-analyzers with known alleles of these genes, which is not always easy.

It was observed that for some hazelnut cultivars, which are alleles of self-incompatibility, S-gene prevented self-fertilization.[1] The cultivars are 'Barcelona', 'Gironell', 'Nocchione', 'Montebello', 'Fertile de Coutard', genotype—S_1S_2; Ennis—S_1S_{11}; 'OSU 20.058' (tester of Oregon University)—S_2S_2; 'Gem'—S_2S_{14}; 'Mortarella'—S_2S_{17}; 'Tonda di Giffoni'— S_2S_{23}; 'Corabel', 'Nonpareil', 'Willamette'—S_3S_1; 'Butler', 'Jemtegaard 5'—S_3S_2; 'OSU 194.001'—S_4S_4; 'Henneman No. 3'—S_6S_{10}; 'Ronde du Piémont'—S_7S_2; 'San Giovanni'—S_8S_2; 'Segorbe'—S_9S_{23}; 'Impériale de Trébizonde'—$S_{10}S_2$; 'Casina'—$S_{10}S_{21}$; 'Negret'—$S_{10}S_{22}$; 'OSU 278.121'— $S_{11}S_4$; 'Tombul'—$S_{12}S_4$; 'OSU 382.026'—$S_{12}S_{23}$; 'USOR 98–83'—$S_{13}S_6$; 'OSU 39.044'—$S_{15}S_{11}$; 'OSU 458.010'—$S_{16}S_{11}$; 'Mortarella'—$S_{17}S_2$; 'Pauetet'—$S_{18}S_{22}$; 'OSU 452.026'—$S_{19}S_4$; 'OSU 455.087'—$S_{20}S_9$; 'Uriase de Valcea'—$S_{20}S_{11}$; 'OSU 168.026'—$S_{21}S_2$; 'OSU 219.133'—$S_{22}S_4$; 'OSU 385.003'—$S_{23}S_4$; 'OSU 54.041'—$S_{24}S_4$; 'OSU 447.015'—$S_{26}S_{26}$, and others.[1,2,37]. In genotypes of cultivars 'Clark', 'Lewis'—S_3S_8; 'Badem', 'Halls Giant', 'Merveille de Bollwiller', 'Tonda Romana'—S_5S_{15} codomination of S-alleles was recorded.[1,2]

The choice and isolation of the branches with female flowers were done at the beginning of spring growth of male inflorescences (before pollen release). Tubular isolators in the form of a sleeve were used; they were tied at the lower part of a branch and also above the branch. They were fixed tightly (with cotton wool layer) and remained there till the end of flowering of male inflorescences of all hazelnut bushes and hazelnut species (*C. avellana* and *C. colurna*), which were planted along the perimeter of a hazelnut orchard as a reserve source of pollen in the years of freezing of male flowers of cultivars-pollinators. The flowering of female flowers was not simultaneous and lasted longer (1.5–2.5 months). The flowering of male catkins usually lasts 5–10 days, and it takes place in the period of extensive flowering of female flowers. The amount and quality of pollen produced by different cultivars depend on many factors. The following ones should be mentioned:

- Genotype (specific and varietal)
- Last year yield of nuts
- Winter resistance of catkins
- Meteorological conditions of last and current years
- Soil conditions and agro-technical factors

A positive correlation between pollen grain size and pollen viability was determined—small pollen is less viable. Pollen viability of wild *C. avellana* forms surpassed the indicators of the studied hazelnut cultivars, which ranged within 0.5–62%. It turned out that the cultivars which produced a lot of pollen were less prolific. On the contrary, the forms which lose their catkins in the second half of summer give better nut yields.

Pollen for the crossing was taken from branches cut with catkins; their development was evoked in room conditions. The branches were put in a container with water and placed on a sheet of paper, leaving them for a night in a cool room (15–16°C). Branches with catkins of one (the same) cultivar were kept in the room to prevent mutual pollen contamination, and they remained there till pollen falling.

Pollen was stored the best in the conditions of higher humidity (not lower than 74%) and at temperature below 0°C, but not in a desiccator under chloride calcium, as it is recommended to store pollen of the majority of plants. Pollination was performed without removing isolators. An upper part of a sleeve was untied, pollen was put on pistil stigma, and an isolator was tied again after pollination.

With help of medical insufflator MO-03, equipped with a medical needle, it is possible to perform pollination, without untying isolators, by injecting pollen into an isolator with insufflator.[1,2]

Pollination was repeated after 2–3 days. When flowering was over, the number of set fruits was revised. The first examination was done 3–4 weeks after mass nut emergence on non-isolated branches, where female flowers freely pollinated, and the second examination was performed 3 weeks before harvesting.

In addition to interspecific crossings, hybridization of the cultivars from the collection of NDP "Sofiyivka" with the representatives of *C. chinensis* (known synonyms—*C. chinensis* var. *macrocarpa* Hu., *C. colurna* var. *chinensis* (Franch.) Burkill and *C. papyracea* Hickel.) was carried out; they were used because of their large fruits and high content of raw protein and oil. The output of kernel in *C. chinensis* nuts was more than 50%.

In the process of studying the ways of growing hybrid seedlings, early fall sowing of nuts (with involucres) proved to be efficient, as it gave 100% germination and remained less labor-consuming.[38]

F_1 seedlings were grown in the containers in controlled conditions in a vegetation facility with small-disperse moistening; later, they were moved to regrow and to be evaluated in a hybrid orchard; there comprehensive evaluation of seedlings and nuts was performed. As a wide spectrum of splitting with respect to main economic-valuable characteristics was recorded in seed progeny from controlled crossings and from free pollination, a number of promising forms were chosen in seedling populations for both successive hybridization cycles and cultivar studies, the aim of which is to further use them in polyclonal plantations where self-pollination is possible.

New hazelnut cultivars 'Sofiyivsky 1', 'Sofiyivsky 2', and 'Sofiyivsky 15', which develop fruits with an early globular shape, characterized by higher winter and drought resistance, compared with Turkish and Azerbaijani cultivars, and lack of fruiting intervals, are prepared to be submitted to the state registration of the State Veterinary and Phytosanitary Service of Ukraine.

'Sofiyivsky 1', developed as a result of artificial clonal selection from hybrid population, received from pollination of Ukraina-50 with *C. chinensis* pollen. The plants of 'Sofiyivsky 1' have a medium vigor and spreading habit, bush is 3.9–4.2 m high, diameter—2.3–2.7 mm, and are branchy. It forms stool shoots. Buds are sharp, tomentous, and brown. Leaves are of average size, 11 cm long, 9 cm wide, short stalked, round, serrate edges, with heart-shaped base, dark green, blade (lamina) is tomentous. Young leaves are green. Catkins are numerous, average in size, 4.1–4.7 cm long, diameter is 0.8–0.9 cm, gathered in panicles— three to five pieces in each, light green. Nuts are of average size— $19.9 \times 19.8 \times 15.9$ mm³, tubular, slightly flat on the sides, with a sharp top, weight range is 1.8–2.4 g (Fig. 13.3).

FIGURE 13.3   (See color insert.) Fruits of hazelnut 'Sofiyivsky 1'.

Kernel is large, tubular, covered with thin, rugose brown husk; some-times, there are small cavities inside the kernel. Shell is thin (0.15 mm), light brown, lighter toward the top, and smooth. Nuts mature in the first decade of September. Fruit involucre is large and opened with serrate edges. While ripening, fruit involucres open and nuts fall freely. Percentage of nut weight that is from the kernel is 55.6%. Oil content is 74.2%. Yield per bush is 6.9 kg. Average yield capacity is 19.1 cwt/ha. It gives the first yield after 3 years of planting; a maximum is reached at the age of 15–16 years.

'Sofiyivsky 2', developed as a result of selection from hybrid population, is received from pollination of cultivar 'Dar Pavlenko' with *C. chinensis* pollen. The bush is highly grown—4.5–5.0 m high, diameter—2.5–2.9 m, branchy, pyramidal shape on the top, spreading, and inclined to thicker crown. It forms stool shoots. Shoots are covered with brown rugose bark, and is rather leafy. Young shoot are green and tomentous. Buds are sharp-ened, woolly, and green. Leaves are of average size, 9 cm long, 7 cm wide; short petiole, round, serrate edges, with heart-shaped basis, light green, and lamina is tomentous. Young leaves are green. Catkins are numerous; average in size, 3.7–4.1 cm long, diameter is 0.6–0.8 cm, gathered in panicles—three to four pieces in each, and light green. Nuts are of average size—20.1 × 17.1 × 17.0 mm³, tubular, slightly flat on the sides, with blunt top, weight range is 2.0–2.5 g (Fig. 13.4).

FIGURE 13.4   **(See color insert.)** Fruits of hazelnut 'Sofiyivsky 2'.

Kernel is large, tubular, covered with thin, rugose brown husk; sometimes, there are small cavities inside the kernel. Shell is thin (0.16 mm), light brown, lighter toward the top, and smooth. Nuts mature in the first decade of September. Fruit involucre is short, wide with serrate edges. While ripening, fruit involucres open and nuts fall freely. Percentage of nut weight that is from the kernel is 53.2%. Oil content is 71.6%. Yield per bush is 6.9 kg. Average yield capacity is 18.5 cwt/ha. It gives the first yield after 3 years of planting; a maximum is reached at the age of 15–16 years.

'Sofiyivsky 15', developed as a result of hybrid population, is received from pollination of cultivar 'Garibaldi' with *C. chinensis* pollen. The bush is average-grown 3.5–4.0 m high, diameter—2.2–2.5 m, branchy, tubular shape, and spreading. It forms stool shoots. Shoots are covered with greenish rugose bark, and are rather leafy. Young shoots are green and tomentous. Buds are sharpened, woolly, and brown. Leaves are of average size—10 cm long, 8 cm wide, short petiole, round, serrate edges, with heart-shaped basis, intense dark green, and lamina is tomentous. Young leaves are green. Catkins are numerous; average in size, 3.5–4.5 cm long, diameter is 0.6–0.8 cm, gathered in panicles—two to three pieces in each and light green. Nuts are large—$20.3 \times 20.2 \times 20.1$ mm³, tubular, slightly flat on the sides, with a sharp top, and weight range is 2.8–3.5 g (Fig. 13.5).

Kernel is large, tubular, covered with thin, rugose brown husk, without cavities inside the kernel. Shell is thin (0.2 mm), with dark brown stripes, lighter toward the top, dim, with visible edges, corrugated, and woolly toward the top. Nuts mature in the first decade of September. Fruit involucre is short, opened, with serrate edges. While ripening, fruit involucres open and nuts fall freely. Percentage of nut weight that is from the kernel is 53.6%. Oil content is 72.4%. Yield per bush is 6.8 kg. Average yield capacity is 17.4 cwt/ha. It gives the first yield after 3 years of planting; a maximum is reached at the age of 14–15 years.

FIGURE 13.5 **(See color insert.)** Fruits of hazelnut 'Sofiyivsky 15'.

The comparison of new hazelnut cultivars 'Sofiyivsky 1', 'Sofiyivsky 2', and 'Sofiyivsky 15' with respect to their economic-valuable characteristics confirms the fact that the difference between them is not very significant. However, as far as the indicators of the nut itself and its kernel are concerned, the advantages of 'Sofiyivsky 15' are quite obvious (Fig. 13.6).

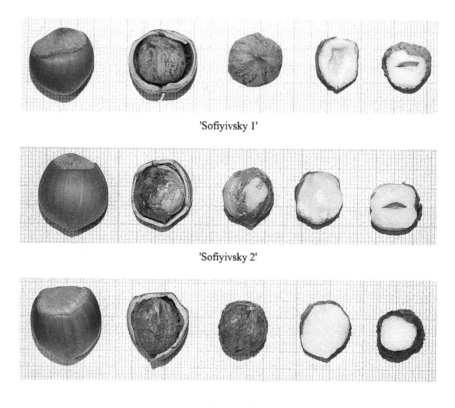

'Sofiyivsky 1'

'Sofiyivsky 2'

'Sofiyivsky 15'

FIGURE 13.6   (See color insert.) Fruits of new hazelnut cultivars: 'Sofiyivsky 1' 'Sofiyivsky 2', 'Sofiyivsky 15'.

The major concerns of confectionary industry are the shape of kernel and the lack of cavity, which are seen in many other cultivars, in particular, 'Sofiyivsky 1' and 'Sofiyivsky 2'.

Better seedlings[39,40] were propagated by the accelerated technology, worked out by us,[41] and which resulted from the comparative evaluation of various hazelnut propagation ways, taking into account the use of growth stimulators, different terms and methods of providing graft materials and choosing the most effective planting material for accelerated cultivation.

13.4 CONCLUSION

The efficiency of a hazelnut breeding scheme, worked out by us, which, in addition to traditional seedling choice from free pollination and interspecific hybridization, includes progeny from controlled crossings of cultivar samples of the collection of NDP "Sofiyivka" of NAS of Ukraine with the representatives of Chinese hazelnut *C. chinensis,* that were used because of their large fruits and high content of raw protein and oil, was confirmed by the development of valuable breeding material. Several new cultivars were chosen from it, namely, 'Sofiyivsky 1', 'Sofiyivsky 2', and 'Sofiyivsky 15', and the application form was submitted to the state registration of the State Veterinary and Phytosanitary Service of Ukraine.

An accelerated passing of the selection material through the stages of the scheme for 3–5 years takes place at the stages of growing seedlings F_1 in a hybrid orchard (1–2 years), speedup propagation of the best seedlings with layerings (1–2 years) using the technology improved by us (1–2 years), and including a tissue culture techniques into hazelnut breeding program (1–3 years). Due to this, hybrid seedling, from which new cultivar 'Sofiyivsky 15' was chosen, started fruiting in the third year after crossing.

ACKNOWLEDGMENT

This study is partly based on the work supported by the National Dendrological Park "Sofiyivka" of NAS of Ukraine (No 115U004248) together with Uman National University of Horticulture (No 101U004495) in compliance with their thematic plans of the research work.

KEYWORDS

- cultivar
- filbert
- genus natural habitat
- hazelnut
- layering
- nut
- screening
- species

REFERENCES

1. Kosenko, I. S.; Opalko, A. I.; Opalko, O. A. Hazelnut: Applied Genetics, Breeding, the Methods of Propagation and Production. Kosenko, I. S., Eds.; Naukova Dumka: Kyiv, 2008; p 256 (In Ukrainian).

2. Kosenko, I. S.; Opalko, A. I.; Balabak, O. A.; Shulga, S. M. *Corylus* spp. Genetic Resources Use in Hazelnuts *Corylus domestica* Kos. et Opal. Improvement. Autochthonous and alien plants, The *Collection of Proceedings of the National Dendrological Park "Sofiyivka" of NAS of Ukraine,* **2016,** *12,* 121–136 (In Ukrainian).

3. Kosenko, I. S. The Genetic Resources Mobilization of *Corylus* L. Genus at the National Dendrological Park "Sofiyivka" of the NAS of Ukraine. News Biosphere Reserve. Askania Nova, **2012,** *14,* 156–160 (In Ukrainian).

4. Kosenko, I. S. Genetic Resources of the Genus *Corylus* l. in the National Dendrological Park "Sofiyivka" of NAS of Ukraine. In *Ecological Consequences of Increasing Crop Productivity: Plant Breeding and Biotic Diversity;* Opalko, A. I.; et al. Eds.; Apple Academic Press: Toronto, New Jersey, 2015; pp 155–166.

5. Molnar, T. J. Corylus. In *Wild Crop Relatives: Genomic and Breeding Resources: Forest Trees;* Kole, C., Ed.; Springer: Berlin, Heidelberg, 2011; pp 15–48.

6. 6. Whitcher, I. N.; Wen, J. Phylogeny and Biogeography of Corylus (Betulaceae): Inferences from ITS Sequences. *Syst. Bot.* **2001,** *26*(2), 283–298.

7. Smolianinova, L. A. *Corylus* (Tourn.) L. Hazelnut. In *Cultured Flora of the USSR;* Vulf, YeV. Ed.; Selhozgiz: Moscow, Leningrad, 1936; Vol. 17, Nut crops. 132–206. (In Russian).

8. Bobrov, YeG. History and Classification *Corylus* Species. *Sovetskaia Botanika* **1936,** *1,* 11–39. (In Russian).

9. Kosenko, I.; Opalko, A. Vegetative Propagation of *Corylus* L. Through Tissue Culture. Monographs of Botanical Gardens: European Botanic Gardens Together Towards the

Implementation of Plant Conservation Strategies. BGCBDCPAS: Warsaw, 2007; Vol. 1, pp 133–136.

10. Hazelnut (*Corylus maxima* Mill.) State Register of Plant Varieties Suitable for Dissemination in Ukraine in 2016. 2016; p 344. http://vet.gov.ua/sites/default/files/Reestr%2022.02.16.pdf (accessed Aug 29, 2016). (In Ukrainian).

11. Kosenko, I. S.; Boyko, A. L.; Opalko, A. I.; et al. Micropropagation of *Corylus Colurna* L. *Acta Hortic.* **2009**, *845*(1), 261–266.

12. Balabak, O. A. Ecological-Biological Peculiarities of Growth, Development and Propagation of Hazelnut (*Corylus domestica* Kosenko et Opalko). Ecology—ways of Harmonization of the Relationship Between Nature and Society: Theses of IV Inter-University Scientific-Practical Conference, Devoted 170-Anniversary of Uman National University of Horticulture (Uman, UNUH, 16–17 October, 2014). UNUH: Uman, 2014; 54–55. (In Ukrainian).

13. Kosenko, I. S.; Opalko, A. I. Dynamics of *Corylus* L. Species as a Proof of the Law of N.I. Vavilov About Homological Lines in Heredity Variability. Plant Introduction at the Beginning of XXI Century: Achievements and Challenges (to the 120th Anniversary of Academician N.I. Vavilov), *Proceedings of the International Scientific Conference,* October, 2–4, 2007), Phytosociocenter, Kyiv, 2007; 70–74 (In Ukrainian).

14. Kosenko, I. S.; Tarasenko, G. A.; Opalko, A. I. Disputable Aspects of *Corylus* L. Genus System. Inspiring Solution in Plant Technology, Horticultural Research and Sustainable Conservation Methods: 2nd World Scientific Congress: Challenges in Botanical Research and Climate Change (Netherlands, Delft, June 29–July 4, 2008). Sieca Repro: Delft, 2008; p 37.

15. Woodworth, R. H. Cytological Studies in the Betulaceae. II. *Corylus* and *Alnus*. *Bot. Gaz.* **1929**, *88*(4), 383–399.

16. Bolkhovskih, Z. V.; Grif, V. G.; Zaharieva, O. I.; et al. *Chromosome Numbers of Flower Plants;* Fedorov, A. A., Ed.;. Nauka (Leningrad Branch): Leningrad, 1969; p 927 (In Russian).

17. Danielsson-Santesson, B. Further Iinvestigations of polyploid hazel. *Sweden Pomol compd. årsskr.* (Fortsatta undersökningar av polyploid hassel. Sveriges pomol förenings årsskr). **1951**, *52*, 38–48 (In Swedish).

18. Thompson, M. M.; Lagerstedt, H. B.; Mehlenbacher, S. A. Hazelnuts. In *Fruit Breeding;* Janick, J., Moore, J. N., Eds.; Wiley: New York, 1996; Vol. 3, pp 125–184.

19. Kubitzki KB. *The Families and Genera of Cascular Plants;* Kubitzki, K. Ed.; *Flowering Plants. Dicotyledons: Magnoliid, Hamamelid and Caryophyllid Families;* Kubitzki, K., Rohwer, J. G., Bittrich, V., Eds.; Springer Science & Business Media: Berlin, Heidelberg, 2013, Vol. 2, pp 152–157.

20. Erdogan, V. Genetic Relationships Among Hazelnut (*Corylus*) Species. Thesis for the degree of Doctor of Philosophy in Horticulture Oregon State University, April 16, 1999; p 218.

21. Botta, R.; Emanuel, E.; Me, G.; Sacerdote, S.; Vallania, R.; Me, G.; Sacerdote, S.; Vallania, R. Survey Karyological in Some Species of the Genus *Corylus*. (Indagine cariologica in alcune specie del genere Corylus.) *Ital. J. Hortic.* (Rivista di ortofloro-frutticoltura italiana). **1986**, *70*(5), 323–329. (In Italian)

22. Wetzel, G. Chromosome Studies in the *Fagales*. (Chromosomenstudien bei den Fagales. Botanisches Archiv). *Bot. Arch.* **1929**, *25*(3/4), 257–283 (In German).

23. Kasapligil, B. Chromosome Studies in Genus *Corylus*. *Sci. Rep. Faculty Sci.* Ege University: Bornova, (Ege Universitesi Matbaasi.) **1968,** *59,* 14.

24. Mehlenbacher, S. A.; Brown, R. N.; Nouhra, E. R.; et al. A Genetic Linkage Map for Hazelnut (*Corylus avellana* L.) Based on RAPD and SSR Markers. *Genome.* **2006,** *49*(2),122–133.

25. Kosenko, I. S.; Opalko, A. I.; Shulga, S. M. Breeding Material for Developing New Hazelnuts Cultivars (*Corylus domestica* Kos. et Opal.) with the Increased Content of Essence Phospholipids in the Nuts. Plant Introduction, Preservation and Enriching of Bio-Diversity in Botanical Gardens, *Proceedings of the International Scientific Conference Devoted the 80th Anniversary of the M.M. Gryshko National Botanic Garden of NAS of Ukraine,* Kyiv, M. M. Gryshko National Botanic Garden of NAS of Ukraine, Sept 15–17, 2015. Phytosociocenter: Kyiv, 2015; 127–129 (In Ukrainian).

26. Hazelnuts, with Shell FAOSTAT Domains Production/Crops: Average. [Electronic Resource]. 2013. http://faostat3.fao.org/browse/Q/QC/E (accessed Dec 2, 2015).

27. Satina, H. M.; Oleshchenko, F. H.; Koshlakova, N. M.; et al. Scientific Principles and Components of the Branch Program of the Development of Nut Production in Ukraine. Logos: Kyiv, 2011; p 100 (In Ukrainian).

28. Lugovskoi, A. P. Hazelnut Breeding. In *Program and Breeding Methodology of Frits, Small Fruits and Nuts;* Sedov, Ye. N., Ed.; VNIISPK: Orel, 1995; pp 436–445 (In Russian).

29. Methods of Doing the Cultivar Expertise of a Group of Fruit Crops, Small Fruits, Nuts, Subtropical and Grapes as to Their Suitability for the Cultivation in Ukraine (PSP). Tkachyk, S. O., Ed.; National and Veterinary and Phytosanitary Agency of Ukraine; Ukrainian Institute of Cultivar Expertise: Kyiv, 2014; p 83 (In Ukrainian).

30. Oil Seeds. Methods of Determining Oiliness/Oil Content. GOST 10857–64 (Instead of GOST 3040–55 in the Part of Determining Raw Oil Content in Oil-Bearing Seeds (p. 76) [Valid from 1964–30–06] GOST Catalog. Standardinform: Moscow, 2010; p 69–74. (Inter-State Standard). (In Russian).

31. Barabat, V. A. *Peroxide Oxidation and Stress;* Nauka: Leningrad, 1992; p 148 (In Russian).

32. Kryshchenko, V. P. *Evaluation Methods of the Quality of Crop Output;* Kolos: Moscow, 1983; p 192 (In Russian).

33. Fisher, R. A. Statistical Methods for Research Workers. Cosmo Publications: New Delhi, **2006;** p 354.

34. Opalko, A. I. Nut Breeding. In *Breeding of Fruit and Vegetable Crops: A Textbook for University Students;* Opalko, A. I.; Zaplichko, F. O., Eds.; Vyshcha Shkola: Kyiv, 2000; pp 386–398 (In Ukrainian).

35. Rahemi, M.; Mojadad, D. Effect of Pollen Source on Nut and Kernel Characteristics of Hazelnut. *Acta Hortic. (ISHS).* **2001,** *556,* 371–376.

36. Anthropoadaptability of Plants as a Basis Component of a New Wave of the "Green Revolution". *Biological Systems, Biodiversity, and Stability of Plant Communities;* Weisfeld, L. I., Opalko, A. I. Bome, N. A.; et al., Eds.; Apple Academic Press: Toronto New Jersey, 2015; 3–17. Part 1: The Optimization of Interaction Anthropogenic Changes with Natural Environmental Variability for Sustainable Land Use.

37. Mehlenbacher, S. A. Geographic Distribution of Incompatibility Alleles in Cultivars and Selections of European Hazelnut. *J. Am. Soc. Hortic. Sci.* **2014,** *139*(2), 191–212.

38. A. C. No.1547733. Ways of sowing the seeds of woody plants. Grodzinskiy, A. M.; Balabushka, V. K.; Balabushka, L. V.; Kosenko, I. S.; Parhomenko, L. I. // Goskomizobretenia. Request No. 4357637 of January 4, 1988. Registered in the National registry of inventions of the USSR November 8,1989 (In Russian)..

39. Kosenko, I. S.; Opalko, A. I.; Balabak, O. A.; Shulga, S. M. Oil-Acid Composition of Nut Oil of New Hazelnut Cultivars (Corylusdomestica Kos. et Opal.) of Domestic Breeding. Protection of Bio-Diversity and Historical-Cultural Heritage in Botanical Gardens and Dendrological Parks: Theses of the International Scientific Conference Devoted to the 60th Anniversary of the National Dendrological Park "Sofiyivka" as a Scientific Institution of NAS of Ukraine (October 6–8, 2015, Uman, NDP "Sofiyivka" of NAS of Ukraine). Visavi: Uman, 2015; pp 91–92 (In Ukrainian).

40. Kosenko, I. S.; Balabak, O. A.; Opalko, A. I. New Hazelnut Cultivar (*Corylus domestica* Kos. et Opal.) Sofiyivsky 15. Plant Introduction, Preservation and Enriching of Bio-Diversity I Botanical Gardens and Dendroparks, *Proceedings of the International Scientific Conference Devoted to the 80th Anniversary of the M.M. Gryshko National Botanic Garden of NAS of Ukraine,* Kyiv, M. M. Gryshko National Botanic Garden of NAS of Ukraine, Sept 15–17, 2015). Phytosociocenter: Kyiv, 2015; pp 124–125 (In Ukrainian).

41. Kosenko, I. S.; Balabak, O. A.; Opalko, A. I.; Tarasenko H. A.; Balabak A. V. Patent on useful model No. 98106. The way of hazelnut propagation // Request No. u 2014 13707 submitted on 22.12.2014; registered in the National patent registry of Ukraine on useful models April 10, 2015. Bull. No. 7. 2015; p 4 (In Ukrainian).

CHAPTER 14

RESULTS OF RARE FRUIT CROP ASSORTMENT IMPROVEMENT

VOLODYMYR MEZHENSKYJ

National University of Life and Environmental Sciences of Ukraine, 15, Heroiv Oborony St., Kyiv, 03041, Ukraine

*Corresponding author. E-mail: mezh1956@gmail.com

CONTENTS

ABSTRACT

The chapter gives a detailed analysis of rare fruit crop breeding in Ukraine. The natural conditions of this country allow cultivating a wide range of fruit plants. In the second half of the 20th century, zoning and registration of rare fruit crops cultivars began. Today, there are 158 registered cultivars of rare fruit crops of Ukrainian breeding, namely Crimean myrobalan and myrobalan—Japanese plum hybrids (20), cornelian cherry (16), hardy kiwi (15), quince (13), persimmon (13), hazelnut (12), mulberry (11), almond (10), blue honeysuckle (8), Japanese quince (8), nectarine (4), jujube (4), pawpaw (4), fig (4), sea buckthorn (3), hawthorn (3), cranberry bush (3), pomegranate (3), blackberry (2), feijoa (2), Chinese magnolia vine (1), olive (1), and rootstocks (20). Twelve institutions and individuals have carried out this fruitful breeding. Breeding work with these and other fruit plants has been going on, ensuring further improvement and expansion of assortment rare fruit crops.

14.1   INTRODUCTION

The development of plant industry, including horticulture, particularly depends on the introduction of new plants. Blueberry, honeysuckle, kiwifruit, sea buckthorn, and other fruit plants were introduced last century. The cultivars of these plants have gained commercial importance around the world. Cultivation of new plant species provides stability of agricultural production, increases productivity, reduces material and energy consumption, and healthier environment. The new raw material base for various manufacturing industries is expanding and the diversity of products is increasing. Growing of crops with a high content of bioactive substances improves the quality and nutritional value of both consumed fresh fruit and processed products and promotes a healthier nation. New fruit crops often fully meet the requirements of organic production that enables the development of organic horticulture. The success of orchard culture largely depends on the cultivars; so creating new rare fruit crops, cultivars promote the development of the commercial and amateur horticulture.

14.2   RARE FRUIT CROPS

The traditional set of fruit crops of Ukraine was formed in the 17th century. These are apples, pears, plums, apricots, peaches, sour and sweet cherries, grapes, strawberries, black and red currants, gooseberries, and walnuts. These crops were included in the first zoned national assortments in the first half of the 20th century. Other fruit plants have local significance. Zoning of fruit plants, called rare fruit crops, began in the second half of the 20th century. They are also often called alternative, exotic, less known, new, nontrivial, nontraditional, promising, unusual crops, and so forth. Over the last decade, cultivars of almonds, blackberries, blue honeysuckles, blueberries, cornelian cherries, cranberry bushes, cinnamon roses, filberts, figs, hardy kiwis, hawthorns, Japanese quinces, jujubes, kiwifruits, magnolia vines, mulberries, nectarines, olives, pawpaws, persimmons, pomegranates, quinces, and sea buckthorns have been registered in the State Register of Plant Cultivars of Ukraine (hereinafter referred as State Register).[1–4] In Ukraine, barberry, black apricot, black raspberry, chokeberry, golden currant, goumi, Manchu cherry, rowan, serviceberry, medlar, and other fruit plants are also growing.[5–8] Breeding work with former fruit plants are carried out, but their cultivars in Ukraine have not registered yet.

In this chapter under name rare fruit crop, we mean cultivated fruit plants that are zoned and registered in the State Register only after the middle of the 20th century, that is, a demarcation point between them and traditional fruit crops. Cultivated fruit plants whose cultivars have not been registered yet are also rare fruit crop. Fruit plants, not having commercially valuable cultivars, are not listed in the State Register. Each rare fruit crop occupies less than 1% of the total area of fruit plantations in Ukraine.

In Ukraine, there are about 1000 species of fruit plants, among them more than three-fourth are introduced. About 110 species and interspecific hybrids are economically important.[9] The national collections of plant genetic resources contain 21,200 accessions of fruit crops, including 14,800 cultivars, and among them, there are 4500 of Ukrainian breeding.[10] All fruit crops can be divided into four groups: pome fruits, stone fruit, berry (small fruit) crops, and nut crops.[11] Both widespread and unusual fruit crops in Ukraine are given in Table 14.1.

TABLE 14.1   Fruit crops in Ukraine.

Pome fruits	Stone fruits	Berry crops	Nut crops
	Traditional fruit crops		
Apple	Apricot	Black current	Persian walnut
Pear	Peach	Gooseberry	
	Prune	Grape	
	Sour cherry and duke	Raspberry	
	Sweet cherry	Red currant	
		Strawberry	
	Fruit crops and rootstocks whose cultivars have been registered since the 1950s		
Hawthorn	Cranberry bush	Blackberry	Almond
Japanese quince	Cornelian cherry	Blue honeysuckle	Hazelnut
Quince	Crimean myrobalan and myrobalan–Japanese plum hybrids	Blueberry	
Pome crop rootstocks	Jujube	Cinnamon rose	
	Nectarine	Feijoa	
	Olive	Fig	
	Stonecrop rootstocks	Hardy kiwi	
		Kiwifruit	
		Lemon	
		Magnolia vine	
		Mulberry	
		Pawpaw	
		Persimmon	
		Pomegranate	
		Sea buckthorn	
		Grapevine	
		Rootstocks	
	Bred fruit plants whose cultivars have not been registered yet		
Checker tree	Chokecherry	Barberry and Mahonia	Pecan
Chokeberry	Bessey cherry	Buffalo berry	Turkish hazelnut
Medlar	Elder	Cudrang	
Nashi	Manchu cherry	Goji	
Rowan	Plumcot	Golden current	
Serviceberry		Goumi	
Service tree		Passionflower	
Whitebeam			

14.3   BREEDING ESTABLISHMENTS

Nikita Botanical Garden, founded in 1812, significantly contributes to the development of the world and Ukrainian horticulture. The breeding work with many fruit crops initiated by the Second Director Nikolai von Hartwiß[12] is of great success, especially over the last century. In the assortment of Ukraine, the cultivars of quinces, myrobalans, pomegranates, jujubes, nectarines, and persimmons of Nikita Botanical Garden breeding occupy leading positions. Owing to subtropical climate, the southern Crimean coast, where the botanical garden is located, is a unique place for research on subtropical fruit plants. Registered cultivars of figs, olives, and almonds are created exclusively in this establishment. The gene pool of rare fruit crops of Southern Fruit Crop Department has 30 species, 392 cultivars of Nikita Botanical Garden breeding, 1004 introduced cultivars, and 1902 elite forms.[13] Moreover, a significant foundation stone hybrid crops, including interspecific hybrids, are studied.

In 1913, the Academician Mykola Kashchenko[14] founded the Acclimatization Garden in Kyiv, and he initiated the introduction and breeding work with many new and rare fruit crops. Further, the collections of his Acclimatization Garden became the basis for breeding of fruit plants in the Department of Fruit Plant Acclimatization at M. M. Hryshko National Botanical Garden of the National Academy of Sciences of Ukraine. Now, collectible-breeding fund cultural and wild plants are about 700 species and cultivars.[15] Many cultivars of hardy kiwi, cornelian cherry, Japanese and common quinces, magnolia vine, nectarines, and plum hybrids have been bred and included in the State Register. The National Botanical Garden is the leading institution in the breeding work on hardy kiwi, cornelian cherry, and both common and Japanese quinces.

In 1981 in Donetsk Experimental Station of Horticulture (nowadays, Bakhmut Experimental Station of Nurseries Cultivation), we started collecting rare fruit plants and began breeding work with them. In 1992, this rare fruit crop collection was included as a component to national collections of newly formed National Centre for Plant Genetic Resources of Ukraine. It consisted of 316 species, subspecies, and interspecific hybrids, and also 730 cultivars and promising forms. The total number of accessories was 1120. After moving to Kyiv, in the National University of Life and Environmental Sciences of Ukraine, we have created a new collection, including practically former collection accessories. This collection is numbering 266 species and interspecific hybrids. The total number of accessories is 981, including 532 cultivars and promising forms of rare fruit crops and ornamental fruits.[8]

The breeding work on rare fruit crops is conducted in other scientific institutions of gardening, forestry, and botanical profile (Table 14.2). In each of them, there are a large collection and hybrid funds. The Ukrainian establishments, in which the registered rare fruit cultivars are created, have been listed in Table 14.3. In 2001, the monopoly of state institutions was terminated and since that time private persons have registered some cultivars.

TABLE 14.2   List of Abbreviations of Scientific Institutions for Tables.

Abbreviations	Full title
Bakhmut Exp. St. Nurs. Cult.	Bakhmut Experimental Station of Nurseries Cultivation of the Institute of Horticulture, Bakhmut, Donetsk region
Donetsk Bot. Gard.	Donetsk Botanical Garden, Donetsk
Inst. Hort.	Institute of Horticulture, Kyiv
Exp. St. Pomol.	L. P. Symyrenko Experimental Station of Pomology of the Institute of Horticulture, Mliiv, Cherkassy region
Sericult. Depart.	Department of Sericulture and Technical Entomology of the National Research Centre "Institute of Experimental and Clinical Veterinary Medicine", Merefa, Kharkiv region
Inst. Viticult. Wine.	National Scientific Centre "Tairov Institute of Viticulture and Winemaking," Odesa
Krasnokutsk Depart.	Krasnokutsk Department of Horticulture of the Institute of Horticulture, Krasnokutsk, Kharkiv region
Nat. Bot. Gard.	M. M. Hryshko National Botanical Garden, Kyiv
Nat. Univ. Life Environm. Sc.	National University of Life and Environmental Sciences of Ukraine, Kyiv
Nikita Bot. Gard.	Nikita Botanical Garden—National Research Centre, Yalta, Autonomous Republic of Crimea
Nova Kakhovka Exp. Farm	Nova Kakhovka Experimental Farm, Nova Kakhovka. Kherson region
Ukr. Res. Inst. Forest. & Forest Melior.	H. M. Vysotskyi Ukrainian Research Institution of Forestry & Forest Melioration, Kharkiv

TABLE 14.3 The Effectiveness of Rare Fruit Crops and Rootstock Breeding in Ukraine*.

Crop	Breeding establishment/Applicant**													
	Nikita Bot. Gard.	Nat. Bot. Gard.	Bakhmut Exp. St. Nurs. Cult.	Inst. Hort.	Ukr. Res. Inst. Forest. & Forest Mellior.	Donetsk Bot. Gard.	Nova Kakhovka Exp. Farm	Sericult. Depart.	Private applicants	Exp. St. Pomol.	Krasnokutsk Depart.	Inst. Viticult. Vine.	Nat. Univ. Life Environm. Sc.	In all applicants
Crimean myrobalan and myrobalan–Japanese plum hybrids	14	1	5***	2***	–	–	–	–	–	–	–	–	–	20
Cornelian cherry	–	14	1	–	–	–	–	–	–	1	–	–	–	16
Hardy kiwi	–	15	–	–	–	–	–	–	–	–	–	–	–	15
Pome crop rootstocks	–	–	2	10***	–	–	–	–	–	1***	2	–	–	14
Quince	8	5	–	–	–	–	–	–	–	–	–	–	–	13
Persimmon	9	–	–	–	–	–	4	–	–	–	–	–	–	13
Hazelnut	–	–	–	–	12	–	–	–	–	–	–	–	–	12
Mulberry	–	–	–	–	–	5	–	6	–	–	–	–	–	11
Almond	10	–	–	–	–	4	–	–	–	1	–	–	–	10
Blue honeysuckle	–	–	–	–	–	4	–	–	–	1	3***	–	–	8
Japanese quince	–	4	4	–	–	–	–	–	–	–	–	–	–	8
Nectarine	3	1	–	–	–	–	–	–	–	–	–	–	–	4
Jujube	3	–	–	–	–	–	1	–	–	–	–	–	–	4
Pawpaw	1	–	–	–	–	–	3	–	–	–	–	–	–	4
Sea buckthorn	–	–	1	–	–	–	–	–	2	–	–	–	–	3

TABLE 14.3 (Continued)

Crop	Breeding establishment/Applicant**													
	Nikita Bot. Gard.	Nat. Bot. Gard.	Bakhmut Exp. St. Nurs. Cult.	Inst. Hort.	Ukr. Res. Inst. Forest & Forest Melior.	Donetsk Bot. Gard.	Nova Kakhovka Exp. Farm	Sericult. Depart.	Private applicants	Exp. St. Pomol.	Krasnokutsk Depart.	Inst. Viticult. Vine.	Nat. Univ. Life Environm. Sc.	In all applicants
Hawthorn	–	–	–	–	–	–	–	–	3	–	–	–	–	3
Cranberry bush	–	–	–	–	–	–	–	–	–	3	–	–	–	3
Pomegranate	3	–	–	–	–	–	–	–	–	–	–	–	–	3
Grapevine rootstocks	–	–	–	–	–	–	–	–	–	–	–	3	–	3
Stone crop rootstocks	–	1	–	2***	–	–	–	–	–	–	–	–	–	3
Blackberry	–	–	–	–	–	–	–	–	–	–	–	–	2	2
Feijoa	2	–	–	–	–	–	–	–	–	–	–	–	2	2
Magnolia vine	–	1	–	–	–	–	–	–	–	–	–	–	–	1
Fig	1	–	–	–	–	–	–	–	–	–	–	–	–	1
Olive	1	–	–	–	–	–	–	–	–	–	–	–	–	1
Blueberry	–	–	–	–	–	–	–	1	–	–	–	–	–	1
Total	55	42	13***	14***	12	9	8	6	6	6***	5	3	2	178

Note: *In the State Register, there are also cultivars of fruit plant with edible/relatively edible fruit which are included to ornamental and forest plant groups, for example, barberry, rose, ornamental peach, ornamental apple, beech, and oak; **short/full research establishment names and their locating; and ***including cultivars of joint breeding more than one research establishment.

14.4 REGISTERED RARE POME FRUITS

14.4.1 QUINCE (AIVA IN UKRAINIAN)

Cydonia oblonga Mill. is one of the oldest fruit crops in the world. Its natural habitat is the Caucasus and northwestern Iran areas, adjacent to the Caspian Sea where quince was domesticated. Then, quince was spread through the Caucasus and Turkey first in Crimea, and then further in the other parts of Ukraine. In Crimea and Trans-Dniester, the secondary diversity centers of quince have been formed.

Formerly, foreign cultivars 'Anger', 'Champion', 'Constantinople', 'Bereczki', 'Portugal', 'Sorocskaia', and 'Turunciukskaia' were recommended for growing in Ukrainian regions with the most favorable climatic conditions (Table 14.4).

TABLE 14.4 List of Rare Pome Fruit Cultivars of Ukrainian Breeding in the State Register of Plant Cultivars of Ukraine.

Crop	Cultivar name (registration year)	Applicant
Quince	'Obilna' (1962)	Nikita Bot. Gard.
	'Krymska Aromatna' (1981)	–
	'Krymska Rannia' (1982)	–
	'Myr' (1981)	–
	'Akademichna' (1999)	Nat. Bot. Gard.
	'Darunok Onuku' (1999)	–
	'Kashchenka No. 18' (1999)	–
	'Mariia' (1999)	–
	'Studentka' (1999)	–
	'Siedobna' (2001)	Nikita Bot. Gard.
	'Skazochna' (2001)	–
	'Novorichna' (2010)	–
	'Oktiabryna' (2010)	–
Japanese quince	'Kalif' (2001)	Bakhmut Exp. St. Nurs. Cult.
	'Nika' (2001)	–
	'Nikolai' (2001)	–
	'Nina' (2001)	–
	'Karavaievskyi' (2001)	Nat. Bot. Gard.

TABLE 14.4   *(Continued)*

Crop	Cultivar name (registration year)	Applicant
	'Pomaranchevyi' (2001)	–
	'Tsytrynovyi' (2001)	–
	'Vitaminnyi' (2001)	–
Hawthorn	'Zbigniew' (2001)	V. Mezhenskyj and L. Mezhenska
	'Liudmyl' (2001)	–
	'Shamil' (2001)	–

The first Ukrainian cultivar Obilna originated from Nikita Botanical Garden was included to the zoned assortment in 1962. It was bred from local Crimean quince forms. Afterward, other cultivars were selected from local forms and seedlings of the best foreign cultivars, especially Krymka 'Aromatna' with its fruits suitable for fresh consumption, and self-fertile Myr and 'Krymska Rannia'.[16] Quince breeding, launched by Kashchenko, led to the creation of winter-hardy assortment suitable for northern and central regions of Ukraine.[17] In 1999, 'Akademichna', 'Darunok Onuku', 'Kashchenko' 'No. 18',' Mariia', and 'Studentka' bred in the National Botanical Garden were included in the State Register. At present, 12 cultivars of the National Botanical Garden breeding have been registered. The new selects 'Hrushopodibna Shaida-rovoi', 'Hrushopodibna Shumskoho', 'Kyivska Aromatna', 'Oranzheva', and 'Shkolnitza' of National Botanical Garden breeding; 'Dniprodzerzhynka', 'Dniprovska Lymonnopodibna', 'Savitri', and 'Slava Donchenka' of Botanical Garden of the Dnepropetrovsk State University breeding; and 'Donetska Hrushopodibna' of Bakhmut Experimental Station of Nurseries Cultivation breeding are also promising for cultivation.[18]

14.4.2   JAPANESE QUINCE (YAPONSKA AIVA)

The natural range of *Chaenomeles* is restricted to East Asia. They have been under cultivation in Ukraine for over two centuries as ornamentals for their abundant flowers. Mykola Kashchenko paid his attention to a new fruit shrub and began breeding in his Acclimatization Garden in 1913. In 1937 near Kyiv, the first commercial plantation was established. Later on, many institutions studied Japanese quince; some other small plantations were developed, but the crop was not spread. That is why in 1981, we started developing cultivars to meet the requirements of modern horticulture. First fruit cultivars were registered in Ukraine in 2001. They are *Ch. × californica* W. B. Clark ex C. Weber Kalif and *Ch. × superba* (Frahm) 'Rehder Nika', 'Nikolai', and 'Nina bred' by us and 'Karavaievskyi', 'Pomaranchevyi' 'Tsytrynovyi', and 'Vitaminnyi', 'Tsytrynovyi', and 'Vitaminnyi' of National Botanical Garden breeding.[16] Japanese quince breeding is continuing. The promising cultivars Amfora, Chudovyi Olhy, Maksym, Sviatkovyi, Vyshukanyi Svitlany, and Yan have been bred.[19] It is advisable to select the adaptive erect and spineless shrubs with bigger fruit and an increased pulp layer. Interspecific hybridization is a way of creating a new interesting combination. The new hybrids have no spines; they have seedless and thick-pulped fruit, different time of ripening, and high yield (Fig. 14.1).

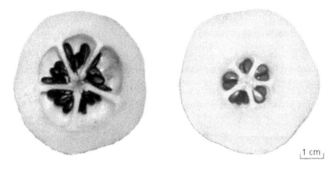

FIGURE 14.1   Cross-sectional view of *Chaenomeles × superba* fruit (left—typical fruit, right—fruit of hybrid '11–21–149').

Japanese quinces of East Asian origin have a secondary European center of diversity with a high-level variability and plants with unique pomological characteristics present in them. The plant domestication in the breeding institutions is similar to the process that once took place in the primary center of origin of cultivated plants, but it is more localized and much faster. Breeding work in institutions significantly accelerates the evolutionary process in plant populations. Original forms that do not exist in nature have been artificially created.[20]

Japanese quinces are valuable for modern horticulture with their early fructification, annual and abundant fruiting, resistant to both biotic and abiotic environmental factors, and suitability for mechanized cultivation. The wide range of adaptability, ecological plasticity, easing in propagation, and high economic efficiency confirms advantages for this pome crop.

Valuable biochemical fruit composition and high resistance to diseases and pests increase the biological value of fruit and make them an indispensable raw material for processing and manufacture of baby food and dietary products.

14.4.3   HAWTHORN (HLID)

Crataegus is a genus of about 250 species spreading in the Old and New Worlds. Almost all hawthorn sp. are ornamentals and suitable for landscaping. Hawthorns have medical properties, a decoction of leaves and fruits is used for treating cardiovascular diseases. The use of hawthorn fruits by a human has a long history. For having large eaten fruit, *C. azarolus* L. has been cultivated through the Mediterranean to Central Asia; *C. pinnatifida* Bunge var. *major* has been cultivated in China; and *C. mexicana* Moc. & Sessé ex DC. has been cultivated in Mexico during ages. In the 20th century,*C. aestivalis* (Walter) Torrey & Gray, *C. opaca* Hook. & Arn., and *C. rufula* Sarg. were domesticated in southeastern states of the United States. In the 20th century, *C. aestivalis* (Walter) Torrey and Gray, *C. opaca* Hook. and Arn., and *C. rufula* Sarg. were domesticated in southeastern states of the United States. In the former USSR, some large-fruited North American and indigenous species are popular as fruit plants.

Although the most native hawthorns in Ukraine are small-fruited, local people collect their fruit for consumption. In Southern Ukraine, there are large-fruited *C. orientalis* Pall. ex M. Bieb., *C. tournefortii* Griseb., and *C. pojarkovae* Kossych. Pojarkova's hawthorn has the greatest value as a fruit plant among indigenous species, combining large yellow fruit with a lack of thorns. This allows to take the best forms of this hawthorn for cultivating directly from the wild nature. We have bred *C. × pojarkovae* Zlat, *C. orientalis* 'Victor' and 'Mark', and *C. × tournefortii* 'Eski Qirim'. Systematically relative to them are Central Asian *C. azarolus* L. var. *pontica* (K. Koch) K. I. Chr. 'Pontii' and *C. × pseudoazarolus* 'Popov Nikita'. Unfortunately, these selects may be damaged in extremely cold winters.[21]

C. pinnatifida Bunge var. *major* N. E. Br. is high winter hardiness and most large-fruited. We have bred 'Kytaiskyi 2', 'Mao Mao', and 'Redflesh Mao'. They have big fruit, but with no pleasant taste for fresh eating, only for processing. Conformably, the large fruits of *C. punctata* Jacq. 'Liudmyl' are suitable for processing, but not for dessert. Instead, other North American species *C. submollis* Sarg. var. *arnoldiana* (Sarg.) Mezhenskyj and *C. pennsylvanica* Ashe are characterized by delicious fruit, 'Zbigniew' and Shamil, respectively.[21] In 2001, we registered 'Liudmyl', 'Zbigniew', and 'Shamil' in the State Register as the first fruit cultivars in this systematic group. We have also bred *C. meyeri* Pojark. 'Vsevolod', *C. chlorocarpa* Lenné & K. Koch. 'Marmeladnyi', *C. rhipidohylla* Gand. 'Lubenskyi', *C. pinnatifida* Bunge 'Donetski Zirochky', and so forth. These cultivars have smaller fruit than the above-mentioned registered cultivars, but they are valuable primarily for the pharmaceutical purposes and as ornamentals.

14.5   REGISTERED RARE STONE FRUIT CROPS

14.5.1   MYROBALAN (ALYCHA)

In Trans-Dniester and Crimea, *Prunus cerasifera* Ehrh. has been cultivated for food at least since ancient times. A specific myrobalan assortment belonging to *P. cerasifera* subsp. *macrocarpa* Eremin & Garkov. with two cultivars—var. *macrocarpa* and var. *taurica* Eremin & Garkov.—has been formed in Crimea. Fruit of var. *macrocarpa* is clingstone, and fruit of var. *taurica* is freestone.[22] The first registered cultivars 'Nikitska Zhovta' and 'Pionerka' were selected from local cultivars and zoned in 1954 (Table 14.5).

TABLE 14.5   List of Rare Stone Fruit Cultivars of Ukrainian Breeding in the State Register of Plant Cultivars of Ukraine.

Crop	Cultivar name (registration year)	Applicant
Crimean myrobalan	'Nikitska Zhovta' (1954)	Nikita Bot. Gard.
	'Pionerka' (1954)	–
	'Kyzyltashska Rannia' (1962)	–
	'Purpurova' (1962)	–
	'Liusha Vyshneva' (1969)	–
	'Krasavytsia' (1978)	–
	'Vasylivska' (1979)	–
Myrobalan–Japanese plum hybrids	'Desertna' (1969)	Nikita Bot. Gard.
	'Obilna' (1969)	–
	'Olenka' (1995)	–
	'Donchanka Rannia' (1997)	Bakhmut Exp. St. Nurs. Cult.
	'Naidionysh' (1998)	–

TABLE 14.5   *(Continued)*

Crop	Cultivar name (registration year)	Applicant
	'Plamenna' (1998)	–
	'Kyivska Hibrydna' (2001)	Nat. Bot. Gard.
	'Heneral' (2007)	Bakhmut Exp. St. Nurs. Cult. and Inst. Hort.
	'Tetiana' (2007)	–
	'Desertna Rannia' (2010)	Nikita Bot. Gard.
	'Femida' (2010)	–
	'Rumiana Zorka' (2010)	–
	'Andromeda' (2012)	–
Nectarine	'Nectaryn Kyivskyi' (1980)	Nat. Bot. Gard.
	'Rubinovyi 8' (2001)	–
	'Krymchanin' (2013)	–
	'Rubinovyi 9' (2013)	–
Olive	'Krymska Prevoskhodna' (1994)	Nikita Bot. Gard.
Cornelian cherry	'Elehantnyi' (1999)	–
	'Olena' (1999)	–
	'Lukianivskyi' (1999)	–
	'Semen' (1999)	–
	'Svitliachok' (1999)	–
	'Volodymyrskyi' (1999)	–
	'Yevheniia' (1999)	–
	'Vydubetskyi' (2000)	–
	'Vavylovets' (2000)	–
	'Hrenader' (2000)	–
	'Mykolka' (2000)	–
	'Radist' (2000)	–
	'Ekzotychnyi' (2001)	–
	'Koralovyi Marka' (2001)	–
	'Bylda' (2001)	Bakhmut Exp. St. Nurs. Cult.
	'Myhailivskyi' (2008)	Exp. St. Pomol.
Cranberry bush	'Koralova' (2001)	–
	'Velykoplidna' (2001)	–
	'Rubinova' (2008)	–
Jujube	'Plodivskyi' (2008)	Nova Kakhovka Exp. Farm
	'Sinit' (2010)	Nikita Bot. Gard.
	'Tsukerkovyi' (2010)	–
	'Koktebel' (2012)	–

During the following decades, other cultivars of var. *macrocarpa* were zoned as well as var. *taurica* 'Liusha Vyshneva' and 'Vasylivska'.

14.5.2 MYROBALAN–JAPANESE PLUM HYBRIDS (VELIKOPLODA ALYCHA, OR HIBRYDNA ALYCHA)

Japanese plum cultivars are very large-fruited, but not hardy under climatic conditions of Ukraine. Claudia Kostina used them for crossing with Crimean myrobalan for the creation of hardy hybrids with larger fruit. In *Prunus salicina* 'Burbank' F_1, she selected two excellent cultivars 'Desertna' and 'Obilna'. They were zoned in 1969. These cultivars and hybrids of the next generations have developed a new fruit crop that has been spread throughout the country. Ukrainian breeders have bred many new cultivars in the 21th century. The cultivars of Russian breeding, for example, 'Kremen', 'Kubanska Kometa', 'Puteshestvennits', 'Sigma', and 'Zhemchuzhina' are registered in the State Register too. All these cultivars are successfully replacing Crimean myrobalan cultivars from orchards.

14.5.3 NECTARINE (NEKTARYNA)

Prunus persica (L.) Batsch var. *nucipersica* (Suckow) C. K. Schneider fruit is appreciated for lack of pubescence, which increases their attractiveness and quality. First, the nectarine 'Blanche' was introduced in Nikita Botanical Garden in 1866. This ancient cultivar retains its value today. Mykola Kashchenko also used nectarine in his hybridization schemes of peach improving. Nowadays in the National Botanical Garden, winter-hardy 'Nectaryn Kyivskyi' was created. It is the first Ukrainian cultivar that was registered in 1980. In Nikita Botanical Garden, the world's nectarine gene pool was introduced. It allowed to start a breeding program for creation of nectarine cultivars of different time of ripening and adapted to the conditions of Ukraine. As a result, 12 cultivars were bred, and 'Rubinovyi 8', 'Rubinovyi 9', and 'Krymchanin' were registered. It allows expanding the industrial plantations of nectarine.

Unfortunately, these cultivars have been created on the narrow genetic base of large-fruited American cultivars strongly affected by fungal diseases. Therefore, the breeding program has been begun for a creation of new cultivars resistant to powdery mildew, leaf curl, shot-hole, and brown rot. This is possible only due to the remote hybridization between peach and nectarine on the one hand and almond on the other hand. As a result, the hybridogenous genotypes with low susceptibility to complex pathogens *Sphaerotheca pannosa, Taphrina deformans, Clasterosporium carpophilum, Monilia cinerea,* and *Monilia fructigena,* which have no analogs in the world, have been created. These hybrids have been crossed with large-fruited male sterility genotypes for increasing the yield and improving fruit quality. It provides new opportunities for creating disease-resistant cultivars that will produce constant offspring by way of seed propagation. Own-rooted orchards are cheaper, long-lived and better suitable to the environmental conditions. Resistant hybrids with large and palatable fruit after 4–5 generations of repeated crossings have been selected.[23]

14.5.4   OLIVE (MASLYNA OR OLYVA)

Olea europea L. was distributed throughout the Mediterranean in ancient times. The evergreen olive tree is hardiest and the most frost-resistant subtropical tree, that can withstand winter temperatures up to −12 to −18°C. In Ukraine, olive tree is cultivated on the southern coast of Crimea, where it was introduced probably in the 13th century. In 1994, the universal cultivar 'Krymska Prevoskhodna' was registered.

14.5.5   CORNELIAN CHERRY (DEREN)

Cornelian cherry is native to Ukraine. It is spread in Crimea and Right-Bank Ukraine in the wild nature, and also throughout the whole country under cultivation. The origin of large-fruited cultivars is unclear, either they were bred from the local forms or they were imported from abroad many years ago. Cornelian cherries are long-lived trees. In many regions of Ukraine, there are the age-old specimens. The cornel cherry is early blooming and flowering and does not suffer from spring frosts. Ripe fruits are eaten out-of-hand and used for preserves, jellies, syrups, compote, and so forth.

In the 1930s, the local large-fruited cultivars were recommended for cultivation according to the first zoned assortment, but only in 1999, the first cultivars, called 'Elehantnyi', 'Olena', 'Lukhianivskyi', 'Semen', 'Svitliachok', 'Volodymyrskyi', and 'Yevheniia' were registered. Later, a lot of new cultivars being resulted by analytical and synthetic breeding were added to the State Register. Moreover, 'Koralovyi', 'Kostia', 'Kozeroh', 'Mriia Shaidarovoi', 'Nespodivanyi', 'Nizhnyi', 'Oryhinalnyi', 'Prior-skyi', 'Prezent', 'Starokyivskyi', 'Suliia', and 'Yantarnyi' were tested. In the National Botanical Garden, there is the world's largest and very diverse cornelian cherry collection. Cultivars are different in the fruit shape from elliptical to pear-shaped, and fruit weighs from 3–4 to 8–10 g. The colors fruit are yellow, pink, orange, red, or almost black; the time of fruit ripening is from late July until late September. There is a hybrid between *Cornus mas* L. and *C. officinalis* Siebold & Zucc. called 'Etiud'.[24]

14.5.6   CRANBERRY BUSH OR GUELDER ROSE (KALYNA)

Viburnum opulus L. is a symbol of Ukraine, the Ukrainian people, Cossacks, homeland, and national unity. It embodies the maiden purity and beauty, eternal love, fidelity, femininity, and motherhood. It is a favorite image of Ukrainian folk poetry and has a sacred meaning. Since the ancient times, the cranberry bush has been planted near the house as an ornamental and fruit plant with medical properties. However, only in the 1980s, the commercial growing of this plant started. Some promising forms have been selected in several scientific institutions, for example, 'Kyivska Sadova 1' and 'Horianka',[16] but there are only three registered, namely 'Koralova', 'Rubinova', and 'Velykoplidna'. Spreading of sweet cultivars contributes to the growth of popularity of the cranberry bush as a fruit crop, because its fruit consists of less glycoside viburnin. The overall objectives of modern cranberry bush breeding are creating adaptive, productive, and resistant to pest cultivars with a high content of viburnin in fruit, which assigns their healing properties. However, there is a requirement to select cultivars without disagreeable smell in processing products and pleasant taste for use as a dessert.

14.5.7   JUJUBE (ZYZYFA)

Ziziphus jujuba Mill. is one of the oldest cultivated fruit crops, which is widely distributed in warm and drought regions throughout the world. It tolerates fairly cold winters, surviving temperatures down to about −25°C, but the tree crown of most cultivars is destroyed by the same frost.

Nevertheless, the crown quickly grows again and the plant bears fruit on 1-year-old wood. Jujube was introduced in Nikita Botanical Gardens in 1814. In 1954, the collection was enlarged a lot with Chinese cultivars.[25] The best Chinese cultivars under Crimean condition are Kytaiskyi 60 registered in 1994 and also 'Kytaiskyi 62', 'Kytaiskyi 93', 'Dabai-zao', 'Tayang-zao' (Lang), 'Xiaobai-zao', 'Ya-zao', and so forth. The new cultivar Yuzhanin of Tajikistan breeding also was proposed for growing. The collection of Chinese cultivars was served as the basis for successful jujube breeding.[26] The first Ukrainian cultivar 'Plodivskyi' was registered in 2006, and later this group was enlarged with new cultivars 'Koktebel', 'Sinit', and 'Tsukerkovyi'. The amateur horticulturists are cultivating foregoing jujube cultivars and seedlings. Most seedlings are small–sized fruit, but harder and less heat-loving plants. It allows jujube to grow in colder areas, for example, Vytryvala is suitable for the northeastern region.

14.6   REGISTERED RARE BERRY CROPS

14.6.1   HARDY KIWI (KTYNIDIIA)

Actinidia sp. became known to Europeans in the 19th century and in the second half of 19th century, they were put into cultivation. Kashchenko's works proved the possibility of cultivation of *A. kolomikta* (Maxim. & Rupr.) Maxim. and *A. arguta* (Siebold & Zucc.) Planch. ex Miq. as fruit plants in Ukraine. In the National Botanical Garden, a lot of fruit cultivars were created by analytical and synthetic breeding. The first of them were registered in 1992.[27] Among them, there are *A. arguta* 'Sentiabrska', *A. arguta* var. *purpurea* (Rehder) C. F. Liang Purpurna Sadova, *A. arguta* var. *arguta* × *purpurea* 'Fihurna', 'Karvaiska urozhaina'. 'Kiyvska Hibrydna', 'Kyivska Krupnoplidna', 'Lasunka', 'Nadia', 'Oryhinalna', 'Perlyna Sadu', 'Rima', 'Rubinova', 'Zahadkova', and pollinator variety 'Don Zhuan', and *A. polygama* (Siebold & Zucc.) Maxim. 'Pomarancheva' (Table 14.6).

The crossing combination *A. arguta* var. *arguta* × *purpurea* has been very productive. Some more promising cultivars, for example, 'Yuvileina' and 'Smarahdova', have been selected out of these hybrids.[28] In Ukraine, *A. arguta* is winter-hardy enough, though it may be damaged by spring frosts. It exceeds *A. kolomikta* in the yield and fruit size because the *A. kolomikta* is less common. Nevertheless, amateurs are cultivating *A. kolo-mikta* cultivars of Russian breeding and various seedlings.

TABLE 14.6 List of Rare Berry Crop Cultivars of Ukrainian Breeding in the State Register of Plant Cultivars of Ukraine.

Crop	Cultivar name (registration year)	Applicant
Hardy kiwi	'Fihurna' (1992)	Nat. Bot. Gard.
	'Kyivska Hibrydna' (1992)	–
	'Kyivska Krupnoplidna' (1992)	–
	'Purpurna Sadova' (1992)	–
	'Sentiabrska' (1992)	–
	'Karavaivska Urozhaina' (2001)	–
	'Oryhinalna' (2001)	–
	'Nadiia' (2001)	–
	'Perlyna Sadu' (2001)	–
	'Rima' (2001)	–
	'Rubinova' (2001)	–
	'Zahadkova' (2001)	–
	'Lasunka' (2006)	–
	'Pomarancheva' (2006)	–
	'Don Zhuan' (2008)	–
Persimmon	'Suputnyk' (1994)	–
	'Hora Hoverla' (2006)	Nova Kakhovka Exp. Farm
	'Hora Kosh' (2006)	–
	Hora Rodzhers' (2006)	–
	'Novynka' (2006)	Nikita Bot. Gard.
	'Mriia' (2010)	–
	'Nikitska Bordova' (2010)	–
	'Pivdenna Krasunia' (2010)	–
	'Rosiianka' (2010)	–
	'Suvenir Oseni' (2010)	–
	'Ukrainka' (2010)	–
	'Zirochka' (2010)	–
	'Zolotysta' (2010)	–
Pomegranate	'Sochnyi' (1994)	–
	'Niutinskyi' (2013)	–
	'Nikitskyi Rannii' (2013)	–
Fig	'Sabrutsiia Rozheva' (1994)	–

TABLE 14.6   *(Continued)*

Crop	Cultivar name (registration year)	Applicant
Magnolia vine	'Sadovyi 1' (1998)	–
Sea buckthorn	'Kyivskyi Yantar' (2000)	V. Dmytriiev
	'Lybid' (2000)	–
	'Solodka Zhinka' (2000)	Bakhmut Exp. St. Nurs. Cult.
Blue honeysuckle	'Bohdana' (2000)	Krasnokutsk Depart. and Inst. Plant Gen. Resour.
	'Donchanka' (2001)	Donetsk Bot. Gard.
	'Skifska' (2001)	–
	'Stepova' (2001)	–
	'Ukrainka' (2001)	–
	'Alisiia' (2010)	Krasnokutsk Depart.
	'Spokusa' (2010)	–
	'Chaika' (2010)	Exp. St. Pomol.
Mulberry	'Bilosnizhka' (2001)	Donetsk Bot. Gard.
	'Dina' (2001)	–
	'Mashenka' (2001)	–
	'Ukrainska 5' (2003)	Sericult. Depart.
	'Ukrainska 6' (2003)	–
	'Ukrainska 7' (2003)	–
	'Merefianska' (2004)	–
	'Nadiia' (2004)	–
	'Slobozhanska 1' (2004)	–
	'Merezhyvo' (2009)	Donetsk Bot. Gard.
	'Pivdenna Nich' (2009)	–
Pawpaw	'Michurinka' (2008)	Nova Kakhovka Exp. Farm
	'Novokakhovchanka' (2008)	–
	'Plodivchanka' (2008)	–
	'Victoriia' (2010)	Nikita Bot. Gard.
Blackberry	'Nasoloda' (2010)	Nat. Univ. Life Environm. Sc.
	'Sadove Chudo' (2013)	–
Feijoa	'Nikitska Aromatna' (2010)	Nikita Bot. Gard.
	'Aromatna Fantaziia' (2014)	–
Blueberry	'Fiolent' (2015)	V. Dmytriiev

14.6.2   KIWIFRUIT (KIVI)

Actinidia chinensis Planch is a subtropical plant; therefore, it should be covered in winter in most territory of Ukraine. The experimental kiwi plantation was established in Crimea in 1986. In amateur culture, it happens in the south of the country. In 2000, two New Zealand cultivars 'Monti' and 'Tomuri' (pollinator) were included in the State Register.

14.6.3   PERSIMMON (KHURMA)

Diospyros lotus L. is native to the Caucasus. This species was naturalized in Crimea a long time ago. Both East Asian *D. kaki* Thunb. and North American *D. virginiana* L. were introduced here as fruits in the 19th century. The first commercial plantation of oriental persimmon was established in Nikita Botanical Garden in 1901.[29] The first Ukrainian variety of *D. kaki* 'Suputnyk' was bred here and entered into the State Register in 1994, as well as some more were well-adapted for local conditions cultivars. The hybridogenous cultivars, namely *D. kaki* × *virginiana* F_1 'Rosiianka', *D. kaki* × *virginiana* F_2 'Nikitska Bordova', and *D. kaki* × *virginiana* F_3 'Hora Hoverla', 'Hora Kosh', and 'Hora Rodzhers', have been broadened the prospects persimmon cultivation in Ukraine. F_2 and F_3 interspecific hybrids can be grown outside the southern coast of Crimea, Crimean steppe and the southern areas such as Odesa, Mykolaiv, Kherson, Zaporizhzhia, and Zakarpattia regions. Extensive testing has proved that some cultivars of *D. kaki* can be grown outside the southern coast of Crimea.[30] 'Rosiianka' attains a weight of 50–60 g, and can be cultivated in the southern and central parts of Ukraine with winter temperatures below −20°C. The hardiest American cultivars of *D. virginiana* 'Meader' and 'John Rick' can be grown in most parts of Ukraine. *D. virginiana* 'Darunok Mamy' hasbeen bred in Nikita Botanical Garden.

14.6.4   POMEGRANATE (HRANAT OR HRANATNYK)

Punica granatum L. needs covering in winter in most of the territory of Ukraine. Therefore, it is rarely cultivated, mostly by amateurs. In outdoor conditions without a winter covering, the pomegranate tree grows only in Crimean subtropics. It was introduced there in ancient times.[31] Azerbaijan and Uzbekistan cultivars 'Achik-anor', 'Bala Miursal', 'Kazake-anor', and so forth are mainly grown. The first cultivar 'Sokovytyi' of Ukrainian breeding was registered in 1994 and later 'Niutinskyi' and 'Nikitskyi Rannii' were also included in the State Register.

14.6.5 FIG (SMOKIVNYTSIA OR INZHYR)

Figs were most likely to have been brought to Crimea by Genoeses. It sometimes runs wild there. Naturalized forms are adapted to local conditions, but they have the fruit of poor quality. In Nikita Botanical Garden, the numerical fig assortment of world breeding has been collected and tested. Foreign cultivars, such as 'Brunswick', 'Dalmatica', 'Datt de Nappl', 'Cadota', 'Lardaro', 'Muasson', and 'Napolitano' are the best under Crimean condition. In the 1920s, the breeding program based on intraspecific and interspecific hybridization, and polyploidy was begun.[32] As a result, some cultivars, including 'Nikitskyi Aromatnyi', 'Podarok Oktiabriu', 'Smena', 'Zhovtoplodyi', and 'Sabrutsiia Rozheva', have been bred.[33] The later cultivar has been listed in the State Register.

14.6.6 MAGNOLIA VINE (LYMONNYK)

In Ukraine, the Far-Eastern climber *Schisandra chinensis* (Turcz.) Baill. has been cultivated as fruit and for medical purposes since 1950s. It is popular as an excellent tonic and restorative. 'Sadovyi 1' is the first and the only cultivar registered in Ukraine.[34] Magnolia vine breeding carried out in the Institute of Horticulture as well[16] in addition to National Botanical Garden.

14.6.7 SEA BUCKTHORN (OBLIPYKHA OR SHCHETS)

Hippophae rhamnoides L. is widespread species of the genus, with the ranges of its eight subspecies. Only one of them, subsp. *carpatica* Rousi native to Ukraine, occurs in the Danube Delta. Some promising genotypes were selected out of natural populations,[16,35] but they were not introduced in horticulture practice. At the same time, the excellent Russian cultivars 'Novost Altaya', 'Prevoskhodnaya', and 'Chuyskaya' belonged to subsp. *mongolica* Rousi were zoned in Ukraine in 1988. In following decades, some cultivars of subsp. *mongolica* and subsp. *rhamnoides* from the Russian Federation began growing in the different regions of Ukraine. However, most of them are susceptible to fungal disease if they grow in chernozem. The first sea buckthorn cultivars 'Kyivskyi Yantar', 'Lybid', and 'Solodka Zhinka' of Ukrainian breeding were registered in 2000.

Sea buckthorn fruit is a rich source of biologically active substances, a valuable raw material for pharmaceutical and cosmetic industry. The fruit is nutritious, although they are very acidic. On the contrary, the fruit of 'Solodka Zhinka' (in the English language "sweet woman") is very palatable with high sugar/acid ratio. Fruit of cultivars of subsp. *mongolica* is of high quality, but the plants have a lower disease resistance level if they grow in firm soil.[36] Therefore, we have been created a hybrid of subsp. *rhamnoides* × *mongolica* 'Orange Revolution' (Fig. 14.2), which combines the best features of parental subspecies. It is very large-fruited, high-yielding, and what is important, resistant to verticillium wilt.

FIGURE 14.2   *Hippophae rhamnoides* 'Orange Revolution' fruit.

14.6.8   BLUE HONEYSUCKLE (SYNIA ZHYMOLOST OR YISTIVNA ZHYMOLOST)

First cultivars of edible honeysuckle were created and zoned in the 1980s in Russia, and then they were spread around the world. Amateurs have been cultivated 'Berel', 'Vasyuganskaya', 'Goluboye Vereteno', 'Lenin-gradskiy Velikan', 'Pavlovskaya', 'Sinyaya Ptitsa', 'Tomichka' and many other cultivars of Russian breeding, including seedlings and local selects. In 2000, *Lonicera caerulea* L. subsp. *kamtschatica* (Sevast.) Gladkova 'Bohdana' of joint breeding of Krasnokutsk Department of Horticulture of the Institute of Horticulture and N. I. Vavilov Institute of Plant Genetic Resources (the Russian Federation) was included in the State Register. Besides, there are registered cultivars 'Donchanka', 'Skifska', 'Ukrainka', and 'Stepova' of Donetsk Botanical Garden breeding; 'Alisiia' and 'Spokusa' of Krasnokutsk Department of Horticulture of the Institute of Horticulture, and 'Chaika' of L. P. Symyrenko Experimental Station of Pomology.[4,37]

14.6.9   MULBERRY (SHOVKOVYTSIA)

In Southern Ukraine, a mulberry was probably introduced during the ancient Greek colonization of the Black Sea Coast. In 1631, the first industrial plantation of mulberry was established in Kyiv.[38] In the following centuries, the mulberry plantations as fodder for feeding silkworm caterpillars were created throughout the country. The plants ran to the wild places from there, and feral plants were multiplicated by birds. Cultivars 'Kharkivska 3', 'Ukrainska 1', 'Ukrainska 9', and 'Ukrainska 107' were zoned in 1960–1970 as fodder for silkworms. The first fruit cultivars were registered in 2001. They are the tetraploid seedlings of Azerbaijan cultivars 'Abşeron-tut' called 'Bilosnizhka', 'Dina', and 'Mashenka' of Donetsk Botanical Garden breeding.[39] Later, there were also bred 'Merezhyvo' and 'Pivdenna Nich'. The cultivar 'Chornobrova' selected by us is a descendant of 'Abşeron-tut' too. The new cultivars for universal purpose as fodder and for fruit consumption 'Merefianska', 'Nadiia', 'Slobozhanska 1', 'Ukrainska 5', 'Ukrainska 6', and 'Ukrainska 7' are resistant to pathogens *Fusarium oxysporium*, *Verticillium dahliae*, *Mycosphaerella mori*, and *Pseudomonas amygdali*. They were bred in the Department of Sericulture and Technical Entomology with the help of interspecific hybridization of *Morus alba* × *bombycis* × *multicaulis*.[40] People grow many named local forms of *M. alba* with black and white fruit. *M. nigra* L. is important as fruit too, but due to poor winter hardiness, it is distributed only in the south regions. *M. nigra* 'Kartatsiia' distributed in Crimea.

14.6.10   PAWPAW (AZYMINA)

Asimina triloba Dunal was introduced in Ukraine in the early 20th century, but it has recently gained an importance as the fruit crop after importation of new cultivars from the United States.[41] Some cultivars 'Michurinka', 'Novokakhovchanka', and 'Plodivchanka' of the Nova Kakhovka Experimental Farm breeding and 'Victoriia' of Nikita Botanical Garden breeding are registered. In the National Botanical Garden, the promising select 'Rannia' is bred.

14.6.11   BLACKBERRY (OZHYNA)

In the natural flora of Ukraine, there are several dozens of species and nothospecies of blackberries. Fruit of native species mainly *Rubus plicatus* Weihe & Nees, *R. sulcatus* Vest, *R. nessensis* W. Hall, and *R. caesius* L. are gathered by people. In the world, a large assortment of blackberry cultivars was bred for industrial growth. In 1937, *Rubus flagellaris* Willd. 'Izobilnaya' ('Riasna') and *Rubus ×loganobaccus* L. H. Bailey 'Tekhas' of Russian breeding were recommended by the zoned assortment for cultivation in some regions of Ukraine, but later blackberries were not included to zoned assortments. In 2010, the first Ukrainian cultivar 'Nasoloda' was only registered. It is originated by crossing 'Triple Crown' × 'Thornfree'. Another cultivar 'Sadove Chudo' is a clone of 'Thornfree'.[42] Nowadays in Ukraine, blackberries and blackberry–raspberry hybrids have been supported mainly by amateurs and in the small farms, although experiments on industrial cultivation have been conducted and some modern cultivars of foreign breeding have been registered. Thorn free cultivars of foreign breeding are mainly spread.

14.6.12   FEIJOA (FEIKHOA)

As *Acca sellowiana* (O. Berg) Barrett is a subtropical plant, the usable area for cultivation is the southern coast of Crimea. The names of registered cultivars 'Nikitska Aromatna', and 'Aromatna Fantaziia' designate fruit aromaticity. 'Buhrysta' 'Krupnoploda', 'Nikitska, 'Rannia Aromatna', 'Pervenets', and 'Svitla' of Nikita Botanical Garden breeding along with American cultivars 'Coolidge' and 'Superb' are recommended for growing.

14.6.13   CINNAMON ROSE (SHYPSHYNA)

Many years ago, the varietal material and seedlings of *Rosa albertii* Regel, *R. iliensis* Chrshan. were used for the establishment of industrial plantations in Ukraine. Now, the industrial culture of this plant, rich with vitamin C, is based on the cultivars of Russian breeding originating mostly from *Rosa majalis* Herrm. Two cultivars 'Vitaminnyj VNIIVI' and 'Pozdno-sozrevayuschij' were entered to the State Register for Crimea. We have been bred the promising forms in the 'Vitaminnyi VNIIVI' progeny, that is derived from crossing *R. majalis* × *R. webbiana* Royle.

14.6.14   BLUEBERRY (LOKHYNA)

The indigenous species *Vaccinium uliginosum* L. and *V. myrtillus* L. are not domesticated in this country. People have gathered fruit in local plant population. In the early 20th century, *V. corymbosum* hort. was introduced into the commercial culture in the United States. Since 1980, its cultivars have been cultivated in Ukraine too. In 2008, 'Amanda', 'Bluestar', 'Carrie', 'Chyck', and 'Jonne' of American breeding were registered in the State Register. The list of introduced cultivars has been enlarged year by year. In 2015, first Ukrainian cultivar 'Fiolent' was registered. Many farmers have established the new plantation of this promising fruit crop.

In amateur cultures, there are also cultivars 'Ben Lear', 'Bergman', 'Franklin', 'Howes', 'MacFarlin', 'Piligrim', 'Stevens', and so forth of allied fruit crop cranberry (zhuravlyna).

14.7   REGISTERED RARE NUT CROPS

14.7.1   ALMOND (MYHDAL)

Prunus dulcis (Mill.) D. A. Webb has been cultivated in Ukraine from the 16th century. According to prewar assortment, it was recommended for growing in some southern areas, but in the zoned assortment of 1948, which took the coldest winters 1939/1940 and 1941/1942 into consideration, this crop was missing. Only in 1954, the first almond cultivars, named 'Krymskyi', 'Nikitskyi 62', and 'Yaltinskyi' bred in Nikita Botanical Garden, were included to the zoned assortment (Table 14.7).

TABLE 14.7   List of Rare Nut Crop Cultivars of Ukrainian Breeding in the State Register of Plant Cultivars of Ukraine.

Crop	Cultivar name (registration year)	Applicant
Almond	'Krymskyi' (1954)	Nikita Bot. Gard.
	'Nikitskyi 62' (1954)	–
	'Yaltynskyi' (1954)	–
	'Desertnyi' (1976)	–
	'Nikitskyi 2240' (1986)	–
	'Pryberezhnyi' (1986)	–
	'Milas' (2000)	–
	'Bospor' (2011)	–
	'Oleksandr' (2011)	–
	'Vitiaz' (2013)	–
Hazelnut	'Bolhradska Novynka' (1981)	Ukr. Res. Inst. Forest. & Forest Melior.
	'Shedevr' (1995)	–
	'Stepovyi 83' (1985)	–
	'Klynovydnyi' (1988)	–
	'Shokoladnyi' (1988)	–
	'Lozovskyi Kuliastyi' (1989)	–
	'Raketnyi' (1990)	–
	'Borovskyi' (1991)	–
	'Dar Pavlenka' (1991)	–
	'Koronchastyi' (1991)	–
	'Sribliastyi' (1991)	–
	'Pyrozhok' (1996)	–

Breeding of late-blooming cultivars has been improved to expand the area of industrial cultivation.[43,44] In Nikita Botanical Garden, more than 50 cultivars have been bred. It allows growing almonds in the whole Crimea and across the southern areas in Odesa, Kherson, Mykolaiv, and Zakarpattya regions. New cultivars combine later flowering with early fruit ripening.[45] 'Bospor', 'Desertnyi', 'Milas', 'Nikitskyi 2240', 'Oleksandr', 'Pryberezhnyi', and 'Vitiaz' have been listed in the State Register.

14.7.2   HAZELNUT (FUNDUK)

Corylus avellana L. is native to the flora of Ukraine and is widely cultivated. At the beginning of the last century, Western European and Turkish
cultivars were encouraged to grow there. In 1930, hazelnut breeding started
in the Ukrainian Research Institute of Forestry & Forest Melioration. As
a result, 'Bolhradska Novynka', 'Borovskyi', 'Dar Pavlenka', 'Klynopo-
dibnyi', 'Koronchastyi', 'Lozovskyi Kuliastui', 'Pyrozhok', 'Raketnyi',
'Shedevr', 'Shokoladnyi', 'Sribliastyi', and 'Stepovyi 83' were selected
among infra- and interspecies hybrids.[46] In the 1980s, they were zoned for
cultivation for all natural zones of Ukraine. Unfortunately, now these cultivars are not practically propagated by the industrial nurseries, although the
demand for planting material of hazelnut is very large. Therefore, propagation of foreign cultivars 'Ata-baba', 'Barcelona', 'Cosford', 'Ganja',
and 'Halle' has recently started.

14.8   REGISTERED ROOTSTOCKS

14.8.1   APPLE ROOTSTOCKS

Modern industrial horticulture is based on the use of clonal rootstocks.
Both dwarf and semidwarf rootstocks are necessary for intensive apple
production. In Ukraine, *Malus pumila* L. 'M 4' and 'M 9' are zoned. They
are able to withstand soil temperature drops up to −10 to −11.5°C and
−9 to −9.5°C, respectively, without damage.[47] Under winter condition of
the Left-Bank Ukraine, where the root system may be damaged, the new
rootstocks were bred and registered, in particular, 'D 471' and 'D 1071'.
They are results of crossing 'M 4' × 'Krasnyj Shtandart' and 'M 9' × 'Anis
Alyj', respectively. Other rootstocks 'KD 4' and 'KD 5' were created by
the way of hybridization between 'Paradizka Budagovskogo' ('PB 9'
= 'M 8' × 'Krasnyj Shtandart') and hybrid forms derived from crosses
'M 9' with *M. baccata* (L.) Borkh. and *M. prunifolia* (Willd.) Borkh.
The group of the newest rootstocks, namely 'Baturynska', 'Konotopska',
'Maliuk', 'Nadiia', 'Nizhynska', 'Sambirska', and 'Slobozhanska', are
progeny of crosses between 'Paradyzka Poltavska' (='M 9' × *M. prunifolia*) and '57–146' (= 'PB 9' F_1 originated from free pollination). These
new rootstocks are tolerant to winter soil temperature drops up to −12 to
−16°C.[48] Thus, the involvement of *M. baccata* and *M. prunifolia* to the

crossing with popular *M. pumila* rootstocks allowed to bred winter-hardy apple rootstocks that have good graft compatibility, providing the highest quality of budded plant material and high-yield grafted trees (Table 14.8).

TABLE 14.8   List of Rootstocks of Ukrainian Breeding in the State Register of Plant Cultivars of Ukraine.

Crop	Cultivar name (registration year)	Applicant
Apple rootstock	'D 471' (2000)	Bakhmut Exp. St. Nurs. Cult.
	'D 1071' (2000)	–
	'KD 4' (2007)	Krasnokutsk Depart.
	'KD 5' (2007)	–
	'Baturynska' (2009)	Inst. Hort.
	'Konotopska' (2009)	–
	'Maliuk' (2009)	–
	'Nadiia' (2009)	–
	'Nizhynska' (2009)	–
	'Sambirska' (2009)	–
	'Slobozhanska' (2009)	–
Pear rootstock	'IS 2–10' (2000)	–
	IS 4–6' (2000)	–
Universal pome crop rootstock	'UUPROZ-6' (2010)	Exp. St. Pomol. and Inst. Hort.
Myrobalan rootstock	'Vesennee Plamya' (2005)	Krymsk Exp. Breed. St. and Inst. Hort.
Sweet Cherry rootstock	'Studenikivska' (2013)	Inst. Hort.
Peach rootstock	'Pidshchepnyi 1' (1990)	Nat. Bot. Gard.
Grapevine rootstocks	'Dobrynia' (2007)	Inst. Viticult. Wine.
	'Harant' (2014)	–
	'4923' (2014)	–

14.8.2   PEAR ROOTSTOCKS

Pear rootstocks 'QA' and 'BA 29' of French breeding have been registered in Ukraine. Ukrainian rootstocks *C. oblonga* Mill. 'IS 2–10' and 'IS 4–6' produce higher sucker output. Pear trees on these stocks have been healthy and productive, less vigorous than those on *Pyrus* seedlings and early bearing.

14.8.3   UNIVERSAL POME CROP ROOTSTOCK

In 2010, × *Cydolus rudenkoana* Mezhenskyj (= *C. oblonga* × *Malus pumila*) 'UUPROZ-6' was registered. It is positioning as a rootstock for apples, pears, quinces, Japanese quinces, hawthorns, and rowans.[49]

14.8.4   SWEET CHERRY ROOTSTOCKS

Clonal rootstocks are an essential component of modern sweet cherry orchards. In 2006, dwarf rootstock 'VSL 2' of Russian breeding is registered in Ukraine. It is bred by crossing between *Prunus fruticosa* Pall. × *P. serrulata* Lindl. 'VSL 2' and is easily propagated by softwood cuttings, but very sensitive to viral infection.[50] In Ukraine, *Prunus* × *eminens* Beck 'Studenikivska' has been selected. The sweet cherry grafted on 'Studeni-kivska' has productivity up to 50 t/ha.[51]

14.8.5   MYROBALAN ROOTSTOCK

Semidwarf clonal rootstock 'Vesennee Plamya' of joint breeding of the Crimean Experimental Breeding Station VIR (Russian Federation) and the Institute of Horticulture in Kyiv was registered in the State Register in 2005. It has a complex hybrid from crossbreeding of several plum species (*Prunus americana* Marsh. × *P.* ×*simonii* Carrière) × *P. cerasifera.*

14.8.6   PEACH ROOTSTOCKS

Usually, peach is grown on peach, apricot, myrobalan or almond seedlings. In 1990, seed rootstock 'Pidshchepnyi 1' of the National Botanical Garden breeding, which is a hybrid between introduced from China *Prunus david-iana* (Carrière) Franch. and *P. kansuensis* Rehder, for the first time was zoned in the steppe. Currently, a number of new rootstocks originating from distant crosses have been tested.

14.8.7 GRAPEVINE ROOTSTOCKS

In Ukraine, three *Phyloxera*-resistant rootstocks of Ukrainian breeding have been registered. Taxonomically, rootstock '4923' belong to the *Vitis ×instabilis* Ardenghi, Galasso, Banfi & Lastrucci (= *V. riparia × rupestris*). 'Dobrynia' is a hybrid*Vitis vinifera* L. × *Vitis rupestris* Scheele, and 'Harant' is a complex hybrid from crossbreeding of *Vitis koberi* Ardenghi, Galasso, Banfi & Lastrucci × [(*Vitis vinifera* × *Vitis riparia* Michx.) + *Vitis × ruggerii* Ardenghi, Galasso, Banfi & Lastrucci + *Vitis riparia* Michx.]. There are five long-standing rootstocks of foreign breeding too. In 1974, *Vitis × koberi* 'SO4', '5BB', and '5C' and *Vitis × instabilis* '101–14' and '3309' were included in the State Register.

14.9 BRED FRUIT PLANTS WHOSE CULTIVARS HAVE NOT BEEN REGISTERED YET

14.9.1 ROWAN (HOROBYNA)

The clones of native to Ukraine *Sorbus aucuparia* L. with non-astringent fruit and the Far-Eastern *Sorbus sambucifolia* (Cham. & Schlecht.) M. Roem. are the most important for cultivation as fruits. Sweet-flavor clones of rowan were cultivated in Ukraine in the late 19th century. They were introduced to Germany from here under the names 'Rossica' and 'Rossica Major'. The variety 'Moravska Vrozhaina' bred by us belongs to another group of sweet-flavor rowan clones, originated from Moravia.[6,8]

14.9.2 SERVICE TREE (VELYKOPLODA HOROBYNA OR DOMASHNYA HOROBYNA)

Cormus domestica (L.) Spach has been cultivated in the Mediterranean and Crimea since ancient times. In Nikita Botanical Garden, the best local forms 'Lymonna', 'Nikitska', 'Nikitska 520', 'Rubinova', 'Solodka', and 'Tavryda' were preserved. They are offered for growing in the coastal areas of Crimea, Odesa, Kherson, and Mykolaiv regions.[52] In the different parts of this country, some genotypes were listed, for example, 'Barvinok 1', 'Barvinok 2', and

'Medvedivska' in Zakarpattia region; 'Zhovtneva' in Kyiv; and 'Rumiana Hrushka' and 'Rumiane Yabluchko' in the Donbas area.[6,8]

14.9.3   CHOKEBERRY (ARONIA OR CHORNOPLODA HOROBYNA)

Aronia mitschurinii A. K. Skvortsov & Maitulina has been grown in commercial plantations in Ukraine since 1958. Michurin's chokeberry propagated for industrial plantations from seed as well as apomictic offspring inherits economic features of mother plants. Since genes of rowan were found out in the genome of Michurin's chokeberry, it was proposed to rename it in *Sorbaronia mitschurinii* (A. K. Skvortsov & Maitulina) Sennikov, although morphologically it is closer to *Aronia*. Genetically, our select 'Vseslava' (Fig. 14.3) is certain to be *Sorbaronia*, but it is similar to Michurin's chokeberry. 'Vseslava' has more fruit in the inflorescences and fruit is bigger.[6,8]

FIGURE 14.3   Chokeberry 'Viking' (above) and 'Vseslava' (below) fruit.

14.9.4 SERVICEBERRY (SADOVA IRHA)

Native to Ukraine, *Amelanchier ovalis* Medik. is not important as fruits crop compared to North American species in this genus. In 1886, serviceberry was first tested in Uman School of Horticulture for the winemaking purpose. Today, serviceberries are common plants in amateur orchards as fruit crop as well as pear dwarf rootstock. We have bred promising forms of *A. alnifolia* and *A. spicata* called 'Bluesun' and 'Bluemoon', respectively.[6,8]

14.9.5 MANCHU CHERRY OR NANKING CHERRY (POVSTYANA VYSHNIA)

Since the 1930s, *Prunus tomentosa* Thunb. has been grown by Ukrainian amateurs. This fruit shrub is very popular among them because of profuse fruiting, early ripening, and ease of growing and propagation. Manchu cherry is important as the dwarf rootstocks for stone fruits. As Manchu cherry is not resistant to *Monilia,* hybrids between *P. tomentosa* and *P. ulmifolia* Franch. are more promising for growing. Some cultivars, for example, 'Oksamytova', 'Temna Rannia', 'Temna Piznia', and 'Yefimka', which are more tolerant to *Monilia* with bigger and darker fruit and with firmer flesh, have been bred.[6,8]

14.9.6 BESSEY CHERRY (BESSEIA)

Prunus besseyi L. H. Bailey is known as rootstock and a component of interspecific crossings in the breeding programs of the clonal rootstocks for stone fruits. We have bred variety 'Sonechko' with yellow edible fruit.[6,8]

14.9.7 PLUMCOTS (ABRYKOSOSLYVA)

In Melitopol Experimental Station of Horticulture, *Prunus ×dasycarpa* Ehrh. 'Melitopolska Chorna' has been bred. It has big palatable fruit that is intermediate between apricot and myrobalan, but the yield is lower. In Nikita Botanical Garden, hybrids between *P. brigántina* Vill. and *P. armeniaca* L. have been created. They are later blooming and more tolerant to biotic and abiotic environmental factors.[53] Out of the F_2 and F_3 hybrids, some promising genotypes with a complex of valuable features have been selected. These include 'Briol', 'Brigmas', 'Krymskyi Samotsvit', 'Nikitski Rossypi', 'Nikitske Sokrovyshche', 'Nikitskyi Novyi', 'Yaltynski Vohni', 'Yaltynskyi Klad', and so forth. A new selects are different from apricots on phenology. They have fruit with an original appearance, taste, and biochemical composition.[54,55]

14.9.8   GOLDEN CURRENT (ZOLOTYSTI PORICHKY)

In Ukraine, *Ribes aureum* Pursh. was introduced in the early 19th century. Seedlings of golden currant, due to tolerance to droughts, are widely used in phytoreclamation plantations. Golden currant promising forms have been selected in Melitopol Experimental Station of Horticulture and the National University of Life and Environmental Sciences of Ukraine. In the latter of them, 'Monastyrska', 'Perlyna Didorivky', 'Pyriatynska Pokrashchena', 'Samorodok', 'Vyshneva', and 'Yantarna' have been bred.[42]

14.9.9   BARBERRY (BARBARYS)

In the flora of Ukraine, there are indigenous species *Berberis vulgaris* L. and more than 100 introduced barberries. Many of them may be fruit crop. The selects of our breeding are promising for fruit growing. These are 'Chervonyi Veleten', 'Likhtaryk', and 'Black Giant' with large fruit; 'Likhtaryk' and 'Tsukerka' with delicious sour-sweet fruit; 'Beznasinnievyi Chervonyi' and 'Beznasinnievyi Zhovtyi' with seedless fruit. Large-fruited *B. aquifolium* Pursh 'Bluegrape' and 'Bluecloud' are fruit plants too.[7,8]

Moreover, scientists and amateurs have bred many interesting genotypes of Nashi (Nashi or Yaponska hrusha)—*Pyrus pyrifolia* (Burm. f.) Nakai; Medlar (Mushmula)—*Mespilus germanica* L.; Whitebeam (Ariia or Mukynia)—*Aria nivea* Host, *A. szovitsii* Decne., and intergeneric hybrids related to them; Checker tree (Bereka)—*Torminalis glaberrima* (Gand.) Sennikov; Chokecherry (Virdzhynska cheremkha)—*Prunus virginiana* L.; Elder (Buzyna)—*Sambucus nigra* L.;Goumi (Gumi)—*Elaeagnus multiflora* Thunb.; Buffalo berry (Sheferdia)—*Shepherdia argentea* (Pursh) Nutt.; Goji (Povii or Dereza)—*Lycium barbatrum* L., Cudrang (Kudraniia)—*Maclura tricuspidata* Carrière; Passionflower (Strastotsvit)—*Passiflora incarnata* L.; Turkish hazelnut (Vedmezha lishchyna)—*Corylus colurna* L.; Pecan (Pekan)—*Carya illinoensis* (Wangenh.) K. Koch, and so forth.[6-8]

In Ukraine, 178 cultivars of rare fruit crops and rootstocks of Ukrainian breeding have totally been registered (see Table 14.1). This fruitful breeding has been carried out by 12 institutions and individuals. The leaders in the rare fruit breeding are Nikita Botanical Garden (55 registered cultivars) and the National Botanical Garden (42 cultivars). The list of Crimean myrobalan and myrobalan–Japanese plums cultivars consists of 20 cultivars. A numerousand diverse assortment of cornelian cherry and hardy kiwi is represented by 16 and 15 cultivars, respectively. Rather large assortment of pome crop rootstock (14 cultivars), persimmon (13), hazelnut (12), quince (13), mulberry (11), almond (10), honeysuckle (8), and Japanese quince (8 cultivars) has been bred. Other crops are represented in the State Register by one to four cultivars.

14.10 CONCLUSION

The natural conditions of Ukraine allow cultivating about 1000 species of fruit plants. Apple, pear, apricot, peach, prune, sour cherry and duke, sweet cherry, gooseberry, black and red currant, raspberry, strawberry, grape, and Persian walnut have been traditionally cultivated. In the second half of the 20th century, the cultivars of new fruit crop were started to bring to the State Register of Plant Cultivars of Ukraine. These are pome fruits: quince, hawthorn, Japanese quince; stone fruits: Crimean myrobalan and myrobalan–Japanese plum hybrids, nectarine, olive, cornelian cherry, cranberry bush, jujube; berry fruit crops: hardy kiwi, persimmon, pomegranate, fig, magnolia vine, cinnamon rose, seabuckthorn, blue honeysuckle, kiwifruit, mulberry, lemon, pawpaw, blackberry, blueberry, feijoa; and nut crops: almond, hazelnut, and rootstocks. In the State Register, cinnamon rose, kiwifruit, and lemon are represented by exclusive cultivars of foreign breeding. Total 158 cultivars of rare plants and 20 rootstocks of Ukrainian selection have been registered. There are also many selects of other fruit plants that are not registered yet.

KEYWORDS

- **horticulture**
- **plant breeding**
- **pome fruits**
- **stone fruits**
- **small fruit crops**
- **nut crops**
- **rootstock breeding**

REFERENCES

1. The Regionalized Cultivars of Agricultural Crops for Ukrainian SSR for. Urozhai: Kyiv, 1991; p 269 (in Ukrainian).
2. The Register of Plant Cultivars of Ukraine for. Alefa: Kyiv, 2001; p 139 (in Ukrainian).
3. State Register of Plant Cultivars Suitable for Dissemination in Ukraine in 2005. Alefa: Kyiv, 2005; p 243 (in Ukrainian).
4. State Register of Plant Cultivars Suitable for Dissemination in Ukraine in 2016, (in Ukrainian). http://vet.gov.ua/sites/default/files/Reestr%2022.02.16.pdf.
5. Andriienko, M. V. Introduction, Cultivation and Use of Rare Fruit and Small Fruit Crops in Woodlands and Forest-Steppes of Ukraine. Abstract of Dr. Agr. Sc. Thesis, Kyiv, 1999; p 50 (in Ukrainian).
6. Mezhenskyj, V. M.; Mezhenska, L. O.; Melnychuk, M. D.; Yakubenko, B.Ye. *Rare Fruit Crops: Recommendations for Breeding and Propagation;* Phytosociocentre: Kyiv, 2012; p 80 (in Ukrainian).
7. Mezhenskyj, V. M.; Mezhenska, L. O.; Yakubenko, BYe. *Rare Small Fruit Crops: Recommendations for Breeding and Propagation;* Comprint: Kyiv, 2014; p 119 (in Ukrainian).
8. Mezhenskyj, V. M.; Mezhenska, L. O. The *Formation of the Collection and Improvement of Plant Breeding Methods of Rare Fruit and Ornamental Crops;* Comprint: Kyiv, 2015; p 480 (in Ukrainian).
9. Mezhenskyj, V. M. Species Composition of Fruit Plants in Ukraine and Prospect for Their Use. *Plant Introd.* **2008,** *1,* 8–19 (in Ukrainian).
10. Riabchun, V. K.; Bohyslavskyi, R. L. *Plant Genetic Recourses and Their Importance for Breeding. Theoretical Principles of Field Crops Breeding;* V. Ja. Yuriev Institute Plant Industry: Kharkiv, 2007; pp 363–398 (in Ukrainian).
11. Mezhenskyj, V. M. Improvement of Economic Botanical Classification of Fruit Plants. Scientific Proceedings of the National University of Life and Environmental

Sciences of Ukraine. **2011,** *4,* (in Ukrainian). http://www.nbuv.gov.ua/e-journals/Nd/2011_4/11mvm.pdf.

12. Klimenko, Z. K.; Rubtsova, E. L.; Zykova, V. K. *Nilolai von Hartviss—A Second Director of the Imperial Nikita Garden;* Ahrarna Nauka; Simferopol: Oreanda: Kyiv, 2012; p 80 (in Russian).

13. Smykov, A. V. Genofond Formation of South Fruit Crops in Nikita Botanical Garden. *Works Nikita Bot. Gard.* **2010,** *132,* pp 5–7 (in Russian).

14. Kashchenko, N. F. The First Step of my Nursery for Plant Acclimatization in the City of Kiev. Rostov-on-Don, 1914; p 24 (in Russian).

15. Klimenko, S. V. Introduction and Breeding of South, New and Non-Traditional Fruit Plants in the National Botanical Garden of Ukraine: History, Results and Prospects. Non-Traditional, New and Forgotten Plant Species: Scientific and Practical Aspects of Cultivation. Knihonosha: Kyiv, 2013; pp 56–64 (in Russian).

16. Kopan, V. P., Ed. *Atlas of Promising Fruit and Small Fruit Crops Cultivars of Ukraine;* Odeks: Kyiv, 1999; p 454 (in Russian).

17. Klimenko, S. V. *Common Quince;* Naukova Dumka: Kyiv, 1993; p 286 (in Russian).

18. Klimenko, S. V.; Skripchenko, N. B. Cultivars of Both Fruit and Small Fruit Crops of the M. M. Hryshko National Botanical Garden breeding. Kyiv, 2013; p 104 (in Russian).

19. Klymenko, S. V.; Mezhenskyj, V. M. Origin of Chaenomeles Cultivars of the Ukrainian Breeding. *Plant Introd.* **2013,** *4,* 25–30 (in Ukrainian).

20. Mezhenskyj, V. M. Evolutionary Changes in Breeding Improvement of Japanese Quinces as Fruit Crop. Indigenous and Introduced Plants, Proceedings of the National Arboretum "Sofoivka," **2009,** *5,* 126–132 (in Ukrainian).

21. Mezhenska, L. O.; Mezhenskyj, V. M. *Genus Hawtorn (Crataegus L.) in Ukraine: Introduction, Breeding, and Eco-biological Features;* Comprint: Kyiv, 2013; p 234, [40 ill.] (in Ukrainian).

22. Eryomin, G. V.; Garkovenko, V. M. Intraspecific Taxonomy of Myrobalan Plum. *Bull. Appl. Bot. Genet. plant Breed.* **1989,** *123,* 9–15 (in Russian).

23. Shoferistov, YeP.; Tsiupka, SYu.; Ivashchenko, YuA. Prospects of Breeding Fungal Diseases Resistant Cultivars of Nectarine (*Prunus persica* (L.) Batsch. subsp. *Nectarina* (Ait.) Shof.). *Plant Var. Stud. and Prot.* **2014,** *1,* 49–53 (in Russian).

24. Klymenko, S. V. *Cornelian Cherry in Ukraine: Growing Biology, and Cultivars;* Phytosociocentre: Kyiv, 2000; p 91 (in Ukrainian).

25. Sinko, L. T. Jujube is One of the Most Important Subtropical Fruit Crops for the South of the USSR. *Works Nikita Bot. Gard.* **1971,** *52,* 31–53 (in Russian).

26. Sinko, L. T.; Litvinova, T. V.; Shevchenko, S. V. Jujube Breeding. *Works Nikita Bot. Gard.* **1999,** *118,* 78–84 (in Russian).

27. Moroz, P. A.; Grynenko, N. S.; Skripchenko, N. V. Introduction and Breeding of Hardy Kiwi: Progress and Prospects of the Research Advancement. *Plant Introd.* **2002,** *2,* 14–23 (in Russian).

28. Skrypchenko, N. V.; Moroz, P. A. *Hardy Kiwi (Cultivars, Growing, Propagation);* Phytosociocentre: Kyiv, 2002; p 43 (in Ukrainian).

29. Kazas, A. I. History of the Persimmon Culture in Ukraine. Progress and Problems of Plant Introduction in Ukraine. Ailant: Kherson, 2007; p 55 (in Russian).

30. Derevianko, V. M. Cultural Prospects of the Japanese Persimmon (*Diospyros* kaki) and It Hybrids with American Persimmon (D. virginiana) in South of Ukraine. In *Progress and Problems of Plant Introduction in Ukraine*. Ailant: Kherson, 2007; p 37–38 (in Ukrainian).

31. Sinko, L. T.; Yadrov, A. A. *Pomeganate. Nut and Subtropical Fruit Crops;* Tauria: Simferopol, 1990; pp 55–69 (in Russian).

32. Arendt, N. K. Breeding of Fig in the Crimea. *Works Nikita Bot. Gard.* **1964**, *37,* 190–213 (in Russian).

33. Shishkina, YeL. The Creation of Fig Collection Fund in Nikita Botanical Garden and Its Using. *Works Nikita Bot. Gard.* **2010**, *132,* 185–189 (in Russian).

34. Shaitan, I. M.; Kleeva, R. F.; Chuprina, L. M.; Tereshchenko, T. P. *Magnolia Vine. Introduction and Breeding of South and New Fruit Crops;* Naukova Dumka: Kyiv, 1983; pp 114–121 (in Russian).

35. Lebeda, A. F.; Dzhurenko, N. I. *Sea Buckthorn in Ukraine;* Naukova Dumka: Kyiv, 1990; p 78 (in Russian).

36. Mezhenskyj, V. N. Some Results of Variety Investigation and Breeding of Sea Buckthorn in the Donbas. *Problems of Stable Development of Horticulture of the Siberia;* Barnaul, 2003; pp 79–81 (in Russian).

37. Glukhov, A. Z.; Kostyrko, D. R.; Osavliuk, S. N. *Species of the Genus Lonicera L. in the South-East of Ukraine. Introduction, Biomorphology and Using;* Lebed: Donetsk, 2002; p 122 (in Russian).

38. Kokhno, N. A.; Kurdiuk, A. M. *Theoretical Principles and Experience of Wood Plants Introduction in Ukraine;* Naukova Dumka: Kyiv, 1994; p 186 (in Russian).

39. Hlukhov, O. Z.; Kostyrko, D. P.; Mitina, L. V. *Fruit Mulberry Morus alba L. in the South-East of Ukraine;* Lebid: Donetsk, 2003; p 138 (in Ukrainian).

40. Oleksiichenko, N. O. *Breeding of Mulberry in Ukraine: Problems, Progress, and Prospects;* Kyiv National Linguistic University: Kyiv, 2007; p 303 (in Ukrainian).

41. Hrabovetska, O. A. Pawpaw (*Asimina triloba* (L.) Dunal) in the Steppe of Ukraine: Introduction, Biology, and Reproduction. Ph. D. Thesis, Kyiv, 2011, p 21 (in Ukrainian).

42. Sherenhovyi, P. Z. *My Life Is in My Cultivars;* Vinnytsia, 2011; p 167 (in Ukrainian).

43. Rikhter, A. A. Almond. *Works Nikita Bot. Gard.* **1972**, *57,* p 111 (in Russian).

44. Yadrov, O. O.; Chernobai, I. H. Main Directions and Progress of Almond Breeding in Ukraine. Genetics and Breeding in Ukraine on the Verge of Millennium. Lohos: Kyiv, 2001; Vol. 3, pp 429–442 (in Ukrainian).

45. Chernobaj, I. G. Almond in the Crimea: Biological Peculiarities and Economical Importance. *Works Nikita Bot. Gard.* **2010**; *132,* pp 196–202 (in Russian).

46. Pavlenko, F. A. *Hazel. Nuts;* Urozhai. P: Kyiv, 1987; pp 147–181 (in Russian).

47. Omelchenko, I. K. *Growing of Apple in Ukraine;* Urozhai: Kyiv, 2006; p 304 (in Ukrainian).

48. Kodratenko, P. V.; Kondratenko, T. Ye., Eds. Pomology. *Apples;* Nilan-Ltd: Vinnytsia, 2013; p 624 (in Ukrainian).

49. Matviienko, M. V.; Babina, R. D.; Kondratenko, P. V. *Pear in Ukraine;* Ahrarna Dumka: Kyiv, 2006; p 315 (in Ukrainian).

50. Mezhenskyj, V. M.; Brusentsov, V. P. Supersensitivity of Cherry and Sweet Cherry Clonal Rootsticks to the Virus Infection. *Horticulture* **2006**, *59,* 177–181 (in Ukrainian).

51. Tretiak, K. D.; Zavhorodnia, V. H.; Turovtsev, M. I. *Sour and Sweet Cherries;* Urozhai: Kyiv, 1990; p 173 (in Ukrainian).
52. Chernobaj, I. G. Service Tree. *Works Nikita Bot. Gard.* **2010,** *132,* 181–184 (in Russian).
53. Kostina, K. F. Hybrids Between Alpine Plum and Myrobalan, also Alpine Plum and Apricot. *Works Nikita Bot. Gard.* **1978,** *76,* 111–122 (in Russian).
54. Komar-Temna, L. D.; Horina, V. M. Promising Hybrids Between Alpine plum (*Prunus brigantiaca* Vill.) and Apricots (*Armeniaca vulgaris* Lam. and *Armeniaca leiocarpa* Kostina). Scientific Herald of the National University of Life and Environmental Sciences of Ukraine. **2009,** *133,* 137–143 (in Ukrainian).
55. Mezhenskyj, V. M. Introduction of *Prunus brigantiaca* Vill. Hybrids with Both Myrobalan and Apricot in South-East of Ukraine. *Bull. Nikita Bot. Gard.* **2009,** *99,* 76–79 (in Ukrainian).

CHAPTER 15

ECOLOGICAL PLASTICITY AND PRODUCTIVITY OF BLACK CURRANT CULTIVARS UNDER INSTABILITY OF ENVIRONMENT

TAT'YANA V. ZHIDYOKHINA

I. V. Michurin All-Russia Research Institute for Horticulture. Russian Academy of Sciences, 30, Michurin St., Michurinsk, Tambov Region, Russia, 393774

`Corresponding author. E-mail: berrys-m@mail.ru

CONTENTS

ABSTRACT

The chapter presents a comprehensive assessment for adaptive and productive potential of black currant cultivars bred in I. V. Michurin All-Russian Scientific Research Institute of Certification [now I. V. Michurin Federal Scientific Center (I. V. Michurin FSC)]. It was stated that 'Zeljonaja dymka', 'Talisman', 'Tamerlan', 'Chernavka', and 'Shalun'ja' cultivars are characterized by standardly good and excellent plant state. The plant state was greatly affected by plant age ($r=-0.63$) but not by sum of the temperatures of the previous period ($r=0.35$). The cultivars were grouped by the initial flowering dates. Early blooming cultivars are 'Divo Zvjaginoj', 'Karmelita', 'Sensej', and 'Shalun'ja'. The medium-blooming cultivars are 'Zeljonaja dymka', 'Izumrudnoe ozherel'je', 'Malen'kij princ', 'Prima', 'Rossiyanka', 'Sozvezdie', 'Talisman', 'Charovnica', 'Chernavka', and 'Chjornyi zhemchug'; medium-late flowering cultivars are 'Bagira', 'Vospominanie', 'Pandora', 'Tamerlan', 'Tat'janin den'', and 'Elevesta'; and late blooming cultivar is 'Amirani'. The complex resistance to fungal diseases was found in the 'Malen'kij princ', 'Pandora', 'Tamerlan', and 'Chernavka' cultivars and to mites in the 'Amirani', 'Izumrudnoe ozherel'je', 'Karmelita', 'Prima', 'Rossiyanka', and 'Sensej' cultivars. The most plastic black currant cultivars are Chernavka, Malen'kij princ', and 'Zeljonaja dymka', on the whole, according to 2000–2016 period of investigations; they are characterized by high productivity, namely 9.7 ± 0.6, 9.2 ± 0.6, and 9.1 ± 0.6 t/ha, respectively.

15.1   INTRODUCTION

Michurinsk, scientific town of Russia Federation, is All-Russian Center of Horticulture. It is located in the territory of Tambov Region, in the western part of Oka-Don plain, on the right bank of the Lesnoy Voronezh river (the basin of the Don river). Geographic coordinates of the town are 52°53′32″ of northern latitude and 40°29′34″ of Eastern longitude. The Central Chernozem Region including Tambov Region is characterized with favorable soil climatic conditions for the cultivation of a great set of horticultural crops.[1] For the last time, we can observe the intensification of complex effect of unfavorable factors on agricultural crops. The main stressors of winter period are long and intensive thaws followed by the sharp decrease of air temperature; in spring, the effect of low temperatures on the background of overmoistured soil, significant differences of day and night temperatures with a high level of solar radiation, in the afternoon are observed; in summer, there is drought on the background of extremely high temperatures with big amounts of rainfalls; in autumn, there is sharp decrease of air temperature up to negative meanings with the absence of snow cover.[2] In regions with variable meteorological conditions, the ecological sustainability of cultivars is becoming increasingly important along with potential productivity. In this context, the important trend of the investigations is a task for development of ecologically plastic genotypes, providing rather high yield under the favorable conditions of cultivation and stability of the yield in stressor situations.

The black currant is the second popular berry-like crop on the territory of Tambov Region. It is a typical mesophytic cold-resistant plant of temperate and temperate-cold climate. In general, it is a highly winter-resistant crop, which can survive the temperature decrease up to -35 to $-40°C$, and in definite winters, the plants are damaged by frosts. Very often generative buds are affected and the first characters of damaged buds are observed after temperature decrease up to -25 to $-30°C$ and after continuous thaws with sharp increase at $-15°C$. In definite years, "scalds of branches" are observed (damage of cambial lay) on the level of snow cover and they are obviously exposed to further withering. The main roots of black currant can survive the temperature decrease up to $-20°C$, but the feeding roots are strongly damaged at soil temperature -3.7 to $-4.7°C$. The recurrent spring frosts promote a great damage of buds-flowers and fruit-sets which begin to die at -2 to $-3°C$. Black currant is a moisture-requiring crop which in case of temperature decrease above $30°C$ and at the low moisture of air loses the fruit-sets and the long-term effect of the given factors promotes defoliation even under the circumstances of decent watering. The unshaded fruits of several cultivars can be scalded in hot and sunny weather.[3]

Breeding and scientific researches of the black currant in All-Russian Research Institute of Horticulture started in 1948 under the leadership of Klavdija D. Sergeyeva. The main task at the first stage was the obtaining of winter-resistant, high-yielding large-fruited cultivars of different terms of maturation with high vitamin C content. The breeding work was based on the intervarietal crossings within *Ribes nigrum* ssp. *europaeum* (Jancz) Pavl. As initial parental forms, the following cultivars were used: Baldwin, Boskoops Goliath, Davison's Eight, and Lees Prolific. As a result of that work, the cultivars 'Krupnoplodnaya', 'Michurinka', and 'Smena' were developed. The comparison of the obtained cultivars with initial forms showed that they are likely to combine the best traits of the parents, and even exceed them in yield, fruit-size, and vitamin C content. But the progeny that has been obtained from *Ribes nigrum* ssp. *europaeum* (Jancz) Pavl. was not resistant to currant bud mite, reversion disease, and anthracnose. Later, the breeders focused on the development of high selfed cultivars with higher resistance to currant bud mite and reversion disease. The breeding was conducted with the use of reciprocal and convergent crossings with inclusion of *Ribes nigrum* ssp. *europaeum* (Jancz) Pavl., *R. nigrum* ssp. *sibiricum* E. Wolf., and *R. dikuscha* Fisch. As initial parental forms, 'Boskoop Goliath', 'Vystavochnaya, 'Golubka', 'Lees Prolific', 'Nadezhda', 'Narodnaya', 'Pamyat' 'Michurina', 'Pobeda', 'Primorsky champion', and 'Smena' cultivars were used. As a result of all these, the cultivars 'Otbornaya', 'Prima', 'Rossiyanka', and 'Smuglyanka' were obtained; such cultivars are characterized with a good selft-fertilization, high resistance to currant bud mite, and high accumulation levels of sugars and vitamin C in berries. In 1969, for the first time in Michurinsk, the *Sphaerotheca mors-uvae* (Schw.) Berk. et Curt. was observed on the black currant. The numerous crossings were held with inclosing of *Ribes nigrum* ssp. *scandinavicum* Jancz. Hybrid fund of the Institute promoted selection of the 'Bagira', 'Vospominanie', 'Zeljonaja dymka', 'Ljubava', 'Sozvezdie', 'Tat'janin den'', 'Charovnica', and 'Chjornyi zhemchug' cultivars. During the experiment in February of warm days equal to 10 was in 2000 (14 days), 10 days in 2001, 24 days in 2002, 11 days in 2008, 12 days in 2013, 12 days in 2014, 12 days in 2015, and 23 days in 2016.[4]

At the beginning of the 1970s, the consecutive enrichment of genetic diversity has been performed. Not only *Ribes nigrum* ssp. *europaeum* (Jancz) Pavl., *R. nigrum* ssp. *sibiricum* E. Wolf., *R. nigrum* ssp. *scandinavicum* Jancz., and *R. dikuscha* Fisch. were included in breeding programs but the offsprings of the currant *R. glutinosum* Benth., *R. procumbens* Pall., *R. bracteosum* Dougl., *R. ussuriensis* Jancz., and *R. petiolare* Dougl were also included. Breeding strategy was based on interspecific and distant hybridization, convergent crossings, backcrosses, sibs crossings, and inbreeding. Involvement in breeding of such genetically diverse and geographically distant initial material enabled to obtain the breeding material with a great range of variation and improved inheritance. The new cultivars with high level of economically important traits were isolated, these were: 'Divo Zvjaginoi', 'Izumrudnoye ozherel'ye', 'Karmelita', 'Malen'kij princ', 'Sensej', 'Tamerlan', 'Gernavka', 'Shalun'ja', and 'Elevesta'.[5]

The modern conception for development of black currant cultivars consists of adaptability to major diseases and pests.[1,6–8] In the cause of adaptive breeding, the most efficient technique of cultivar improving is in the construction of genotypes on the base of the known oligogenes, controlling as a rule resistance to diseases and pests, and polygenes providing the combination of adaptability and productivity components.[9,10] The recent generation includes the black currant cultivars sent to the state cultivar trials, which were 'Talisman', 'Pandora', and 'Amirani' after 2010.

On May 28, 2017, the 110 anniversary of birth of the outstanding breeder, an honorary active worker of science RF, and doctor of agricultural sciences K. D. Sergeyeva will be celebrated. On the eve of the jubilee, we devote our study to the evaluation of black currant cultivars released in different years for complex of agronomic traits in unstable conditions of the environment.

15.2 MATERIALS AND METHODOLOGY

The black currant adaptive and productive potential was studied on the experimental plot of berry-like crops at I. V. Michurin All-Russian Scientific Research Institute of Certification in 2000–2016. The plots were designed as 4.0×0.75 m. The plants of 21 black currant cultivars of were observed as experimental objects bred at the institute (Table 15.1) 30 plants per cultivar.

Investigations were held using standard methodical recommendations.[11] General plant state was estimated in points: 1—very weak plants and dying; 2—weak state, bushes with a lack of foliage and have a bad increment, and so forth; 3—medium: slightly weakened, medium increment, undeveloped leaves; 4—good: healthy bushes with a good foliage and increment, show insignificant injuries by frosts, diseases, and pests; and 5—excellent: healthy bushes with typical well-developed leaves. While estimating the resistance to fungal diseases [American mildew (*Sphaerotheca mors-uvae* (Schw.) Berk. et Curt), anthracnose (*Pseudopeziza ribis* Kleb.), septoria leaf spot (*Septoria ribis* Desm.)], and European red mite (*Tetranychus*

TABLE 15.1   Black Currant Cultivars Bred in I. V. Michurin All-Russian Scientific Research Institute of Certification (now I. V. Michurin FSC).

Cultivar	Color of berries	Year		Genetic origin*	Term of maturation
		Sent for state trials	Including in the list of the state register of breeding achievement		
Amirani	Black	2015	–	ESUScGR	Medium late
Bagira	–	1985	1994	ESDSc	–
Vospominanie	–	1995	1997	–	Medium
Divo Zvjaginoi	–	2009	–	–	Medium early
Zeljonaja dymka	–	1985	1994	–	Medium
Izumrudnoe ozherel'je	Green	2008	–	–	–
Karmelita	Black	2005	–	–	Very early
Malen'kij princ	–	2001	2004	–	Medium early
Pandora	–	2013	–	ESDU	Early
Prima	–	1969	?	ES	Medium early
Rossiyanka	–	1969	?	ES	Medium late
Sensej	–	2008	2011	ESDSc	Medium early
Sozvezdie	–	1985	1997	–	–
Talisman	–	2011	–	–	Medium late
Tamerlan	–	2001	2004	–	–
Tat'janin den'	–	1995	1997	–	Late
Charovnica	–	1999	2006	–	Medium early
Chernavka	–	2004	2006	–	Medium late
Chjornyi zhemchug	–	1983	1992	–	Medium
Shalun'ja	–	2004	2006	ESD	Very early
Elevesta	–	1999	2001	ESDSc	Medium

Note: D—*Ribes dikuscha* Fisch., E—*R. nigrum* ssp. *europaeum* (Jancz) Pavl., G—*R. glutinosum* Benth., R—*Grossularia reclinata* (L.) Mill., S—*R. nigrum* ssp. *sibiricum* E. Wolf, Sc—*R. nigrum* ssp. *scandinavicum* Jancz., and U—*R. ussuriensis* Jancz.

urticae Koch.)], visual estimation per bush was applied. It was done according to the scale: 0—absence of injuries; 1—very weak, injuries were observed on single leave; 2—weak, up to 25% leaves for mildew and up to 10% for anthracnose, septoria leaf spot, and spider mite; 3—medium up to 50 and 30%; 4—severe up to 70 and 50%, and 5—very severe over 70% and above 50% of leaves, respectively. Resistance to currant bud mite (*Cecidophyopsis ribis* Westw.) evaluated in spring before bud breading according to the 6-point scale: 0—absence of injuring; 1—very weak, single bud; 2—weak up to 10% of buds; 3—medium up to 30%; 4—severe up to 50%; and 5—very severe over 50%. The stage of flowering was observed at the period of mass bud breaking and degree of fructification before the beginning of maturation; when the fruit-set became maximal, the 6-point scale was used: 0—the absence of florescence and fructification; 1—very weak, single flower and fruit-set were present; 2—weak, florescence and fructification on separate shoots and terminal shoots; 3—medium, no less than on one-half part of the shoot length; 4—very, good from one-half till three-fourth shoot length; 5—abundant, along the whole length of the shoot. The average mass of berry was determined by weighing 100 berries in 3 replicates. The result of weighing was divided by 100 and the average mass of berries was determined in grams. While analyzing the experimental data, the group with very small berries included the cultivars with values <0.50 g, small berries group 0.51–0.70 g, medium berries group 0.71–1.00 g, large berries group 1.01–1.50 g, and very large berries group >1.50 g. Weighing registration was held on the whole by a plot and then the average yield per bush was recalculated, and with the use of multiplication of the received value per number of bushes, the output per hectare was determined. Summing up of the indices for some years provides the characteristics of the average meaning. The statistical processing of the data was held according to the "methods of field experiment,"[12] with the use of computer program Microsoft Excel.

15.3 RESULTS AND DISCUSSION

Climatic conditions for the years of the investigation were not always favorable for growth and cultivation of black currant plants. The environment was characterized taking into account the data of agro-meteostantion in Michurinsk (Table 15.2).

TABLE 15.2   Characterization of Climatic Conditions for the Period Investigations in 2000–2016 Years.

Years	Winter (December–February)		Spring (March–May)		Summer (June–August)		Autumn (September–November)	
	ΣT°C	ΣPn, mm*	ΣT°C	ΣPn, mm	ΣT°C	ΣPn, mm	ΣT°C	ΣPn, mm
2000	−430.1	105.0	628.8	95.3	1720.2	350.3	505.2	75.0
2001	−374.7	164.7	676.1	161.5	1818.7	266.4	537.5	127.2
2002	−600.1	151.2	763.1	54.5	1856.8	67.6	542.6	163.7
2003	−951.7	48.7	486.8	61.1	1617.7	287.2	595.7	100.7
2004	−490.2	119.9	626.7	133.6	1710.5	168.0	610.7	177.7
2005	−499.8	160.2	545.8	113.6	1746.0	195.6	699.5	133.3
2006	−928.0	149.8	494.9	118.2	1769.4	253.5	638.1	223.9
2007	−330.0	125.9	796.6	61.2	1895.9	236.2	590.5	229.1
2008	−605.5	84.1	838.9	129.3	1797.9	183.2	696.7	81.6
2009	−510.6	110.5	594.7	86.2	1767.6	189.2	762.6	118.5
2010	−968.9	99.8	723.9	68.2	2285.7	41.0	708.8	90.7
2011	−795.3	159.0	560.4	87.7	1967.8	256.2	544.2	107.9
2012	−673.7	144.7	739.9	103.3	1868.7	239.8	741.8	173.8
2013	−600.6	120.0	681.2	112.2	1857.7	239.9	659.0	173.4
2014	−521.8	93.1	827.3	87.8	1858.3	91.6	507.3	59.3
2015	−462.1	142.2	707.2	102.4	1769.3	235.7	639.7	90.9
2016	−359.7	176.1	776.6	272.7	1903.1	218.3	503.9	129.0

*Note: ΣT°C—sum of temperatures, in °C, ΣPn, mm—sum of precipitations, in mm.

Analysis of meteorological data showed that there were the frost-iest winters in 2002/2003, 2005/2006, 2009/2010, and 2010/2011 with minimum thaws 10, 9, 18, and 14 days, respectively. It was stated that average daily temperature on the investigation period was 1.8°C higher than average long-term indices in December, 2.0°C in January and February and its average index per winter was 7.0°C. The maximum number of 52 thaw days was in winter of 2015/2016 and -42 thaw days in 2013/2014. In the other winters, the number of such days varied from 24 (2007/2008) till 39 days (2000/2001 and 2014/2015) per season. The scientific publi-cations report the effect of thaws in different periods of winters as the unequal one on frost resistance of tissues. Therefore, the frost resistance of wood and bark in December is medium and for cambium, it is rela-tively high. After breakage of deep dormancy, the cambium response to thaws increases sharply, but for the wood and bark, their response is retardant. During the whole experiment, there were ≥ 10 February warm days in i.e. 14 days in 2000; 10 in 2001; 24 in 2002; 11 in 2008; 12 in 2013, 12 in 2014; 12 in 2015, and 23 in 2016, but an absolute minimum of temperatures ($t°C_{min}$) $-23.5°C$ for the whole winter was registered just in February where an average daily temperature was $-6,5°C$. The coldest winter month (except 2001) was February in 2004 ($\sum T°C = -206.8°C$, $t°C_{min} = -22.5°C$), 2005 ($\sum T°C = -265.6°C$, $t°C_{min} = -24.5°C$), 2006 ($\sum T°C = -397.6°C$, $t°C_{min} = -34.7°C$), 2007 ($\sum T°C = -278.0°C$, $t°C_{min} = -25.0°C$), 2011 ($\sum T°C = -349.8°C$, $t°C_{min} = -30.1°C$), and 2012 ($\sum T°C = -373.6°C$, $t°C_{min} = -27.1°C$) with a small quantity of thaws and at the rest years, January was the frostiest month. The effect of negative temperatures on black currant plants is often overthrown by the amount of precipitation. Our experiment revealed that there was a few less precipi-tation than a month norm (by 1.7%) in February 2005. In other years, the February precipitation exceeded average many years index by 3.3% (2006), 7.7% (2012), 9.3% (2011), 23.3% (2004), 47.3% (2007), and 83.7% (2001). However, we need to take into consideration that deep snow and soft weather—all at once—very often cause the asphyxiation of plants and development of mold fungi. On the whole, the precipitation was on the level of climatic norm in 2003/2004 (95.9%), 2006/2007 (100.7%), and 2012/2013 (96.0%), and there were less precipitation in 1999/2000 (84.0%), 2002/2003 (39.0%), 2007/2008 (67.3%), 2008/2009 (88.4%), 2009/2010 (79.8%), and 2013/2014 (74.5%). In subsequent winter seasons, the precipitation was from 113.8% (2014/2015) till 140.9% (2015/2016) from long-term level.

Thus, weather conditions of winter periods over the years of investigations were different from each other. However, black currant cultivars, synthesized by I.V. Michurin FSC at different years began vegetating approximately at the equal terms independently on the genetic origin (Table 15.3).

It was stated that under Michurinsk conditions, beginning of vegetation was considered to be constantly early in following cultivars: 'Divo Zvjaginoj', 'Karmelita', 'Pandora', 'Chernavka', and 'Shalun'ja'. Obviously, to begin the vegetation, the cultivars required 46.6–53.3°C of positive temperatures or 1°C of warmth per 16.6–20.2°C frosty winters. The black currant varieties Bagira and Elevesta shower the latest bud-breaking and they demanded 66.6...70.0 degrees of C warmth per 12.2...12.5 degrees of frosty winters.

Adaptability of black currant cultivars is characterized by general plant state which directly depends on winter resistance, draught resistance, regenerative ability, and resistance to pests and diseases. Analysis of general plant state after winters showed that local currant cultivars had rather high resistance to unfavorable environmental factors (Table 15. 4).

Stably good and excellent state was in 'Zeljonaja dymka', 'Talisman', 'Tamerlan', 'Chernavka', and 'Shalun'ja'. It was determined that plant state greatly depended on the age of plantations ($r=-0.63$) and significantly less on the sum of temperatures of previous vegetation period ($r=-0.35$) and precipitation ($r=0.22$). Other scientific publications inform that some factors are decreasing the growth intensity, for example, drought can stimulate the flowering.

The flowering of black currant is one of the fundamental stages in plant growth and development. Under the circumstances of the Central Chernozem Region in different years of experiments, the black currant began blooming from the second 10 days period of April till the second 10 days period of May, taking into account average long-term dates which occur at the third 10 days period of April to first 10 days period of May (Table 15.5).

The minimum number of days (18) from initial vegetation to initial blossom was spent in 2000, 2001, 2011, and 2013 years with a sum of positive temperatures ($\sum T°C>0$) equal to 233.1, 205.6, 232.1, and 221.9°C, respectively. The longest periods were in 2002 (42 days), 2004 (34 days), and 2007 (36 days) in spite of the fact that initial blossom demanded $\sum T°C>0$ equal 249.6, 216.3, and 229.0°C. It was found that there was

TABLE 15.3 Terms of Vegetation (Beginning) in Black Currant Cultivars in 2000–2016 Years

Cultivar	Date of vegetation			Sum of positive temperatures at the beginning of vegetation (°C)		
	Medium	Early	Late	Average many years	Min	Max
Amirani	9.04	16.03	19.04	63.4	19.2	104.4
Bagira	9.04	17.03	20.04	70.0	18.2	106.3
Vospominanie	8.04	16.03	18.04	65.6	26.8	106.5
Divo Zvjaginoi	7.04	13.03	18.04	53.3	8.8	99.7
Zeljonaja dymka	7.04	15.03	17.04	54.3	29.9	84.5
Izumrudnoe ozherel'je	8.04	17.03	19.04	63.9	26.8	106.5
Karmelita	6.04	14.03	17.04	52.3	15.8	89.9
Malen'kij princ	7.04	14.03	20.04	57.0	18.2	95.0
Pandora	6.04	12.03	17.04	46.6	12.6	83.8
Prima	7.04	13.03	21.04	62.2	8.8	106.3
Rossiyanka	6.04	14.03	19.04	59.9	15.8	96.3
Sensej	7.04	18.03	18.04	56.4	15.8	95.0
Sozvezdie	7.04	16.03	19.04	57.7	18.2	95.0
Talisman	7.04	13.03	18.04	53.6	15.8	84.5
Tamerlan	7.04	11.03	19.04	56.1	7.0	99.7
Tat'janin den'	8.04	14.03	20.04	62.9	26.8	106.5
Charovnica	8.04	17.03	20.04	59.8	18.2	95.0
Chernavka	6.04	12.03	17.04	51.0	12.6	89.9
Chjornyi zhemchug	7.04	14.03	18.04	56.6	15.8	89.9
Shalun'ja	6.04	13.03	17.04	49.5	8.8	89.9
Elevesta	9.04	18.03	19.04	66.6	33.7	106.5

TABLE 15.4 Biological Features of Black Currant on Average in 2000–2016 Years

Cultivar	Plant state after winter (point)				Degree of flowering (point)				Degree of fructification (point)			
	X	Max	Min	V%	X	Max	Min	V%	X	Max	Min	V%
Amirani	4.7	5.0	4.0	10.2	4.6	5.0	4.0	11.0	4.4	5.0	4.0	11.6
Bagira	4.5	5.0	4.0	10.7	4.9	5.0	4.0	6.8	4.5	5.0	3.0	12.6
Vospominanie	4.4	5.0	3.5	13.1	4.8	5.0	3.0	10.5	4.5	5.0	3.0	13.8
Divo Zvjaginoi	4.5	5.0	3.0	15.6	4.6	5.0	2.0	18.6	4.2	5.0	1.0	24.4
Zeljonaja dymka	4.8	5.0	4.0	7.1	4.9	5.0	4.0	5.4	4.8	5.0	4.0	7.5
Izumrudnoe ozherel'je	4.4	5.0	3.0	16.9	4.3	5.0	2.0	21.4	3.6	5.0	1.0	29.1
Karmelita	4.6	5.0	4.0	10.6	4.6	5.0	4.0	10.5	4.1	5.0	3.0	15.2
Malen'kij princ	4.6	5.0	3.0	13.4	4.7	5.0	3.0	14.0	4.4	5.0	2.0	17.3
Pandora	4.4	5.0	3.0	13.7	4.5	5.0	3.0	14.0	4.3	5.0	2.0	25.1
Prima	4.0	5.0	3.0	17.4	4.4	5.0	3.0	15.1	3.3	5.0	2.0	30.9
Rossiyanka	4.6	5.0	4.0	10.7	4.9	5.0	4.0	6.1	4.3	5.0	2.0	21.8
Sensej	4.3	5.0	3.0	17.8	4.6	5.0	4.0	10.5	4.0	5.0	3.0	12.8
Sozvezdie	4.5	5.0	4.0	10.7	4.7	5.0	3.0	11.9	4.3	5.0	3.0	17.2
Talisman	4.9	5.0	4.5	4.8	5.0	5.0	5.0	0.0	4.6	5.0	3.0	12.4
Tamerlan	4.7	5.0	4.0	9.2	4.9	5.0	3.0	9.9	4.4	5.0	3.0	14.3
Tat'janin den'	4.5	5.0	3.5	12.6	4.9	5.0	4.0	5.8	4.3	5.0	3.0	13.1
Charovnica	4.3	5.0	3.0	14.3	4.6	5.0	2.0	16.5	4.1	5.0	2.0	22.5
Chernavka	4.7	5.0	4.0	9.2	5.0	5.0	4.5	2.4	4.7	5.0	3.5	10.0
Chjornyi zhemchug	4.3	5.0	3.0	16.3	4.9	5.0	4.0	7.1	4.5	5.0	3.0	15.2
Shalun'ja	4.7	5.0	4.0	9.2	4.7	5.0	3.0	12.5	4.4	5.0	2.0	17.1
Elevesta	4.4	5.0	3.0	20.6	4.7	5.0	3.0	14.6	4.4	5.0	3.0	15.9

TABLE 15.5 Features of Blossom Phase in Black Currant Cultivars in 2000–2016 Years.

Cultivar	Initial blossom, date			X±m	Duration of blossom, days			ΣT at initial blossom, °C	
	Medium	Early	Late		Min	Max	V%	>0	ef.>5*
Amirani	3.05	27.04	1.05	10±0.8	5	18	32.4	249.5	135.9
Bagira	1.05	20.04	11.05	11±0.8	7	19	30.8	223.5	116.3
Vospominanie	1.05	20.04	11.05	11±0.8	6	19	30.2	223.7	116.3
Divo Zvjaginoi	28.04	16.04	7.05	12±0.7	8	18	24.5	196.1	96.5
Zeljonaja dymka	30.04	17.04	11.05	12±0.9	8	21	29.9	219.6	111.0
Izumrudnoe ozherel'je	30.04	18.04	9.05	11±0.8	8	18	28.0	215.0	111.8
Karmelita	28.04	15.04	6.05	11±0.7	7	17	26.2	196.3	96.5
Malen'kij princ	30.04	18.04	8.05	11±0.8	6	18	30.1	221.0	112.2
Pandora	1.05	23.04	8.05	11±1.0	6	20	38.6	231.3	116.1
Prima	1.05	18.04	10.05	10±0.6	7	14	20.5	234.4	120.7
Rossiyanka	30.04	18.04	12.05	10±0.5	8	14	16.9	233.3	119.3
Sensej	28.04	16.04	8.05	11±0.8	8	18	27.2	200.0	99.6
Sozvezdie	30.04	17.04	9.05	11±1.0	7	23	36.1	212.3	106.8
Talisman	30.04	17.04	9.05	12±0.6	8	18	22.0	217.8	109.4
Tamerlan	1.05	23.04	8.05	11±0.7	7	17	25.6	235.0	120.2
Tat'janin den'	30.04	18.04	10.05	11±0.7	7	18	24.7	218.4	112.1
Charovnica	29.04	15.04	8.05	11±0.9	8	22	31.8	207.3	104.7
Chernavka	30.04	17.04	10.05	11±0.7	7	17	25.8	229.1	115.7
Chjornyi zhemchug	30.04	17.04	9.05	12±0.9	7	21	31.6	219.0	111.6
Shalun'ja	27.04	15.04	6.05	11±0.7	7	17	24.7	196.8	97.0
Elevesta	1.05	18.04	12.05	11±0.7	8	19	27.2	227.1	118.8

Note: ∑Tem. > 5, total effective temperatures above 5°C.

inverse and strong relationship between terms of initial blossom, and the average daily temperature was −0.80 within cultivars, and there was fluctuation from −0.72 (Talisman) up to −0.94 (Prima). Despite the fact that winter and spring periods of different years had different weather conditions, initial blossom demanded 1°C of warmth per −2.5°C ('Amirani') and −3.0°C ('Divo Zvjaginoj', 'Karmelita', and 'Shalun'ja') of winter frost.

According to the results of long-term evaluation, the black currant cultivars are divided into three groups of initial blossoms:

- Early blooming—'Divo Zvjaginoj', 'Karmelita', 'Sensej', and 'Shalun'ja'
- Medium blooming—'Zelenaya dymka', 'Izumrudnoye ozherel'je', 'Malen'kij princ', 'Prima', 'Rossiyanka', 'Sozvezdie', 'Talisman', 'Charovnica', 'Cherrnavka', and 'Chjornyi zhemchug'
- Medium late blooming—'Bagira', 'Vospominanie', 'Pandora', 'Tamerlan', 'Tat'janin den', and 'Elevesta'
- Late blooming—'Amirani'

On average, black currant cultivars showed 10 days ('Amirani', 'Prima', and 'Rossjanka') to 12 days ('Divo Zvjaginoj', 'Zeljonaja dymka', 'Talisman', and 'Chjornyi zhemchug') of blossom. Actually, the duration of blossom was determined by temperature regime and the presence of close and inverse interrelationship: r = − 0.71 was found between them. For the whole experimental period, the minimum average daily temperature of air during the blossom was in 2000 (9.6°C), 2007 (10.2°C), 2014 (12.2°C), and 2016 (12.8°C). In those seasons, subfrosts were recorded when the t°Cmin was −0.9 to −1.2°C (2000), −0.8°C (2007), −2.2°C (2014), and −0.3°C (2016). However, in 2014 at the period of flowering, the absolute maximum temperature (t°Cmax) 31.4°C was recorded. On the whole, in spite of significant fluctuations of temperature, black currant cultivars showed a good blooming, average long-term degree of blossom ranged from 4.3 ("Iz'Izumrudnoye ozherel'je'") up to 5.0 ('Talisman' and 'Chernavka') (see Table 15.4).The genotypic differences in black currant cultivars determine their unequal response to change in environment. Usually, it is very hot in summer in the Chernozem Region (Table 15.2). On average for the years of investigations, the hottest summer seasons were in 2010, 2011, and 2016 years and the coolest one in 2003 year. Analyzing the distribution of precipitations during a year, it was found that there were from 13.7% (2010) precipitation up to 57.7% (2003) from the yearly amount.

One of the soil moisture indices of location is hydrothermal coefficient (HTC) which is in itself proportion ratio of rainfalls sum over the period within the temperatures above 10°C to the sum of average daily temperatures 10 times decreased. It is considered that moisture deficit is observed when the HTC is lower than 1.0 and its shortage at 1.0–1.3 (HTC), favorable conditions at 1.4–2.0 (HTC), and the higher HTC values testify to the surplus of moisture. From the observations of Michurinsk agro-meteostation, it was concluded that moisture deficit was recorded in summer of 2002, 2010, and 2014 and the shortage in the following years: 2004, 2005, 2007, 2008, 2009, 2011, 2012, 2013, 2015, and 2016. Summer seasons in 2000, 2001, 2003, and 2006 were characterized with the optimal amount of rainfalls.

During the summer, the maximum average daily temperature (°C) was recorded (according to the years) in June 2006, 19.8; 2013, 20.8, and 2015, 19.6; in July 2000, 19.8; 2001, 23.8; 2002, 23.8; 2003, 20.6; 2005, 20.2; 2009, 21.2; 2010, 27.3; 2011, 24.0; 2012, 21.7; 2014, 22.0, and 2016, 21.9. The highest August index was in 2004 (19.9°C), 2007 (22.7°C), and 2008 (21.0°C). Moreover, $t°C_{max}$ was very often recorded in August in 2000 (30.9°C), 2004 (34.5°C), 2005 (32.2°C), 2007 (35.6°C), 2008 (36.7°C), 2010 (40.6°C), 2012 (36.6°C), 2013 (32.6°C), 2014 (36.5°C), 2015 (33.5°C), and 2016 (34.7°C). High levels of $t°C_{max}$ were found in July 2001 (35.6°C), 2002 (34.3°C), 2003 (29.6°C), 2009 (33.7°C), 2011 (35.0°C), and in June 2006 (34.1°C).

In the unstable conditions of the environment, the beginning of maturation in black currant berries was recorded on average in 40 days from finishing of blossom with fluctuations from 35 days (2014) to 46 days (2008). The black currant berries begin maturing at the second 10 days period and third 10 days period of June, with T°C>0°C equal to 908.1°C ('Shalun'ja') and 1104.1°C ('Amirani') (Table 15.6). The presence of close negative correlation between the terms early maturation and average daily temperature of air was stated and the correlation on average was $r=-0.71$ by the cultivars, with fluctuations from −0.19 ('Izumrudnoe ozherel'je') up to −0.75 ('Talisman').

The quickest maturation of black currant berries was recorded in 2000 (19 days), 2006 (20 days), and 2014 (20 days), and the longest period was both in 2008 and 2013 years (30 days). On average, according to the years of observation, the interval between initial vegetation and full maturation of berries demands 95 days with fluctuations from 87 ('Karmelita and

TABLE 15.6 Features of Phenological Phase for Maturation of Berries in Black Currant Cultivars in 2000–2016 Years.

Cultivar	Beginning of berries maturation, date			Duration of berries maturation, days				$\sum T$ at the beginning of maturation of berries, °C	
	Medium	Early	Late	X±m	Min	Max	V%	>0	ef.>5
Amirani	22.06	15.06	30.06	22±1.0	16	27	16.5	1104.1	738.6
Bagira	21.06	12.06	28.06	22±1.1	17	33	19.9	1064.2	706.4
Vospominanie	20.06	10.06	28.06	23±1.0	19	33	17.5	1048.6	693.6
Divo Zvjaginoi	14.06	6.06	25.06	25±1.1	18	36	17.8	943.5	611.3
Zeljonaja dymka	19.06	11.06	26.06	23±1.0	17	32	17.4	1046.0	688.2
Izumrudnoe ozherel'je	19.06	6.06	30.06	23±2.0	17	40	34.6	1057.8	704.0
Karmelita	12.06	3.06	21.06	20±0.7	15	25	14.4	912.6	588.7
Malen'kij princ	18.06	9.06	26.06	22±1.3	12	32	23.6	1013.0	663.9
Pandora	16.06	6.06	22.06	26±0.9	19	31	14.7	992.2	647.2
Prima	17.06	10.06	22.06	25±1.8	16	33	23.1	998.5	655.2
Rossiyanka	16.06	10.06	25.06	25±1.5	18	33	18.9	988.0	645.8
Sensej	14.06	5.06	24.06	25±1.3	13	37	21.7	942.5	611.8
Sozvezdie	19.06	12.06	27.06	23±1.0	19	34	18.5	1042.8	686.9
Talisman	17.06	11.06	26.06	24±1.1	19	36	17.9	1012.9	663.6
Tamerlan	20.06	13.06	29.06	24±1.1	17	32	18.9	1066.9	703.5
Tat'janin den'	21.06	15.06	29.06	24±0.8	18	30	14.0	1076.4	713.2
Charovnica	19.06	10.06	28.06	22±1.1	16	30	18.3	1041.3	686.5
Chernavka	18.06	10.06	26.06	24±1.1	15	32	18.9	1037.8	680.7
Chjornyi zhemchug	19.06	10.06	28.06	24±1.0	19	33	16.9	1039.8	683.2
Shalun'ja	11.06	3.06	21.06	21±0.7	15	27	15.0	908.1	585.1
Elevesta	20.06	14.06	26.06	23±1.0	19	31	16.5	1063.5	706.3

Shalun'ja') till 98 days ('Tamerlan and Tat'janin den'). The speed of berry maturation depended mainly on biological features of the cultivar, and the temperature sum effect was not high, $r = 0.35$. The fructification degree of black currant cultivar varied from 3.3 (Prima) till 4.8 points ('Zeljonaja dymka'), for the whole period of the investigations (see Table 15.4). The stably high degree of fructification was found in 'Zeljonaja dymka', 'Chernavka', 'Talisman', 'Bagira', 'Vospominanije', and 'Chjornyi zhemchug' cultivars. 'Prima' and green-colored cultivar 'Izumrudnoe ozherel'je' were exhibited to almost all stress factors. The green color of berries in black currant is regulated by recessive allele rb, which is linked to a lethal gene l in some forms. That is why very often the green-colored fruit of black currant cultivars are inferior to black colored fruit cultivars in viability.

Therefore, while selecting the important forms, the breeders pay a great attention to cultivars with optimum economical important traits and high potential of productivity; a comprehensive evaluation of genotypes by the elements of the yield capacity is necessary. The evaluation of new black currant cultivars was carried out on the industrial plots of fruit variety trials with cultivation of the plants on the bogharic agricultural system. It was found that very severe Michurinsk conditions of cultivation provided high levels for elements of black currant productivity component (Table 15.7).

All components of the productivity are variable, in spite of the fact of their determination by cultivar genotype, but they depend on the plant age, agro-technical factors, meteorological conditions, disease and pest resistance, and so forth. The evaluation of the studied cultivars by the number of racemes per running meter of the fruiting wood showed that 57.1% of the total number formed more than 50 racemes.

Initiation of two to three buds in fruiting knot, capable of the normal development of inflorescences, is the character allowing increase in the yielding capacity of the cultivar. There was difference among cultivars by the number of racemes per running meter of fruit-bearing wood 1.9 times, from 33 pieces ('Prima') and 38 pieces ('Shalun'ja') (Fig. 15.1a) up to 63 pieces ('Divo Zvjaginoj') (Fig. 15.1b).

One of the indices characterizing the cultivar importance is the ability of variety to form the long polycarpic racemes. Our experiment demonstrated that on average for the years of investigation, six and more fruits in raceme were formed in 'Divo Zvjaginoj', 'Prima', 'Pandora' (Fig. 15.2a), and 'Izumrudnoe ozherel'je' (Fig. 15.2b) cultivars. Malen'kij princ', 'Talisman', and 'Shalun'ja' cultivars showed a lower maturation of berries in a raceme than other cultivars.

TABLE 15.7 The Main Components of Productivity in Black Currant, on Average for the 2000–2016 Years.

Cultivar	Number of, pieces				Length of raceme, cm		Mean fruit weight, g	
	Racemes per running meter		Berries in a raceme					
	X±m	V%	X±m	V%	X±m	V%	X±m	V%
Amirani	53±5.1	32.1	5±0.5	29.9	4.7±0.4	26.7	1.1±0.1	26.8
Bagira	54±3.8	29.1	5±0.4	33.7	4.0±0.2	19.8	1.1±0.1	32.9
Vospominanie	55±4.1	25.4	5±0.4	26.6	3.9±0.2	18.8	1.1±0.1	22.6
Divo Zvjaginoi	63±9.2	58.3	6±0.5	34.1	3.9±0.3	25.5	1.2±0.1	29.1
Zeljonaja dymka	48±3.4	29.9	5±0.3	26.9	4.6±0.2	18.8	1.2±0.1	17.2
Izumrudnoe ozherel'je	44±3.4	28.8	8±0.3	13.8	4.6±0.2	15.9	0.8±0.1	26.8
Karmelita	53±4.7	35.5	5±0.5	40.6	3.5±0.2	27.5	1.5±0.1	26.5
Malen'kij princ	50±4.6	38.0	4±0.2	22.8	3.8±0.2	20.4	1.3±0.1	26.1
Pandora	51±5.3	37.3	7±0.3	16.4	4.7±0.4	27.0	1.1±0.1	22.3
Prima	33±5.7	35.3	7±0.5	22.2	2.9±0.3	19.8	0.7±0.1	10.1
Rossiyanka	35±7.4	42.8	5±0.6	20.1	4.0±0.5	21.7	0.8±0.1	20.7
Sensej	44±3.6	33.0	5±0.3	21.1	3.8±0.2	22.5	1.2±0.1	34.0
Sozvezdie	51±6.3	48.4	5±0.4	29.9	4.3±0.2	19.4	1.2±0.1	25.2
Talisman	50±4.3	33.5	4±0.3	28.2	3.5±0.2	18.0	1.2±0.1	16.5
Tamerlan	48±2.9	24.2	4±0.4	34.6	4.8±0.3	22.4	1.3±0.1	18.9
Tat'janin den'	53±5.6	36.5	5±0.3	19.4	4.7±0.2	16.8	1.1±0.1	25.1
Charovnica	50±3.2	25.7	5±0.2	20.1	3.6±0.2	20.0	1.1±0.1	18.8
Chernavka	48±3.3	27.5	5±0.3	25.1	4.4±0.2	17.0	1.2±0.1	27.4
Chjornyi zhemchug	45±4.7	38.5	5±0.3	23.0	4.8±0.3	23.0	1.3±0.1	26.1
Shalun'ja	38±3.1	32.6	4±0.3	30.8	3.9±0.2	21.7	1.5±0.1	27.1
Elevesta	54±5.2	40.0	5±0.3	23.3	3.8±0.2	21.3	1.0±0.1	21.9
LSD$_{05}$	5.3	–	0.4	–	0.2	–	$F_r < F_{05}$	–

(a)   (b)

FIGURE 15.1 (See color insert.) Raceme formation in black currant; (*a*) 'Shalun'ja' and (*b*) 'Divo Zvjaginoj'.

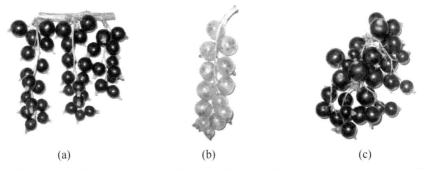

(a)   (b)   (c)

FIGURE 15.2 (See color insert.) Biological features of berry formation in a raceme of the black currant cultivars; (a) Pandora, (b) 'Izumrudnoe ozherel'je', and (c) 'Talisman'.

Such index indicates that the length of racemes is closely connected with a number of berries in a raceme. It was stated that the raceme length varied from 3.3 cm in 2015 till 4.9 cm on average in 2012. The longest racemes for the years of investigations were formed in 'Tamerlan', 'Chjornyi zhemchug' (Fig. 15.3a), 'Amirani', 'Pandora', 'Tat'janin den', and 'Zeljonaja dymka' cultivars. The following cultivars 'Prima', 'Karmelita' (Fig. 15.3b), 'Talisman', and 'Charovnitsa' had short racemes

As a result of the investigations held, the presence of negative dependence between the berry amount in raceme and medium mass of berries was stated as $r = -0.64$. Among the indices of the potential productivity, in cultivar, the medium mass of berries is the main one and it characterizes the adaptive potential. The phenotypic manifestation of large size fruit in black currant is influenced by the age of plantations, the location of the plot, and environment. It is known that on average taking into account the cultivars, the largest berries were in 2005 (1.6 g) and the smallest berries in 2010 (0.9 g). 'Shalun'ja' cultivar has very large berries and 'Izumrudnoe ozherel'je',' Prima', ''Rossiyanka', and 'Elevesta' have medium size berries, and the other studied cultivars have large berries.

(a)   (b)

FIGURE 15.3 **(See color insert.)** Differences in cluster length of black currant cultivars; (a) Chjornyi zhemchug and (b) Karmelita.

Disease and pest resistance has a significant impact on the realization of productivity potential in black currant. Fungal disease resistance was always one of the most important tasks in black currant breeding. The varieties were evaluated for resistance to those diseases which were economical important for the industrial plantations in Chernozem Region—American mildew, antracnose, and septoria leaf spot.

Warm and wet weather with temperature 17–28°C and air moisture 90–100% promotes dispersion of American mildew. The first symptoms of disease can appear on the young shoots and leaves at the end of May. It was stated that insignificant damage of black currant leaves by *Sphaerotheca mors-uvae* (Schw.) Berk. et Curt. was observed annually within 0.1–1.1. Dispersion of the fungus on plantation was in inverse dependence from the sum of temperatures for May and June—$r = -0.51$ and in direct dependence with the amount of precipitation for this period $r = -0.64$.

It was revealed that among the black currant cultivars bred at I. V. Michurin FSC, the following cultivars had immunity to *Sphaerotheca mors-uvae* (Schw.) Berk. et Curt.: 'Amirani' and 'Izumrudnoe ozherel'je', 'Divo Zvjaginoj', 'Karmelita', 'Malen'kij princ', 'Pandora', 'Prima', 'Sensej', 'Talisman', 'Tamerlan', 'Chernavka', and 'Shalun'ja' were tolerant (Table 15.8).

TABLE 15.8 The Resistance of Black Currant Cultivars in 2000–2016 Years.

Cultivar	Degree of leaf injury by disease, point								
	Sphaerotheca mors-uvae (Schw.) Berk. et Curt			Pseudopeziza ribis Kleb.			Septoria ribis Desm.		
	X±m	Max	Min	X±m	Max	Min	X±m	Max	Min
Amirani	0.0±0.0	0.0	0.0	1.3±0.2	3.0	0.0	1.2±0.2	3.0	0.0
Bagira	1.0±0.2	2.0	0.0	1.5±0.2	3.0	0.0	1.0±0.2	3.0	0.0
Vospominanie	0.7±0.2	2.5	0.0	1.6±0.2	3.0	0.0	1.1±0.2	2.0	0.0
Divo Zvjaginoi	0.2±0.1	1.0	0.0	1.4±0.2	3.0	0.0	0.9±0.2	3.0	0.0
Zeljonaja dymka	0.9±0.2	2.1	0.0	1.3±0.2	2.0	0.0	0.8±0.2	2.0	0.0
Izumrudnoe ozherel'je	0.0±0.0	0.0	0.0	1.1±0.3	4.0	0.0	0.7±0.2	2.0	0.0
Karmelita	0.1±0.1	1.0	0.0	1.3±0.2	3.0	0.0	0.9±0.2	3.0	0.0
Malen'kij princ	0.1±0.1	1.0	0.0	1.1±0.1	2.0	0.0	0.5±0.1	1.0	0.0
Pandora	0.1±0.1	1.0	0.0	0.8±0.1	2.0	0.0	0.6±0.1	1.0	0.0
Prima	0.4±0.1	1.0	0.0	1.8±0.1	3.0	1.0	0.3±0.1	1.0	0.0
Rossiyanka	1.1±0.2	3.0	0.0	2.0±0.1	3.0	1.0	0.6±0.1	1.7	0.0
Sensej	0.1±0.1	1.0	0.0	1.5±0.2	3.0	0.0	0.5±0.1	2.0	0.0
Sozvezdie	1.1±0.2	3.0	0.0	1.3±0.2	3.0	0.0	0.9±0.2	3.0	0.0
Talisman	0.3±0.1	1.0	0.0	1.8±0.2	4.0	0.0	0.8±0.2	3.0	0.0
Tamerlan	0.2±0.1	1.0	0.0	1.1±0.2	2.0	0.0	0.5±0.2	2.0	0.0
Tat'janin den'	1.4±0.2	3.0	0.0	1.2±0.1	2.0	0.0	0.9±0.2	3.0	0.0
Charovnica	0.8±0.1	1.3	0.0	1.5±0.2	3.0	0.0	0.9±0.1	2.0	0.0
Chernavka	0.4±0.1	1.0	0.0	1.3±0.1	2.0	0.0	0.6±0.2	2.0	0.0
Chjornyi zhemchug	1.7±0.2	3.0	0.0	1.2±0.1	2.0	0.0	0.7±0.2	3.0	0.0
Shalun'ja	0.1±0.1	0.1	0.0	1.7±0.2	3.5	0.0	0.7±0.2	2.0	0.0
Elevesta	0.9±0.2	2.0	0.0	1.2±0.1	2.0	0.0	0.9±0.1	2.0	0.0

Pseudopeziza ribis Kleb. and *Septoria ribis* Desm. on black currant plantation were especially spread at the end of harvesting period. Candle snuff fungus (*Pseudopeziza ribis* Kleb.) is well developed at 16–20°C and with the presents of drop-like moisture. Our experiment revealed the presence of direct positive dependence between the age of plantation and degree of injuring of plant by the fungus $r=0.53$. The high resistance to *Pseudopeziza ribis* Kleb. is observed in 'Zeljonaja dymka', 'Malen'kij princ', 'Pandora', 'Tamerlan', 'Tat'janin den', 'Chernavka', 'Chjornyi zhemchug', and 'Elevesta'.

The black currant leaves are damaged by *Septoria ribis* Desm. in the middle of May, because the dispersion of ascospores is observed at the end of April and at the beginning of May at the temperature 11°C and the shoots are infected at the second half of summer. The years of the investigations showed the minimum infestation by *Septoria ribis* Desm. of plants in 2007 (0.1 point), 2009 (0.3 point), and 2011 (0.1 point) and maximum in 2000 (1.6 point), 2015 (1.5 points), and in 2016 (1.9 points). The presence of positive dependence between the sum of the May–June precipitation and dispersion of the conditions of the fungus $r=0.58$ was found. In the conditions of the Chernozem Region, fungus infestation depended partly on the age of plantations $r=0.39$. Under unstable conditions of the environment, average *Septoria ribis* Desm. resistance occurs in 'Vospominanie', 'Zeljonaja dymka', 'Izumrudnoe ozherel'je', 'Malen'kij princ', 'Pandora', 'Prima', 'Rossiyanka', 'Sensej', 'Tamerlan', 'Charovnica', 'Chernavka', 'Shalun'ja', and 'Elevesta' cultivars.

'Malen'kij princ', 'Pandora', 'Tamerlan', and 'Charovnica' have complex resistance to fungal diseases.

In connection with climatic changes, the resistance of black currant to European red mite (*Tetranychus urticae* Koch.) and currant bud mite (*Cecidophyopsis ribis* Westw.) is more and more significant (Table 15.9).

European red mite was greatly dispersed in 2001 (1.9 points) and 2014 (1.9 points); there was no pest invasion in 2015 and 2016. There was a negative dependence on harmfulness of *Tetranychus urticae* Koch. and the sum of the precipitations has fallen in May–June, which was $r=-0.41$. The black currant cultivars such as 'Amirani', 'Zeljonaja dymka', 'Izumrudnoe ozherel'je', 'Malen'kij princ', 'Pandora', 'Rossiyanka', 'Sensej', 'Talisman', 'Tamerlan', 'Tat'janin den', 'Chernavka', and 'Elevesta' are resistant to *Tetranychus urticae* Koch.

TABLE 15.9 Pest Resistance in Black Currant Cultivars in 2000–2016 Years.

Cultivar	Degree of injury by pests, point					
	Tetranychus urticae Koch.			Cecidophyopsis ribis Westw.		
	X±m	Max	Min	X±m	Max	Min
Amirani	0.4±0.1	1.0	0.0	0.0±0.0	0.0	0.0
Bagira	1.0±0.2	4.0	0.0	0.6±0.1	1.0	0.0
Vospominanie	0.7±0.3	4.0	0.0	1.1±0.3	3.0	0.0
Divo Zvjaginoi	1.1±0.3	5.0	0.0	0.3±0.1	1.2	0.0
Zeljonaja dymka	0.6±0.1	1.0	0.0	0.7±0.2	2.6	0.0
Izumrudnoe ozherel'je	0.6±0.2	2.0	0.0	0.5±0.1	1.0	0.0
Karmelita	1.0±0.3	4.0	0.0	0.1±0.1	1.0	0.0
Malen'kij princ	0.7±0.1	2.0	0.0	1.2±0.3	3.6	0.0
Pandora	0.5±0.1	2.0	0.0	0.7±0.2	3.0	0.0
Prima	0.8±0.2	3.0	0.0	0.3±0.1	1.0	0.0
Rossiyanka	0.6±0.1	1.0	0.0	0.1±0.1	1.0	0.0
Sensej	0.8±0.2	2.0	0.0	0.1±0.1	1.0	0.0
Sozvezdie	1.0±0.2	4.0	0.0	1.3±0.2	3.0	0.0
Talisman	0.5±0.2	2.0	0.0	0.5±0.1	1.0	0.0
Tamerlan	0.6±0.1	1.0	0.0	1.1±0.2	3.0	0.0
Tat'janin den'	0.9±0.1	2.0	0.0	0.8±0.2	2.0	0.0
Charovnica	1.0±0.2	4.0	0.0	0.9±0.2	4.0	0.0
Chernavka	0.6±0.1	1.0	0.0	0.5±0.1	1.5	0.0
Chjornyi zhemchug	1.0±0.2	3.0	0.0	0.6±0.1	1.0	0.0
Shalun'ja	1.1±0.3	4.0	0.0	0.0±0.0	0.0	0.0
Elevesta	0.8±0.2	2.0	0.0	0.5±0.1	2.0	0.0

The currant bud mite (*Cecidophyopsis ribis* Westw.) is one of the most harmful pests on the black currant. In 2013, the cultivars were not damaged by *Cecidophyopsis ribis* Westw. 'Amirani' and 'Shalun'ja' cultivars occurred to be immune to the pest, but 'Bagira', 'Izumrudnoe ozherel'je', 'Karmelita', 'Prima', 'Rossiyanka', 'Sensej', 'Talisman', and 'Chjornyi zhemchug' are resistant to the pest.

'Amirani', 'Izumrudnoe ozherel'je', 'Rossiyanka', 'Sensej', and 'Talisman' have complex resistance to the mites.

The economical importance of cultivar is defined by the yielding capacity, which depends on its biological features and weather conditions. Analysis of the experimental material showed that black currant cultivars were characterized by the high productivity in Central Chernozem Region. On average, for the years of investigations, the output per hectare of the cultivars on the fruit-bearing plantations differed 3.9 times. The maximum productivity was observed in 'Chernavka' (9.7 ± 0.6; $b_1=1.03$), 'Malen'kij princ' (9.2 ± 0.6; $b_1=1.01$), and 'Zeljonaja dymka' (9.1 ± 0.6; $b_1=0.89$) (Fig. 15.4).

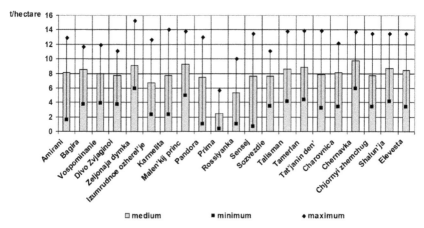

FIGURE 15.4   Characteristic of black currant cultivars for average output (t/ha) for the years of investigations.

As a result of the mathematical analysis of experimental data, the correlation dependence was found between the yielding capacity and the sum of precipitations fallen out during the year previous to fructification, —$r=0.41$. Among the investigated diseases and pests, antracnose is considered to be the most harmful and decrease the productivity of black currant plantations, because the infected bushes lose $\geq70\%$ of leaves before the end of July and beginning of August. In the middle of August, the bushes can be totally leafless. The presence of inverse dependence between the yielding capacity of cultivars and damage of plants by *Pseudopeziza ribis* Kleb. where $r=-0.45$ was found.

15.4 CONCLUSIONS

As a result of the investigations held, it was stated that the black currant cultivars bred in I. V. Michurin FSC (formerly I. V. Michurin All-Russian Scientific Research Institute of Certification) had high resistance to the complex of unfavorable factors of the environment. 'Zeljonaja dymka', 'Talisman', 'Tamerlan', 'Chernavka', and 'Shalun'ja' cultivars were characterized by stably good and excellent plant state.

'Divo Zvjaginoi' and 'Pandora' cultivars were characterized by the high levels of elements of the black currant productivity component.

'Malen'kij princ', 'Pandora', 'Tamerlan', and 'Chernavka' cultivars have complex resistance to fungal diseases.

'Amirani', 'Izumrudnoe ozherel'je', ''Rossiyanka', 'Sensej', and 'Talisman' cultivars are resistant to *Tetranychus urticae* Koch. and *Cecidophyopsis ribis* Westw.

The new black currant cultivars bred in Michurinsk form high productivity when cultivated in Chernozem Region of Russia. The cultivation of some cultivars allows obtaining more than 8 t of berries per hectare: 'Amirani' (8.2 t/ha), 'Charovnica' (8.2 t/ha), 'Elevesta' (8.4 t/ha), 'Bagira' (8.5 t/ha), 'Talisman' (8.6 t/ha), 'Shalun'ja' (8.7 t/ha), 'Tamerlan' (8.9 t/ha), 'Zeljonaja dymka' (9.1 t/ha), 'Malen'kij princ' (9.2 t/ha), and 'Chernavka' '(9.7 t/ha).

KEYWORDS

- American mildew
- anthracnose
- septoria leaf spot
- European red mite
- currant bud mite

REFERENCES

1. Zhidyokhina, T. V. Modern Cultivars of Black Currant: Advantages and Disadvantages. Scientific Basis for Development of Modern Horticulture in the Conditions of Import Substitution. *Materials of International Scientific and Practical Conference,* Michurinsk, June 1–3, 2016; Quart Publishing House (Kvarta): Voronezh, 2016; pp 54–64 (in Russian).

2. Tsukanova, E. M. Changes in Water-Thermal Status in Central Chernozem Region and Apple Trees Response on the Effect of Certain Abiotic Stressors. Present State of Nursery Practice and Innovation Bases of its Development. *Materials of International Scientific and Practical Conference in Connection with the 100th Anniversary of S. N. Stepanov,* Michurinsk, April 21–23, 2015; Quart Publishing House (Kvarta): Voronezh, 2015; pp 169–174 (in Russian).

3. Pomology. In *Currant and Gooseberry;* Sedov, E. N., Ed.; All-Russian Research Institute for Selection of Fruit Crops: Orel, 2009; Vol. IV, pp 30–292 (in Russian).

4. Zhidyokhina, T. V. Creation of Highly Productive Black Currant Cultivars with Complex Disease Resistance. Innovation Technologies for Production, Storage and Processing of Fruit and Berries. *Materials of Scientific and Practical Conference,* Michurinsk, Sept 5–6, 2009; Michurinsk-naukograd, 2009; pp 58–65 (in Russian).

5. Zhidyokhina, T. V. Results of Black Currant at I. V. Michurin All-Russian Scientific Research Institute of Certification. Present State of Currant and Gooseberry Crop. *Proceeding of I. V. Michurin All-Russian Scientific Research Institute of Certification,* Michurinsk 2007; pp 41–59 (in Russian).

6. Knyazev, S. D.; Nikolaev, A. V. Frost Resistance of Modern Black Currant Cultivars. *Herald of Orel State Agrarian University* **2008,** *13*(4), 14–17 (in Russian).

7. Łabonowska-Bury, D.; Dabrowski, Z. T.; Łabanovwska, B. H. Survey of Currant Crop and Pest Management Practices on Black Currant Plantations in Poland. *J. Fruit Ornam. Plant Res.* **2005,** *13,* 91–100.

8. Gajek, D.; Olszak, R. W.; Łabanovwska, B. H. Perspectives of Integrated Production of Black Currant in Poland. Integrated Plant Production in Orchards "Soft fruits". *Int. O. Biol. Integr. Control/West Palearct. Reg. Sect. Bull. (IOBC/WPRS Bull)* **1998,** *21*(10), 39–43.

9. Markelova, N. V. Possibilities of Black Currant Adaptive Breeding. *Herald of Bryansk State University* **2008,** *4,* 134–137 (in Russian).

10. Knyazev, S. D.; Pikunova, A. V.; Chekalin, E. N.; Bakhotskay, AYu.; Shavyrkina, M. A. Innovation Trends of Breeding Researches of Black Currant. Breeding and Cultivar Investigation of Horticultural Crops. *Proceedings of Innovation Techniques in Breeding and Improvement of Fruit and Berry-Like Assortment.* All-Russian Research Institute for Selection of Fruit Crops: Orel, 2014; Vol. 1, pp 192–211 (in Russian).

11. Program and Methodology for the Variety Research of Fruit, Berry and Nut Crops/ Under the General; Sedova, E. N., Ogoltsova, T. P., Eds.; VNIISPK Publishing House: Orel, 1999; pp 351–373.

12. Dospekhov, B. A. *Methodology of Field Experience (with the Basics of Statistical Processing of Research Results),* 4th ed.; Revised and Additional. Kolos: Moscow, 1979; p 416 (in Russian).

GLOSSARY

Aerohydroponic module
Module for accelerated reproduction of potato minitubers.

Aerohydroponics
Aerohydroponics is one of the methods of plant cultivation, which is based on oxygenation of water by passing the air through it. To do this, there are many ways with respect to the use of air and water pumps or whirlpools.

American mildew (*Sphaerotheca mors-uvae* (Schw.) Berk. et Curt)
One of the most common diseases is peronosporosis, or downy mildew, a disease caused by the lower fungus *Peronospora*; the causative agent of the disease can persist in plant remains and seeds and activate in wet conditions: during rains or just in wet weather.

Ammetric method
This is an electrochemical method for measuring the *antioxidant activity* of substances or the total content of antioxidants in them. The method consists of measurement of the electric current arising at oxidation of investigated substance on a surface of a working electrode at certain electric potential.

Anthracnose of cucurbits (*Colletotrichum orbiculare* (Berk. and Mont.) Arx)
A widespread pathogen of cucurbits (Cucurbitaceae), in particular, the cucumber (*Cucumis sativus* L.), watermelon (*Citrullus lanatus* (Thunb.) Matsum. and Nakai), muskmelon and cantaloupe (*Cucumis melo* L.), pumpkin (*Cucurbita pepo* L.), and other cucurbits as well as several noncucurbit hosts. Anthracnose is a fairly common problem on cucurbits in all humid growing regions. This infection was born from the tropics; in the temperate latitudes, it has appeared relatively recently. Residents of Russia first encountered anthracnose in 1901 when it devastated the cucumber and pumpkin beds in the Novgorod province. The attack was described under the name "bitter rot", and took the "honorable" place in the list of the main enemies of open ground cultures.

Anthracnose of currant and gooseberry (*Pseudopeziza ribis* Kleb.)
This is a fungal disease of currant and gooseberry. This is a disease which occurs to some extent every year, and only occasionally becomes epiphytotic. Cool and rainy weather creates perfect conditions for the spores to spread. Dry and hot weather stops the progression of the disease that may begin at the optimal weather conditions. Anthracnose tends to attack plants in the spring, primarily on leaves and twigs. The fungi overwinter in dead twigs and fallen leaves. The fungus is carried through the winter by either the conidial or the sexual stages and possibly as mycelium in the canes. Anthracnose was discovered on several species of currants—both red and white. Gooseberries are also subject to the disease. The disease is primarily a leaf-trouble, although it also appears on the petioles, young canes, fruit stalks, and fruits. The injury from anthracnose, also known as "leaf spot" and "leaf blight," comes from its effect upon the leaves and fruit. Old plantations suffer more than young ones; yet, the disease is of considerable importance in the nursery. In the nursery, older bushes usually suffer more than the first-year cuttings perhaps due to the fact that the young cuttings are planted on ground, which is not ordinarily used for currants and gooseberries.

Antioxidant activity (AOA)
AOA of enzymes and water- and fat-soluble substances inhibits the development of free radical oxidation in the body; in the process of evolution, an antioxidant defense system was developed. There exist medicinal plants that have AOA; they inhibit the degree of development of free radical oxidation.

Antioxidants
Antioxidants donate their electrons to unsaturated free radicals, while stabilizing themselves as the stable compounds; as a result, the continuous chain of destruction of molecules ceases. Natural dyes are concentrated in generative organs of plants (pollen, flowers), vegetative parts (leaves, roots, shoots), fruits, and seeds. Their amount in the product depends on the energy of photosynthesis and the climate. This is another broad term that refers to any molecule that inhibits the oxidation of another molecule and neutralizes the electron-snatching free radicals that damage cells. When a molecule donates an electron, it becomes unstable, because it is lacking the crucial electron. This is the chain reaction of oxidation. However, antioxidants have an extra electron; therefore, they can donate it

to destruct the chain reaction, yet remain stable, without becoming a free radical. Certain antioxidants become unstable, but it is only transient, due to a network of cooperating antioxidants.

Arctic bramble

Arctic bramble is a group of flowering plants of the Ericaceous family, uniting evergreen creeping shrubs growing on marshes in the Northern Hemisphere.

Arctic raspberry

It is the species of slow-growing bramble belonging to the Rosacea family, found in Arctic and alpine regions in the Northern Hemisphere. In conditions of cultivation is an important source of edible and medicinal raw material. Arctic raspberry is a promising species for the cultivation in conditions of Kirov region of Russia.

Basic seed

This is seed that has been produced by or under the responsibility of the breeder and is intended for the production of certified seed. It is called "basic seed" because it is the base for certified seed, and its production is the last stage that the breeder would normally be expected to closely supervise.

Berries

Berries are very productive and bear fruit annually. For northern horticulture, berry crops are especially promising since they are winter-hardy, early-ripening, and high-yielding. Fruits of wild-growing berries contain significant quantities of biologically active substances, vitamins, and antioxidants.

Berry bushes

Berry bushes are hardy—they can bear fruit in the harshest climatic conditions, even in the Far North, as they tolerate frosts well, especially under the cover of snow. In berry plantations, pest control is less complex than in a fruit garden.

Bilberries

Bilberries are one of several primarily Eurasian species of low-growing shrubs in the genus *Vaccinium* (family Ericaceae), bearing edible and nearly black berries. The species is most often referred to as *Vaccinium myrtillus* L., and there are several other closely related species.

Bioflavonoids or flavonoids
These are natural plant compounds and are responsible for the pigmentation of bright color fruits and vegetables. They are abundant in the pulp and rinds of citrus fruits and in other foods containing vitamin C, soybeans and root vegetables. They are polyphenols and belong to a large class of antioxidants that are named for the Latin word for yellow (golden yellow, golden)—"flavus." There are over 5000 subgroups of flavonoids that are organized in basic groups: anthoxanthins, flavanones, flavanonols, and flavans. These are further divided into more categories and groups, such as catechins.

Biological stock
Biological stock is plant material (phytomass)—volume of raw phytomasses formed by all species in areas both suitable and unsuitable for commercial collection.

Black currant (*Ribes nigrum* L.)
A species of the genus *Ribes* L. family of Grossulariaceae DC., deciduous berry bushes. The genus *Ribes* sp. is distributed in the temperate regions of the Northern Hemisphere and South America. A few of the *Ribes* spp. occur in Central America at high elevations, and some European species are found in western North Africa. At present, the wild, black currant grows throughout Europe, the Urals, Siberia to the Yenisei and Baikal, Kazakhstan, Mongolia, and China. It is also distributed in North America. In culture, it is grown all over the world. In Russia, I. V. Michurin Federal Scientific Center (I. V. Michurin FSC) is breeding black currant cultivars.

Blueberries
Blueberries are perennial flowering plants with indigo-colored berries from the section *Cyanococcus* within the genus *Vaccinium* (a genus that also includes cranberries, bilberries, and grouseberries). The species in the section *Cyanococcus* are the most common fruits sold as "blueberries" and are native to North America (commercially cultivated highbush blueberries were not introduced into Europe until the 1930s). They are very productive and bear fruit annually. In addition, they are fast-breeding: berry bushes yield a harvest 2–3 years after planting; they grow in the taiga regions of Russia, contain vitamins, active substances, and antioxidants.

Boron
Boron is an element of the 13th group (according to the obsolete classification—the main subgroup of group III), the second period of the periodic system of chemical elements with atomic number 5.

Cadmium
Cadmium is a chemical element with symbol Cd and atomic number 48. This soft, bluish-white metal is chemically similar to the two other stable metals in group 12, zinc and mercury. Like zinc, it demonstrates oxidation state $+2$ in most of its compounds, and like mercury, it has a lower melting point than other transition metals. Cadmium and its congeners are not always considered transition metals; the average concentration of cadmium in Earth's crust is between 0.1 and 0.5 parts per million (ppm).

Calcium
Calcium is a chemical element with symbol Ca and atomic number 20. Calcium is a soft grayish-yellow alkaline earth metal, fifth most abundant element by mass in the Earth's crust. The ion Ca^{2+} is also the fifth most abundant dissolved ion in seawater by both molarity and mass after sodium, chloride, magnesium, and sulfate. Calcium is an essential trace element in living organisms. It is the most abundant metal by mass in many animals, and it is an important constituent of bone, teeth, and shells. In cell biology, the movement of the calcium ion into and out of the cytoplasm functions as a signal for many cellular processes. Calcium carbonate and calcium citrate are often taken as dietary supplements. Calcium is in the World Health Organization's List of Essential Medicines.

Carbohydrates
Carbohydrates are organic substances containing a carbonyl group and several hydroxyl groups.

Carotenoids
Carotenoids are any of a class of mainly yellow, orange, or red fat-soluble pigments, including carotene.

Certified seed
This is the first generation of multiplication from basic seed and is intended for the production of ware crops as distinct from a further seed generation. In some agricultural crops, there may be more than one generation between basic and certified seed, in which the number of generations of multiplication after basic seed is stated, for example, first- or second-generation agricultural crops such as cereals. It is usually the multiplication rate of

a species which determines whether, or not, further generations can be produced beyond the first multiplication from basic seed; for example, some grain legume species have a relatively low multiplication rate and a second multiplication may be allowed.

Chemical bond

A chemical bond is a strong attraction between two or more atoms. There are many types of chemical bonds, but all involve electrons, which are either shared or transferred between the bonded atoms.

Chemiluminescence method

This method (also *chemoluminescence*) is the emission of light (luminescence), as the result of a chemical reaction. There may also be limited emission of heat. It is the method of determination of the antioxidant activity in the oxidation system "hemoglobin–hydrogen peroxide–luminol."

Chloride salinization

Chloride salinization studied in experiments with germination of seeds and growth of wheat, rape, decorative plants, and other crops exerts an inhibiting effect. These studies are related to the problem of increasing salinity in the soil used in agriculture.

Cloudberry (*Rubus chamaemorus* L.)

This is a species of perennial herbaceous plants of the Rosaceae family. This berry grows almost exclusively in circumpolar regions and is prized in cultures and cuisines in Scandinavia, Russia, Canada, and throughout Alaska. Though related botanically to red raspberry (*Rubus idaeus* L.) and salmonberry (*Rubus spectabilis* Pursh), the cloudberry plant most closely resembles another highly sought arctic berry, *Rubus arcticus* L. In contrast to the prickly, tall canes of raspberries and salmonberries, cloudberries grow on very slender stems of not more than 2–8 in high in boggy, open tundra and forest. Each stem has two to three circular leaves with rounded lobes and toothed edges and a single white, five-petaled flower. The berry is composed of six to eight drupelets, forming a small roundish berry. Unripe cloudberries are hard, sour, and red; as the berries ripen, they soften, sweeten, and lighten to a rosy peach or amber hue.

Cranberry (*Vaccinium oxycoccos* L.; syn. *Oxycoccus quadripetalus* Gilib., *Oxycoccus palustris* Pers.)

A species of dwarf shrub of the Ericaceae Juss. family is a perennial trailing woody vine that is native to bogs and swamps. Cranberry blooms during

June and July, bearing ranks of solitary flowers along short upright shoots. They are widely wild-harvested throughout much of its natural range and cultivated only in Russia and Estonia. This species, the European cranberry, is a dwarf, woody, evergreen clonal shrub with slender, rooting stems, occasionally up to 0.8–1.0 m tall, with short, usually erect flowering shoots. The leaves are leathery, dark, glossy green dorsally, glaucous ventrally, and frequently revolute with an entire blade margin. Racemes of 1–5, white, pink, or red, protandrous flowers are pollinated mostly by solitary or social bees and high fruit production frequently occurs following autogamy. The fruit is an overwintering, edible berry (the cranberry). Although fruit-set in natural populations may be high, the plant mostly reproduces vegetatively, forming large clones, some hundreds of years old. This plant has three (or four, depending on taxonomic treatment) ploidy levels—mainly tetra- and hexaploid populations are found, but pentaploids are also reported in the Czech Republic and Sweden. Diploids are usually treated as a separate species, namely *Vaccinium microcarpum* (Turcz. ex Rupr.) Schmalh.

Crop rotation
Crop rotation is the successive cultivation of different crops in a specified order on the same fields, in contrast to a one-crop system or to haphazard crop successions, for example, alternation of potato crop in the first rotation (1979–1989): potatoes, barley, winter rye, oats, pea–oats mixture, winter wheat, and perennial grass of first and second year use; the second (1996–2001) and third (1990–1995) rotations: potatoes, barley, perennial grass of first and second year use, winter wheat, and oats.

Cultivar
A cultivar is a taxon that has been selected for a particular attribute or combination of attributes, and that is clearly distinct, uniform, and stable in its characteristics and that, when propagated by appropriate means, retains those characteristics. A contraction of "cultivar" (abbreviated cv.); after International Code of Nomenclature for Cultivated Plants (International Code of Nomenclature for Cultivated Plants – ICNCP); "cultivar" is synonymous with "Sorte" (German), "variety" (English), or "variété" (French).

Cultivation of plants
Cultivation of plants is the organization of special conditions for growth of plants, including sowing, growing, growing technology development, new cultivar breeding, increase of productivity, and resistance to unfavorable meteorological conditions, pests, and so forth.

Cuprum (Latin)
It is denoted by the symbol Cu. Copper (English) is an element of the eleventh group of the fourth period (an indirect subgroup of the first group) of the periodic system of chemical elements of D. I. Mendeleyev, with atomic number 29.

Currant mite (currant bud mite)
Currant mite spends its entire life on a bush of currant. It is small, about 0.15–0.3 mm in size. Insects are wintered in the kidneys, with the onset of heat, and the females lay eggs in young kidneys. When they grow up, it become cramped in one kidney, and they spread to the neighboring ones and populate the entire bush, as well as all growing near it.

Dietary fiber (cellulose)
Food components are not digested by the digestive enzymes of the human body; however, they are processed by the beneficial microflora of the intestine. They possess water-binding ability.

Disease
A bacterial or fungal infection, which has a detrimental effect on a plant.

Dwarfing rootstock
A part of a tree, which has the roots onto which scions of other trees are grafted in order to produce a much smaller version of the original tree.

Electrical conductivity
Electrical conductivity (EC) is a measure of a material's ability to conduct an electric current.

Electrochemical oxidation
Electrochemical oxidation is proceeding under scheme $R–OH\ OH \rightarrow R–O^{\bullet} + e^- + H^+$, and can be used as a model for measurement of free radical absorption activity.

Electrochemistry
Electrochemistry is the branch of physical chemistry that studies the relationship between electricity, as a measurable and quantitative phenomenon. Electrochemistry deals with the interaction between electrical energy and chemical change.

Environment
The environment is a generalized concept that characterizes the natural conditions of a certain locality and its ecological state; physical and

biological factors affect the body or group of organisms. When they say "environment," they usually mean nature. Nature is a natural material system, including objects that are not created by human hands. This is soil, water, air, flora, and fauna. In a more general concept, nature is the entire universe, including planets and solar systems. The environment is not just the things around a person—it depends on the health of people, as well as the opportunity to live on this planet for future generations. If there is an irresponsible approach toward its preservation, it is likely that the destruction of the entire human race will occur.

European cranberrybush (*Viburnum opulus* L.)
The species of genus *Viburnum* L. such as *Viburnum opulus* L. (European cranberrybush), *Viburnum trilobum* Marsh. (American cranberrybush), as well as *Viburnum sargentii* Koehne are widely used in traditional and folk medicine. The European cranberrybush (*Viburnum opulus* L.) is a native plant in Lithuania, which is widely used in traditional and folk medicine. Its flowers, bark, and leaves are an important medicinal raw material because it possesses large amounts of the tannic substances, carotenoids, isovalerianic acid, saponines, and glycosides. The seeds contain up to 21% of fatty oil. Fruits of this species accumulate significant amounts of biologically active substances.

European red mite (*Panonychus ulmi* (Koch))
This is a species of mite of a major agricultural pest of fruit trees. It is found on fruit trees, shade trees, and shrubs worldwide. It is the most common mite occurring on apple trees in the Southeast.

Evapotranspiration (total evaporation)
It is the amount of moisture transferred to the atmosphere in the form of steam as a result of transpiration of plants and physical evaporation from the soil. It is expressed in millimeters of mercury. It directly depends on the air temperature, air humidity, and relative humidity.

Exploitation stock of phytomasses
This is the volume of raw material phytomass formed by marketable individuals of medicinal or berry plants on an area suitable for commercial collection.

Federal Scientific Center
I. V. Michurin FSC is the recent name of I. V. Michurin All-Russian Scientific Research Institute of Certification. Michurinsk, the scientific town of

Russia Federation, is All-Russian Center of Horticulture. It is the Central Chernozem region, situated on the territory of Tambov region, in the western part of Oka-Don plain, on the right bank of the Lesnoy Voronezh river (the basin of the Don river).

Ferrum
Iron (Latin Ferrum), Fe, is a chemical element of the group VIII of the Mendeleyev periodic system. Its atomic number is 26 and atomic weight is 55.847.

Fiber
The fiber is a dietary material containing substances such as cellulose, lignin, and pectin that are resistant to the action of digestive enzymes.

Filbert (hazelnut, cobnut)
One of several cultivated hazel shrubs and trees that bears edible oval or rounded nuts belongs to genus *Corylus* L., family Betulaceae Gray.

Flavonoids
Flavonoids (or bioflavonoids; Latin word *flavus*—meaning yellow, their color in nature) are a class of plant and fungus secondary metabolites. Chemically, flavonoids have the general structure of a 15-carbon skeleton, which consists of 2 phenyl rings (A and B) and heterocyclic ring (C). This carbon structure can be abbreviated C6–C3–C6. Flavonoids are natural antioxidants.

Flavonols
It is a kind of flavonoid (and thus a polyphenol as well). Their phenol structure groups are found in many different positions, making this a very diverse group of flavonoids.

Free radicals
Free radicals, in chemistry, are the molecules (usually unstable) that contain one or more unpaired electrons on the outer electron shell. These are active, and in free movement, they can attach electrons, are able to disrupt the permeability of cell membranes, and cause various diseases (oncological as well). For reducing the damage caused by them, use *antioxidants* (see).

Freezing resistance (freeze hardiness)
The ability of plants to survive temperatures below 0°C. It is the ability of cells to survive frost by means of avoidance or tolerance of ice formation in plant tissues and the ability of plants to withstand the complex of

environmental influences during the winter and early spring periods. This includes the ability to withstand spring burns, low temperatures, trapping, soaking, and the ability to withstand frequent temperature changes.

Frost survival
This is an ability of a plant (individual, population) to survive frost events by prevention from freezing, by freezing resistance, and by recovery after partial damage.

Fructose
Fructose, or fruit sugar, is a simple ketonic monosaccharide (arabino-hexulose, levulose), ketone–alcohol, ketohexose; in living organisms, there is only D-isomer; found in many plants, where it is often bonded to glucose to form the disaccharide sucrose.

Fruit
Broadly, the botanical term "fruit" refers to the ripened ovary of a flowering plant, including its seeds, covering, and any closely connected tissue, without any consideration of whether these are edible. Fruits can be dry or fleshy. Berries, nuts, grains, pods, and drupes are fruits. Fruits that consist of ripened ovaries alone, such as the tomato and pea pod, are called true fruits. Fruits that consist of ripened ovaries and other parts such as the receptacle or bracts, as in the apple, are called accessory fruits or false fruits. With respect to food, the botanical term "fruit" refers to the edible part of a plant that consists of the seeds and surrounding tissues. This includes fleshy fruits (such as blueberries, cantaloupe, peach, pumpkin, tomato) and dry fruits, where the ripened ovary wall becomes papery, leathery, or woody as with cereal grains, pulses (mature beans and peas), and nuts.

Fruit and vegetables
These are edible plant foods excluding cereal grains, nuts, seeds, tea leaves, coffee beans, cacao beans, herbs, and spices.

Fruit growing
A branch of horticulture that is related to the growth of fruit plants over isolated areas of land or in orchards for their fruits (including berries and nuts). Fruit growing involves the raising of pip, drupe, nut, and berry crops, as well as nursery cultivation. Subtropical fruit growing and the raising of citrus plants are usually considered to be independent branches of horticulture.

Garden
Garden is an area of land cultivated as a hobby or for pleasure, rather than to produce an income.

Gardening
Gardening is the science or art of growing flowers, fruits, vegetables, or ornamental plants (e.g., in a garden, orchard, or nursery).

Generative shoot
It is shoot bearing of reproductive organs—flowers or fruits.

Glucose
Glucose ($C_6H_{12}O_6$) is simple sugar, grape sugar, or dextrose (D-glucose), which is found in fruit and berry juices. It is a monosaccharide and a hexatomic sugar (hexose). Simple sugar, grape sugar, or dextrose (D-glucose) is found in fruit and berry juices.

Glucose unit
It is a part of polysaccharides (cellulose, starch, glycogen) and a number of disaccharides (maltose, lactose, and sucrose), which, for example, are quickly broken down into glucose and fructose in the digestive tract.

Heteroauxin
Heteroauxin (indolylacetic acid) is a fertilizer from the group of auxins, which is used as a root stimulant.

Horticulture
The science or art of cultivating fruits, vegetables, flowers, or ornamental plants for a garden, orchard, or nursery of fruit trees.

Hybrid
The offspring of two plants of different species or variety created when the pollen from one plant is used to pollinate a different variety resulting in a plant which has characteristics of both parent plants.

Hybrid F_1
It is the hybrid of first generation (F_1) is a homogeneous set of individuals, which is restored each time by crossing two or more selected offspring, lines, clones, and simple hybrids of the first generation F_1.

Immunity
Immunity is the ability of living beings to withstand the action of damaging agents, preserving their integrity and biological individuality. It is the protective reaction of the body.

Immunostimulation

Immunostimulation is (1) a complex of measures aimed at activating the patient's immune system in the treatment of chronic bacterial, viral, and fungal infections and (2) stimulation of an immune response.

Indole-3-acetic acid (IAA)

A substance ($C_{10}H_9NO_2$) that acts as a growth hormone or auxin in plants, where it controls cell growth, and through interactions with other plant hormones, also influences cytokinesis.

Indole-3-butyric acid (IBA)

It is 2,4-(indolyl-3) butyric acid ($C_{12}H_{13}NO_2$). It is a growth regulator of plants and used to accelerate the rooting and growth of seedlings of trees and garden crops.

Indoor garden

Indoor garden (German: wintergarten, French: jardin d'hiver) is heated room with natural light, designed to accommodate exotic and nonresistant, as well as indoor plants.

Iodum

Iodum is chemical element—Iodine (atomic weight 126.53).

Iron (Fe)

In blood, iron has an important biological function such as transport and activation of molecular oxygen; two-third of all iron in the organism is located in hemoglobin. It is also found in myoglobin and in enzymes: catalase, peroxidase, and cytochrome oxidase; the daily requirement is 15 mg. Iron from vegetable products is digested for just 10%. Thus, the daily requirement of the element can be provided by 3 kg of cloudberry. The majority of wild fruits remain effective sources of iron in diets.

Isoflavones

A type of polyphenol with estrogen-like activity and it is sometimes referred to as phytoestrogens. They are found primarily in legumes.

Kalanchoe

Kalanchoe spp. belong to the two genera: *Kalanchoe* Adans. and *Bryophyllum* Salisb. (Crassulaceae J. St.-Hil. family). Kalanchoe have significant antioxidant activity in their leaves juice, two most active species (*Kalanchoe pinnata* (Lam.) Pers. (This name is a synonym of *Bryophyllum pinnatum* (Lam.) Oken.) and *Kalanchoe integra* (Medik.) Kuntze) were revealed.

Kalanchoe *Bryophyllum pinnatum* (Lam.) Oken

Bryophyllum pinnatum is also known as *Kalanchoe pinnata* (Lam.) Pers., *Bryophyllum calycinum* Salisb., and under the common names: *Zakhm-e-hayat*, life plant, air or maternity plant, love plant, Canterbury bells, Cathedral bells, parnabija, and so forth. It is a perennial herb growing widely and used in folkloric medicine in tropical Africa, tropical America, India, China, and Australia, classified as a weed. The plant flourishes throughout the southern part of Nigeria. This is the only *Kalanchoe* sp. found in South America; however, 200 other species are found in Africa, Madagascar, China, and Java. A number of species are cultivated as ornamentals and are popular tropical house plants. In Brazil, the plant goes by the common names of *saiao* or *coirama* and in Peru, it is called *hoja del aire* (air plant) or kalanchoe.

Kalanchoe *Kalanchoe* Adans.

Kalanchoe refers to succulent plants of the Crassulaceae J.St.-Hil. family. It has long shoots strewn with dense foliage and retains its green color both in summer and in winter. The coloring of the leaves depends on the stem: from the top, they are a saturated green hue, and from the bottom greenish-violet. The plant blooms irregularly, with pinkish-green flowers.

Kornevin

Kornevin root is a growth stimulant and a hormonal drug. It includes *indolylbutyric acid* (see *Indole-3-butyric acid* in Glossary), potassium, molybdenum, phosphorus, manganese, and microelements.

Leskenit

Leskenit is a part of the natural zeolite clays of the North Caucasus, has an alkaline reaction (pH 8.46), high calcium content (30–35%), macro- and microelements in forms accessible to plants as well as iodine, which is absent into other natural clays. It is used as an additive when processing potato planting stock. It is used as a mineral supplement in the ration of young cattle and increases the productivity of laying hens.

Lipids

Lipids are a class of hydrocarbon-containing organic compounds essential for the structure and function of living cells. Lipids are insoluble in water and soluble in nonpolar organic solvents.

Lutrasil

Lutrasil is a special covering material to shelter plants from external adverse factors.

Macroelements
These are chemical elements or their compounds used by organisms in relatively large quantities: oxygen, hydrogen, carbon, nitrogen, iron, phosphorus, potassium, calcium, sulfur, magnesium, sodium, and chlorine. Their content in living organisms is more than 0.01%.

Management
Management (or managing) is a set of principles, forms, methods, and controls for the production and working personnel using the achievements of management science: (1) the main goal of the management is to achieve high production efficiency, as efficient use of the resource potential of the enterprise, firm, company; (2) it is the management of an organization, whether it is a business, nonprofit organization, or a governmental body.

Manganese
It is a chemical element with symbol Mn and atomic number 25. It is not found as a free element in nature. It is often found in minerals in combination with iron. Manganese is a metal with important industrial metal alloy applications, particularly in stainless steels.

Manure
This is organic fertilizer, constituted from solid and liquid excrement of animals. It is the result of enzymatic and microbiological processing of plant food and litter residues. It is used as an organic fertilizer for soil and contains water, nitrogen, phosphorus, potassium, calcium, and magnesium. It is applied on different soils, under various agricultural crops. There are different types of organic fertilizers: litter manure, chicken manure, compost from cattle manure, compost from sewage sludge, and peat and also organic mineral fertilizer based on sewage sludge on yield and quality of potatoes. Particularly, it is favorable for potatoes, corn, cabbage, and cucumber.

Marketing
Marketing is the study and management of exchange relationships; social and management process aimed (1) at meeting the needs and needs of individuals and social groups through the creation, supply, and exchange of goods and services; (2) this is the extraction of profit from the satisfaction of the consumer; and (3) the market concept of managing the company's production and marketing activities, aimed at studying the market and specific customer.

Mesophytes

Mesophytes (from the Greek mesos—intermediate). Plants of this ecological group grow in the conditions of sufficient moisture. Osmotic pressure of cell juice in mesophytes is 1000–1500 KPa. They wilt easily. Mesophytes include most of the meadow grasses and legumes such as vacuum creeping, meadow fox, timothy grass, alfalfa blue, and so forth; and field crops such as hard and soft wheat, maize, oats, peas, soybeans, sugar beet, hemp, almost all fruits (almonds, grapes), and many vegetable crops (carrots, tomatoes, cabbage, and so forth).

Mesotrophic swamp

It is a transitional type of swamp between lowland (eutrophic) bog and highland (oligotrophic) bog, on which usually "blend" wetland plants of both types grow.

Michurin I. V.

See "Federal Scientific Center."

Microelements

These are chemical elements that are biologically significant(opposite of biologically inert elements) necessary for living organisms to ensure normal vital activity. Biologically significant elements are classified into macroelements (content in living organisms is more than 0.01%) and trace elements (content less than 0.001%). They participate in biochemical processes and are necessary for living organisms.

Milligram equivalent

The number of milligrams of a substance equal to its equivalent. Equivalent is the amount of a substance that reacts with 1 g of hydrogen or displaces 1 g of hydrogen from its compounds.

Minitubers

Minitubers of potatoes are obtained in vitro. For the improvement of cultivars, several cells from the seedling are planted in a nutrient medium. A plant grows, which is known to be free from viruses and diseases accumulated in the mother tuber. Then they cut it, plant the seedlings in the greenhouse, which results in minitubers. Of these, in the first year, a crop is called the first generation, next year—the second generation (super-super elite), then—super elite, elite, first reproduction, and the second. In subsequent years, ordinary commercial potatoes of so-called mass reproduction are grown.

Myrobalan
Myrobalan is the name of a large group of Himalayan fruit trees. A special kind of myrobalan is held in the Buddha of Medicine, is called "arura" in Tibetan.

Millimolar
Millimolar (mM) is the expression of the molar concentration (the number of particles in solution, whether or not they carry an electric charge). 1 molar is 1000 millimolar.

Monoecious plants
They represent a population or species having functional male and female organs in separate places on the same plant.

Monoploid chromosome number (x)
The basic chromosome number is the lowest haploid chromosome number in a polyploid series (symbolized by x). The number of chromosomes found in a single complete set of chromosomes is called the monoploid number (x). The haploid number (n) is unique to gametes (sperm or egg cells) and refers to the total number of chromosomes found in a gamete, which under normal conditions is half the total number of chromosomes in a somatic cell.

Morphometric parameters
Morphometric (or morphological) parameters of plants are as measurable characteristics that are supposed to reflect evolutionary processes and include a quantitative analysis of the growth of a plant.

Mycorrhiza
Mycorrhiza is a symbiosis of mushroom, which grows in association with the roots of a plant in a symbiotic or mildly pathogenic relationship. This symbiosis can be a normal interaction between the bend and the roots, without which most plants cannot normally live and develop. It improves the root nutrition of plants and is the most powerful source of water for plants. Moreover, it secretes large amounts of antibiotics inhibiting pathogenic organisms. Mycorrhizas supply plants with mineral salts, vitamins, enzymes, biostimulants, hormones, and other active substances. They also provide the basic supply of phosphorus and potassium to the plants deficit with them. It has been established that even widely used agricultural crops, such as grain and fodder cereals, beans, potatoes, sunflower, are also mycotrophic. If the roots of these plants have mycorrhizal fungi, their productivity may increase from 10 to 15 times. Symbiotic fungi have

a strong protective effect on plants, releasing a large number of antibiotics that inhibit pathogenic organisms.

National Dendrological Park "Sofiyivka"

National Dendrological Park "Sofiyivka" is a park and research institute of the National Academy of Sciences of Ukraine, located in the northern part of Uman city in Cherkassy region on the banks of Kamenka river.

Ontogeny (ontogenesis)

It is the process of an individual organism development.

Orchard

Orchard is an area of land used for growing fruit trees.

Organic acid

Organic acid is an organic compound with acidic properties. The most common organic acids are the carboxylic acids.

P-active substances

P-active substances include anthocyanins, leucoanthocyanins, catechols, chlorogenic acids, and flavonols, differing by chemical composition but having similar activity in a human organism. P-active substances have hypotensive and sclerotic activity. The total content of P-active compounds is 0.02–0.6% in fruits and berries of light-red color. Intensively colored and black-fruited cultures such as bilberry, black currant, cranberry, blueberry, honeysuckle, blackthorn, and Amur grape contain 1–1.5% of P-active substances.

Para-aminobenzoic acid (PABA)

It is 4-amino-2-hydroxybezoic acid that has a molecular mass of 137.1 g/mol. It is an organic compound with the formula $H_2NC_6H_4CO_2H$. It is a white-gray crystalline substance that is only slightly soluble in water, soluble in hot water (80–90°C), and well soluble in benzene, ethanol, and acetic acid. It is classified as nontoxic vitamin-like compound of group B, known also as vitamin H_1 or vitamin B_{10}. Microorganisms use it as a precursor of folic acid synthesis. It participates in the synthesis of purines and pyrimidines—ultimately in the synthesis of deoxyribonucleic acid and ribonucleic acid. PABA as chemical compound is known since 1863, but its high biological activity in low concentrations was first discovered in 1939 by well-known geneticist I. A. Rapoport on *Drosophila*. He showed that the positive effect on living systems is based on the previously unknown phenomenon of its interaction with ferments. This interaction results in the restoration of the ferments activity and decreased in some cases at the genetic level (e.g., because of excess of recessive genes or damaging environmental factors). In subsequent studies, the ranges of PABA suitable for different objects were determined. It was proved that PABA is a promoter of phenotypic activity and it also increases immunity. It has virucidal and antimicrobial action and shows bioxidic functions. There are data about PABA effect decreasing harmful mutagens action. PABA positively affects all characters determining yield structure and increasing adaptive plant properties, including the resistance to a series of diseases.

Partial bush
Partial bush is incomplete bush. It is relatively autonomous young plant that is still connected with maternal root or aboveground shoot.

Peatland
Peatland is a biological object, an ecosystem that includes a complex of plants and their remains, which forms an interdependent community in conditions of high humidity. It is the highest type of existence of living organisms, similar to coral reefs, forests, and urban megacities. Peatlands are the most important source of peat.

Pectic substances
Pectic substances are a group of polysaccharides in plant cell walls, which are endowed with multifunctional properties such as the control of cell wall integrity and porosity, the protection of plants against phytopathogens, and gelling and complexing ability with ions of heavy metals and radionuclides. They are also used in food production.

Pectin
Pectin is a structural heteropolysaccharide contained in the primary cell walls of terrestrial plants.

Perennial
It is a nonwoody plant, which lives for more than 2 years.

Periodicity
It is a repetition of events at fairly regular intervals. The tendency of individuals, cultivars, or species to produce fruit crops at long but often more or less regular intervals (in years).

Pesticides

They represent chemical preparations used to control weeds (herbicides), with pests (insecticides, acaricides, zoocides, etc.) and diseases (fungicides, bactericides, etc.) of agricultural plants, trees and shrubs, grains, and so forth.

Phenological observations

These are regular observations of seasonal changes in nature or at description of the development of a plant or agricultural crop on different physiological development stages. All received data are structured in a single system.

Phenols

A class of chemical compounds made up of an oxygen–hydrogen bond (a hydroxyl group) bonded to an aromatic compound of hydrogen and carbon (a hydrocarbon).

Phosphates

Phosphates and phosphorus organic compounds have various functions in the organism: plastic, maintenance of alkaline–acidic balance, synthesis of phospholipids, nucleotides and nucleic acids, enzyme catalyzes, and so forth. Human muscular and intellectual activity depends on phosphorus intake.

Phosphorus

Phosphorus is a chemical element with symbol P and atomic number 15; metalloid is an important microelement in terms of magnesium and calcium utilization.

Phytocenosis

This is any group of plants belonging to a number of different species that co-occur in the same habitat or area and interact through trophic and spatial relationships. It is typically characterized by reference to one or more dominant species.

Phytohormones

Phytohormones are physiologically active compounds of natural or synthetic origin, capable to induce changes of the growth and development of plants in small quantities.

Phytoncides

Phytoncides are plant-derived biologically active substances that kill or inhibit the growth and development of bacteria, microscopic fungi, and protozoa. The term was proposed by B. P. Tokin in 1928.

Phytotron

This is an automated device in which microclimatic conditions for growing plants are recreated; in modern phytotrons, the light regime and water supply with the solution dosage and (often) the temperature are automatically regulated; there you can get several biomass yields annually, as a result of which the selection process can be accelerated.

Plagiotropic sprouts

It is the shoot, which is more or less horizontal in orientation.

Pollination

The transfer of pollen from the stamen (male part of the flower) to the pistil (female part of the flower), which results in the formation of a seed that eventually becomes a fruit.

Polyphenols

Compounds found in abundance in natural plant food sources that have antioxidant properties. They are distinguished by their molecular structure, which is characterized by many multiple conglomerations of phenol structure units, and hence the prefix "poly," which means "many."

Potassium

Potassium is a chemical element with symbol K (derived from Neo-Latin, kalium) and atomic number 19. It was first isolated from potash, the ashes of plants, from which its name is derived. In the periodic table, potassium is one of the alkali metals. All alkali metals have a single valence electron in the outer electron shell, which is easily removed and create an ion with a positive charge—a cation.

Prebasic seed

This is seed material at any generation between the parental material and basic seed.

Propagation

Methods of starting new plants, for example, by sowing seed, division, cuttings, layering, reproduction, procreation, generation, transmission of an inherited trait, and increase in extension.

Protopectin

Protopectin is a water-insoluble natural plant pectin, consisting mainly of chains of polygalacturonic acids, connected by ether bridges through phosphoric acid and ions of polyvalent metals through unesterified carbonyl groups. They are found in the rind of citrus fruits or in apple peels and that are hydrolyzed to pectin or pectic acid.

Provitamin
It is a substance that may be converted within the body to a vitamin. The term "previtamin" is a synonym.

Radionuclides
Radionuclides are a set of atoms characterized by a specific mass number, the energy state of the nuclei, the atomic number, the nuclei of which are unstable, and undergo radioactive decay.

Reproductive organs
Reproductive organs of plants are the most specialized formations that perform the function of asexual or sexual reproduction. These include sporangia, antheridia, archegonia, and flower.

Retinol
Retinol is an active form of vitamin A. It is found in the liver of animals, whole milk, and other enriched foods.

Rootstock
The part of a plant (usually a tree), which has the roots and which will influence certain growing characteristics of the plant such as height.

Rutin
A combination of the glycosides quercetin and rutinose. (Glycosides are compounds made up of a carbohydrate and a noncarbohydrate component.) It is a biologically active flavonoid, and its bioavailability is influenced by the presence or absence of glucosides. (Glucosides are glycosides in which the sugar component is glucose.)

Saccharose
Saccharose is an obsolete term for sugars in general, especially sucrose.

***Septoria* (*Septoria* Sacc.)**
Septoria leaf spot, also called Septoria blight, is a very common disease in the form of a spot on leaves of currant and gooseberry as white spotting (*Septoria ribis* Desm.). It is a widespread disease on the leaves of many cultures.

Species
A species in biology (abbreviated sp., the plural form species abbreviated spp.) is the basic unit of biological classification and a taxonomic rank.

Standard seed
This category contains seed, which is declared by the supplier to be true to cultivar and purity, but is outside the certification scheme.

Sucrose
The disaccharide molecule: glucose + fructose, $C_{12}H_{22}O_{11}$ detectable in many plants and plant parts.

Sulfur
Sulfur or sulphur is a chemical element with symbol S and atomic number 16. It is abundant, multivalent, and nonmetallic.

Super elite
Super elite of potatoes is the material obtained from the super-super elite and corresponds to the modern notion of certified seed.

Super-super elite
The super-super elite of potatoes is the seeded (or planting) material obtained from the first field generation grown from the microtubers.

Taiga
Taiga, also known as boreal forest or snow forest, is a biome characterized by coniferous forests consisting mostly of pines, spruces, and larches.

Trolox
Trolox (6-hydroxy-2,5,7,8-tetramethylchroman-2-carboxylic acid) is the most common antioxidant that is an analogue of water-soluble vitamins E and C. Currently, it is accepted as a standard for assessing antioxidant activity.

Vegetable
In the broadest sense, the botanical term vegetable refers to any plant, edible or not, including trees, bushes, vines, and vascular plants, and distinguishes plant material from animal material and from inorganic matter. There are two slightly different botanical definitions for the term vegetable as it relates to food. According to one, a vegetable is a plant cultivated for its edible part(s); according to the other, a vegetable is the edible part(s) of a plant, such as the stems and stalk (celery), root (carrot), tuber (potato), bulb (onion), leaves (spinach, lettuce), flower (globe artichoke), fruit (apple, cucumber, pumpkin, strawberries, tomato), or seeds (beans, peas). The latter definition includes fruits as a subset of vegetables.

Vegetative shoots

Vegetative shoots are the vegetative organs of a plant without generative organs, consisting of stem, leaves, and kidneys. They can be apical and lateral, shortened and elongated.

Vitamins

It is a group of low-molecular organic compounds with relatively simple structure and diverse chemical nature. These are organic substances in the chemical nature, which are united on the basis of their absolute necessity for the heterotrophic organism as an integral part of the food. Autotrophic organisms also require vitamins, either by synthesis or from the environment.

Winter resistance (winter hardiness) of plants

This is the ability to withstand a complex of environmental influences during the winter and early spring periods. Winter hardiness is a very changeable concept. It varies with the age of plants, depends on the wind regime, the microclimate, the type, and humidity of the soil of the growing area. Winter hardiness and frost resistance of plants develop by the beginning of winter in the process of hardening of plants. Plants can tolerate frost: winter rye up to $-30°C$, winter wheat up to $-25°C$, some apple varieties up to $-40°C$. The stability of plants to the euthanization is ensured by the accumulation in them, by the beginning of winter, of a large number of sugars and other reserve substances, economical consumption by plants (at a temperature of about $0°C$) of reserve substances for respiration and growth, and protection of plants against fungal diseases. The stability of plants to bulging is determined by the thickness and elongation of the roots. Pushing-out is observed more often on dense, humus, and moist soils with their repeated freezing and thawing. Danger of autumn water stagnation; At it the hardening of plants worsens, and they are more easily damaged by frosts. Even more damaging is the stagnation of water in the spring. Weakened and damaged plants in winter die with a lack of aeration.

Yield

It is the total collection of crop products obtained as a result of growing a certain crop from the whole area of its sowing (planting) in the holding, in the region, or in the country.

Zinc

Zinc (Zn) is element of the periodic system of chemical elements of D. I. Mendeleyev, with atomic number 30, and is a brittle transition metal of bluish-white color. Fruits of berries have high zinc content—cloudberry fruits contain about 40 mg/kg, air-dry weight. Maximum zinc concentration is detected in ashberry.

Zircon

Zircon is immunostimulator, and has functional and anti-stress effect. This preparation is prepared from the medicinal plant *Echinacea purpurea* (L.) Moench (eastern purple coneflower or purple coneflower) and represents a solution of hydroxycinnamic acids in ethanol.

INDEX

9 781774 631560